Plumbing Design and Installation Details

Other McGraw-Hill Books in Mechanical Engineering

AVALLONE & BAUMEISTER *Marks' Standard Handbook for Mechanical Engineers, 9/e* (1987)

BEATY *Sourcebook of HVAC Details* (1987)

BEATY *Sourcebook of HVAC Specifications* (1987)

EDWARDS *Estimating Control Systems for HVAC* (1986)

GLADSTONE, HUMPHREYS, & LUNDE *Mechanical Estimating Guidebook for Building Construction, 5/e* (1987)

HANSEN *Hydronic System Design and Operation* (1985)

KHASHAB *Heating, Ventilating and Air-Conditioning Systems Estimating Manual, 2/e* (1983)

MUELLER *Standard Application of Electrical Details* (1984)

MUELLER *Standard Application of Mechanical Details* (1985)

MUELLER *Standard Mechanical and Electrical Details* (1980)

Plumbing Design and Installation Details

Jerome F. Mueller, P.E.

McGRAW-HILL BOOK COMPANY
New York St. Louis San Francisco Auckland Bogotá
Hamburg Johannesburg London Madrid Mexico
Milan Montreal New Delhi Panama
Paris São Paulo Singapore
Sydney Tokyo Toronto

Library of Congress Cataloging-in-Publication Data

Mueller, Jerome F.
    Plumbing design and installation details.

    Bibliography: p.
    Includes index.
    1. Plumbing. I. Title.
TH6122.M84   1987   696′.1   86-18584
ISBN 0-07-043963-X

1234567890   HAL/HAL   8932109876

ISBN 0-07-043963-X

The editors for this book were Betty Sun and Nancy Young,
the designer was Naomi Auerbach, and the production supervisor was
Thomas G. Kowalczyk. It was set in Times Roman by Techna Type, Inc.

Printed and bound by Arcata Graphics/Halliday.

Contents

NOV   1994

Chapter 7. Drainage Systems

131

Chapter 8. Sewage Disposal

165

Chapter 9. Fire Protection

175

Chapter 10. Special Items

193

Chapter 11. Schedules and Symbols

233

Preface

Of all the various disciplines involved in building, process, and utility engineering, the plumbing system is the one that is most commonly underemphasized. The design and detailing of a plumbing and sprinkler system is frequently presumed to be a simple, straightforward area of design that requires only a limited amount of expertise. In today's competitive society no area of engineering, simple or complicated, can be taken for granted. The universal standard of all engineering design is that if it is not clearly shown, it is not part of the contract.

The nature of the plumbing design inherently requires depiction of large amounts of piping and its connections to equipment in very confined areas. The result of this confinement is to require enlarged-scale plans and details. Frequently the enlarged-scale plans require a number of sectional elevations to further clarify the already enlarged plans. As a result, such detail presentation, sometimes three-dimensional, sometimes one-dimensional, has become an almost mandatory requirement.

This book presents, frequently in condensed form, the elements of the design of a plumbing system. And, in considerable volume and variety, it presents the related details. There will never be a book that has all the details for any given engineering discipline. The constantly changing, steadily evolving state of the art precludes that possibility. The book does, however, include most of the details currently applicable to the design of a plumbing system.

There are many areas of plumbing which are described in this book as "gray" areas. These are areas of design and detail which are technically not plumbing and yet, when the application is limited in scope, are frequently included in the work of the plumbing designer. These applications should be thoughtfully considered by the designer before any attempt is made to include them in his or her work. Under no circumstances should any design be attempted that is beyond the capability, knowledge, and expertise of the designer. Where these applications are involved in this book, references to other, more definitive texts are noted.

As in all texts of this nature, the author has benefited from the efforts of those who have previously published information on the same subject. The author is particularly indebted to the Building Officials and Code Administrators International, Inc., who are the publishers of the BOCA *Basic and*

National Plumbing Code. This is the code referred to throughout this book, and it is the code standard adopted by states, cities, and towns. In various chapters of this book certain portions of the code are, by permission, repeated verbatim. The manufacturer's data on hot water heating design was reprinted by permission of PVI Industries of Fort Worth, Texas. Much of the information on sprinklers and fire protection came from the National Fire Protection Association publications. The author further benefited from many texts on the subject that go back to the World War II era. These were published by McGraw-Hill and are noted where applicable in the various chapters of the book.

In the final analysis all engineers and designers benefit from the exchange of information among their peers as well as the manufacturers, suppliers, their representatives, and the installing contractors. The author of this book would be the first to agree that the single, best source of information for any new or strange-to-the-designer piece of equipment is the manufacturer of that equipment and/or the manufacturer's field representative. Many of the details in this book began in that fashion. Overtime, experience, input from installing contractors, and data from others was added, and the details, as depicted in this book, slowly evolved.

List of Drawings

Plumbing
Design
and
Installation
Details

ONE

Basic Design Principles

In this chapter we will be dealing with one of the facets of plumbing system detailing which is unique to plumbing design. On the one hand the plumbing system itself may seem to be one of the easier areas of design to grasp. On the other hand the plumbing detail, even when diagrammatic, is usually a continuation of the design itself and is frequently difficult to grasp. The detailing is further complicated by the need to fit the resultant detail in a restricted space. While the space limitation factor is important in all areas of detailing, it is not merely important but also vital in the application of the plumbing detail.

This chapter also brings out the requirement of product knowledge. Product knowledge is important in all areas of design, but again, because of the space constraints of the plumbing equipment and related piping, the plumbing equipment and piping knowledge requirement is much more stringent for the plumbing detail designer.

The difficulties facing the plumbing detailer do not end with merely a knowledge of system design and space requirements. Invariably the constraints of piping connections create demands for locations that commonly are needed by other trades as well as by various segments of the plumbing system itself. The plumbing detailer is not a person of limited knowledge and ability. As is noted at the end of this chapter, the skills involved are considerable.

Understanding Space

The common assumption of every detailer is that the space surrounding the equipment is sufficient for the details that he or she is preparing. While this is a reasonable assumption, there are many occasions in the preparation of plumbing details when this is not a reasonable assumption. Specifically it is frequently not true in the situation created in the plumbing waste and vent piping detail accompanied by the fixture water piping. In this instance, as in many others, we will see as we proceed with this discussion that available space is extremely important. The detailer needs to understand the space that is actually available. The plumbing detailer approaches much more closely to the rank of a designer than does a detailer in any other area of engineering detail design because the preparation of his or her details mandate those that will fit within the existing space.

There are many situations in which the necessity of being able to think in three dimensions in truly vital for the plumbing detail designer. In Figs. 1.1 and 1.2 it is important to note that these are not completed details and cannot be used directly as a detail in your work. These partially completed details serve to illustrate the point we are making about the ability to think of available space.

Figure 1.1 represents typical men's and women's toilet rooms with the usual fixtures found in these rooms. The men's toilet room consists of two water closets, two urinals, two lavatories, and a service sink. The women's toilet room consists of three water closets and three lavatories. This is a very common set of toilet rooms that one would encounter frequently in office buildings.

The lower half of Fig. 1.1 illustrates the waste and vent riser detail. The lines are so close together that it becomes difficult to follow. This lack of clarity creates the need to provide an elevation plan of the waste and vent piping. Please note that our detail is incomplete. Note also the vent piping at the crucial crossover point is 2 inches (in) in diameter. Commonly the line that

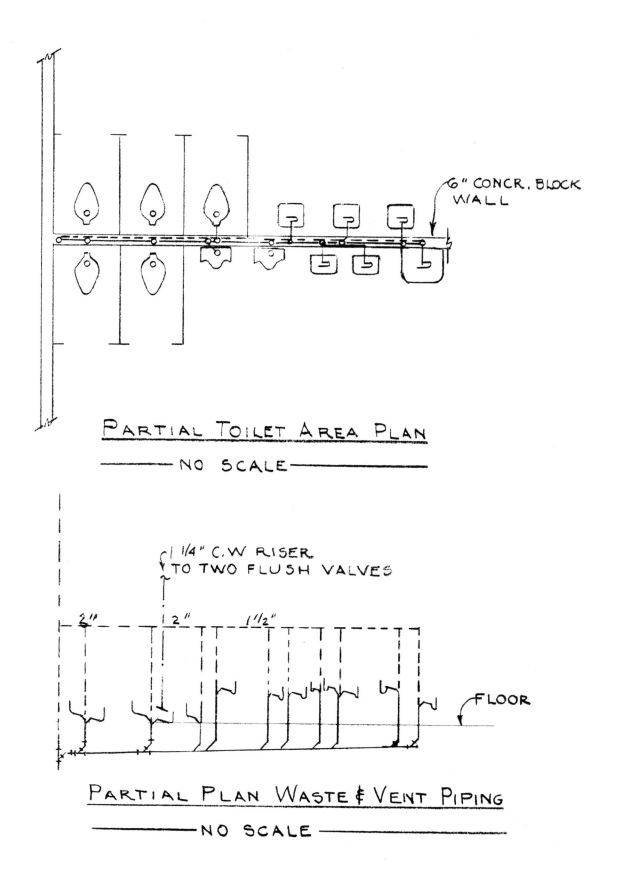

PARTIAL TOILET AREA PLAN

— NO SCALE —

6" CONCR. BLOCK WALL

1 1/4" C.W RISER TO TWO FLUSH VALVES

2"   2"   1 1/2"

FLOOR

PARTIAL PLAN WASTE & VENT PIPING

— NO SCALE —

FIGURE 1-1

must pass by this 2-in vent is a 1¼-in cold water riser line which would drop down to feed two flush valves on opposite sides of the wall. We will agree that a series of single 1-in lines could have been dropped from the ceiling separately into each flush valve. However, it is more economical to drop the 1¼-in line down and tap off of it with two 1-in runouts to feed each of two flush valves.

It may be noted that nothing very unusual has been presented here. But, there is an extraordinarily important point. In Fig. 1.2 we have drawn, to a larger scale, the typical 6-in concrete block wall which was noted on the top portion of Fig. 1.1. This 6-in block wall is actually 5⅝ in thick. It has a series of nearly square holes within the wall that are approximately 3½ by 3½ in. We have drawn in Fig. 1.2 the horizontal vent line superimposed upon a section through the wall along with the 1¼-in water riser. It should be obvious that one cannot chop up the block wall and get the piping in as we have shown

it. The piping simply does not fit. Why does it not fit? It does not fit because the 2-in pipe actually has a 2⅜-in exterior diameter, and the 1¼-in pipe actually has a 1⅜-in exterior diameter. The total space required for the two pipes as they cross each other is slightly more than 3¾ in.

Structural Damage. It might be argued that by judiciously locating the crossing point of the 3¾-in pipe it could be made to fit within the 5⅝-in block space by shaving only a slight amount of the wall thickness on the inside of the block since the space is already 3½ in and we only need ¼ in. This requires the plumbing contractor to perfectly position the 2-in horizontal pipe and the individual 1-in or the combined 1¼-in pipe so that the sum of the two, pressed tightly together as they pass each other, would more or less fit within the block. This is highly impractical and frequently creates a structural weakness. The wall has already been weakened by

NOTES

1. A 6" BLOCK IS NORMALLY 15 5/8" LONG BY 7 5/8" HIGH BY 5 5/8" WIDE

2. A 2" STANDARD WALL STEEL PIPE IS 2 3/8" OUTSIDE DIAMETER

3. A 1 1/4" TYPE L COPPER WATER TUBE IS 1 3/8" OUTSIDE DIAMETER

4. EACH OF THE THREE HOLES IN A STANDARD 6" CONCRETE BLOCK IS 3 3/4" LONG BY 3 1/2" WIDE

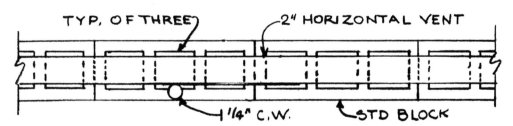

VERTICAL SECTION OF WALL
—— NO SCALE ——

FIGURE 1-2

cutting out a section of the block, which provides structural support for the wall itself, in order to install the horizontal 2-in line. The additional shaving of the block would further weaken the wall and invite cracking.

Most likely neither the architect nor the structural engineer will permit this to occur. If one had casually assumed that a 1¼-in pipe and a 2-in pipe added up to 3¼ in and could easily have been made to fit within the block opening, one would have created a problem and made a serious mistake. It is important that plumbing designers know their products and the exact sizes of them. It is equally important that detailers have similar knowledge. Mistakes such as these occur all too frequently in plumbing details. The question might be raised as to why not make the 2-in line slightly smaller. The answer, of course, is that this size is a code requirement.

Code Requirements. There are many many instances in which the plumbing code plays a very important role in the application of plumbing details. The detailer must know the code since it is highly unlikely that every time one is handed a detail it will exactly correspond to the code requirements. Knowing the code does not mean that one has the code memorized. It means that one has available the applicable code. Usually the code book required is the Building Officials and Code Administrators (BOCA) *Basic Plumbing Code.* By available we mean that the detailer has ready access to this code and preferably has his or her own copy. Merely having it someplace in the office where no one knows exactly where it is when one needs it is not very satisfactory.

Product Knowledge

In addition to knowing the code the detailer as well as the designer must have instant or easy access to applicable manufacturer's literature. This again is extremely important in plumbing detail design since the casual assumption of size of any given item can very easily become a serious design and detail error. Ideally, the detailer's spare time should be used to become familiar with the various and sundry manufactured equipment and materials in office catalogs which are commonly specified by the office in which he or she is employed.

The detailer must also have a knowledge of working clearances, or he or she must be specifically given this information, of many of the plumbing details which are to be located in constricted locations. It is important to realize that one cannot specify a valve, a union, or a fitting of any sort with the presumption that somehow or other it will be installed when in fact the plumber cannot even get a wrench in position to tighten the fittings as detailed. This knowledge and understanding extends even further into at least a simple understanding of structural plans, especially beams and columns and their connections, as well as foundations and footings. The detailer, for example, of the vent piping that we have shown in Figs. 1.1 and 1.2 may very well be con-

fronted with the fact that at each floor level there is a horizontal beam which absolutely precludes the vertical rise of the plumbing pipe. In these cases the detailer cannot casually show vent piping and vent lines continuing up through the building, floor after floor, and on through the roof when in fact one would have to cut apart the connecting beams in order to get this vent piping installed.

Related Trades. Frequently, the plumbing designer in his or her effort to provide carefully detailed plans ends up with an extraordinarily complex installation. While the lengths of piping are not great, the number of fittings may be considerable. In plumbing work the cost of the fitting installation represents a very significant portion of the total plumbing installation cost. In addition to which the probably locations and paths of mechanical and electrical piping to equipment and services play a large role in not only plumbing details but in all related details. Recessed fixtures, or pipes within walls, designed in a position that absolutely precludes the installation of other mechanical or electrical work are a constant related-trade installation problem.

The plumbing detailer cannot resolve every problem in the entire building. However, it is a very great advantage to the firm to have a detailer who recognizes when there is an incipient problem. If he or she cannot solve the problem, at least the problem can be pointed out and a solution requested. As we go into the problem of details in this book, we will invariably come into the design area of the plumbing system since the two are so intimately related. In the toilet room that we depicted in Fig. 1.1 there is a problem of water hammer and the requirement of the location of the water hammer arrestor is not shown in the plans. There could easily have been a roof drain over the toilet room with a diameter of some 4 in that would be scheduled to come down through the same wall in which our vent piping is located. This is a problem of chase requirements for both waste and vent piping as well as for storm water piping.

In Fig. 1.1 you will note that we depicted only the crucial sizes. This approach of locating, early in the detailing effort, the crucial sizes where one pipe passes another is extremely important. This directly relates to a knowledge of exactly how big a pipe actually is as well as physically which way the pipe will be required to run.

Even if the detailer does not know how to size and design the entire plumbing system, he or she should know how to size the vent piping details. Without the proper sizes on the detail, the information depicted is of little practical value.

Pipe Sizing. This emphasis on sizing brings up another serious point in plumbing design details. Plumbing detailing takes in many situations that require a higher order of knowledge than does any other type of detailing because the detailer really has to know how to calculate the hot water, cold water, waste, and vent piping fixture loads and how to apply these fixture loads to tables in the various design manuals and in the applicable codes in order to select actual size of the line involved. The 2-in vent line shown in Fig. 1.1 came from the code, and the 1¼-in water line shown in Fig. 1.1 came from a standard table on sizing of hot and cold water piping. Neither is difficult to comprehend and both are vital in order to solve an incipient problem which we illustrated in Figs. 1.1 and 1.2.

Special Requirements

In addition to the ability to calculate the size of the pipe shown on the detail, the plumbing detailer must recognize and understand the special requirements of unusual fixtures such as those contained in hospitals, factories, and other areas wherein the application is nonstandard. Ideally, the detailer should be able to go to the manufacturer's catalog and suggest a different type of fixture that will perform a similar function whose space requirements may more readily fit the space condition available. We are aware that we are asking for a great deal of knowledge and experience on the part of the detailer. We are also hoping that in situations in which the detailer is a newcomer the supervisors of the detailer spend sufficient time with the detailer to acquaint the detailer with all of the facts that we have presented. The firm should conduct brief programs in the office on these subjects so that the detailer can get a better understanding of job requirements. Far too many details state "comply with the codes" or "comply with all of the design criteria for the various pipes involved." Unfortunately these details frequently cannot be installed as they are drawn. And what makes the problem even more difficult is the fact that in many cases the detail literally must be installed as drawn or as nearly drawn. The tragedy of this situation is that no one pointed out this problem to the building designers and no one in the building design group picked up the problem of the lack of space for the installation of the work designed. Certainly the detailer cannot be all things to all people nor can the detailer be expected to know everything instantly.

Design Responsibility

The detailer, the supervisor, the head of the consulting firm, or someone on the building design team side must be able to come up with alternative solutions to the problem when the required solution simply cannot be made to fit in the space allocated. One of the important points, which is not really the detailer's responsibility but rather that of his or her supervisor, is the ability to know and understand how fittings are installed, the space that they take, and the ways to reduce the quantity and complexity of the various and sundry fittings that are required to install the detail as depicted. While we are on the subject of fittings, there is another area in which

the detailer with experience may be of considerable help. Certainly the firm's supervisor should note the connections, fittings, and miscellaneous parts that are required to make the plumbing fixture completely operable. One of the classic mistakes in the supplying of special sinks, such as stainless steel sinks, occurs when the fixture noted in the specification arrives on the job with only the hot and cold water faucets (sometimes not even those items) and no trap. The plumbing detail shows a trap and the question becomes, "Who supplies the trap?"

This brings us to the final part of the detailer's problem which of course also is the supervisor's problem and the problem of the firm. The final completed design consists of a plan, its details, and an accompanying specification. Frequently, the firm presumes that when in doubt, it should be covered in the specification. They try to make the specification cover everything by the use of words such as "as shown/are required." If the firm can get away with such phraseology, the previously noted missing trap is solved by the words "and/or required." Frequently, this is not acceptable to the installing contractor. The contractor simply will not supply the parts that are needed to do the job in any quantity merely because you are saying, "If I forgot anything, you install it anyhow." This sort of premise normally will not be accepted by the installing contractor. It is entirely possible that if the sink that we mentioned is the one and only sink on the job that does not have a trap, the contractor will not argue over one missing trap. However, if the sink is actually 20 or 30 sinks with missing traps, you can rest assured that missing traps will show up on the job only when the contractor is paid an additional sum.

Qualifications

We have made a great many points in this opening chapter of our book. It easily might appear that we have described almost a miracle worker instead of the poor beginning detailer. Even the detailer with 3 or 4 years experience may have some difficulty in meeting all of the criteria we have listed here. For the benefit of the detailer and the owner of the firm, we will list the ideal qualifications of the detailer. It should be understood that no one can come into the office in a starting position and know all of the things we are listing. However, the list represents a reasonable expectation of what the detailer's knowledge, over time, should become. And perhaps it is a way for the owner of the business to judge the competency of any detailer in his or her employ. The basic ideal qualifications of the plumbing design detailer are as follows:

1. Understands architectural and structural building plans
2. Is able to think in three dimensions
3. Is familiar with plumbing products
4. Knows and understands applicable building codes

5. Understands working clearances and installation requirements

6. Can propose alternative piping arrangements

7. Recognizes incipient piping and installation problems

8. Knows how to size all types of piping related to the detail

9. Recognizes special requirements of special fixtures in institutional and process applications

10. Can offer suggestions on alternative fixtures

11. Understands fittings and their installation space requirements

12. Can relate and understand applicable specifications and point out shortcomings

There is no doubt that a joking remark could be made at this point that if the detailer knows all of this, we ought to make him or her a partner. That is a little excessive, but plumbing detail design is far more difficult than any other form of detail design. The ideal detailer is one who came from some other section of the firm, preferably doing heating or air conditioning piping details. As such, the general approach and the plumbing detailing is not as serious a problem, and bringing this detailer along to do acceptable plumbing details is relatively easy. An alternative probable ideal candidate could also be one who did not know anything but was an avid and bright student of details with an ability to learn rapidly and to reason logically.

TWO

Materials and Methods

Our primary concern in designing the plumbing detail is the connections to specified equipment. Thus when we speak of materials, we are primarily describing piping materials. In order to select a reasonable source of data of piping materials we have used as our basic frame of reference the Building Officials and Code Administrators' (BOCA) code as the basic determinant for materials in any given area of the United States. There may be some further local restrictions or other materials that maybe allowed or be precluded. We suggest that you check these local requirements first before assuming that what we are saying is totally applicable to your local situation.

The following is an abbreviated summary of pipe materials by classification as contained in the BOCA code:

Water service pipes: asbestos cement, brass, cast iron, copper, plastic, galvanized iron, steel

Water distribution pipe: brass, copper, plastic, galvanized steel

Above ground drainage and vent pipe: brass, cast iron, type K, L, M, or DWV copper, galvanized steel, plastic

Underground building drainage pipe: plastic, cast iron, copper

Building sewer pipe: asbestos cement, bituminized fiber, cast iron, type K or L copper, concrete, plastic, vitrified clay

Building storm sewer pipe: asbestos cement, bituminized fiber, cast iron, concrete, type K, L, M, or DWV copper, vitrified tile

Subsoil drain pipe: asbestos cement, bituminized fiber, cast iron, plastic, styrene rubber, vitrified clay

In all of these descriptions of pipe materials, note that wherever we used the term "plastic pipe," we strongly recommend that you consult your local code and the BOCA code requirements for the exact types of pipe. We did not list all the possible types of plastic pipe.

Corrosion

Corrosion is a very complex subject of a highly specialized nature and is not of pressing concern to the detail designer. There are, however, a number of types of corrosion which occur in plumbing piping systems and for which the designer should be aware of in imposing his or her details. In essence corrosion is an electrochemical reaction. By this we mean that both chemical and electrical changes are involved. Among the various types of corrosion are the following:

1. *Concentration cell corrosion.* The electrical effect caused by differences in composition of a solution in contact with a single metal or alloy at different points promotes an accelerated form of corrosion known as concentration cell corrosion.

2. *Dezincification corrosion.* This form of corrosion occurs principally with alloys of copper and zinc in which the zinc content is above 15 percent. In this form of corrosion the zinc is chemically leached out of the copper-zinc compound.

3. *Graphite type corrosion.* This form of corrosion is commonly related to cast iron and resembles zincification. In many cases of contact with corrosive materials the iron in gray cast iron may be leached out, leaving behind a porous layer of graphite over the remaining unattacked iron.

4. *Stress—accelerated corrosion.* This form of corrosion is limited to a number of alloys of which the yellow brass alloys are more susceptible. Under high stress some of these alloys may have a quantity of metal dissolve and create a considerable mechanical weakening.

5. *Galvanic corrosion.* In the case of dissimilar metals, or in the case of various sources creating electrochemical rather than electrical differences, you can create a condition of electro-chemical corrosion.

6. *Pitting and local corrosion.* These are usually caused by breakdown in the passive films which a metal or alloy may build up for itself in corroding solution. Or, as a result of an accelerated corrosion, it may have a surface blemish or nonmetallic inclusion.

Corrosion Protection

Of the more common materials used in plumbing piping, cast-iron pipe is subject to both self-corrosion and galvanic corrosion. The usual pipe protection is tar coating on the outside and asphalt or cement lining on the inside. The most common form of rusting is called tuberculation. Wrought-iron pipe and steel pipe usually rust by tuberculation. Lead and lead-lined pipe are quite low in the electrical-chemical series and resist corrosion very well. Tin and tin-lined pipe serve as good corrosion resisting materials since tin has excellent corrosion resisting properties. Pure tin is commonly used as a lining for other types of pipe. Brass and copper pipe are particularly suspectible to galvanic corrosion when the distribution of zinc and copper crystals is not even. Brass pipe used for water distribution should contain at least two-thirds copper. Galvanic currents may also be set up when fittings are made up of a different composition from the pipe, such as bronze fittings with brass pipe and particularly brass fittings with iron pipe. This is why you always need a special isolating connection when connecting brass to iron. Of all of the pipe materials the plastic pipes of the proper types are the only ones that are simultaneously light, flexible, fairly tough, and exceptionally corrosion resistant.

Materials

So far as the materials for the plumbing system go, the selection undoubtedly has already been made before the detailer becomes involved. The piping in the detail is most commonly an extension of the material selection made for the systems that are supplying the various equipment items in the plumbing contract. However, there should be an understanding by both the detailer and the system designer of the types of materials used and how and why they are selected. The proper type of material for the given application has a great deal to do with the proper functioning of not only the system but of the detail and its connections.

Selection of water pipe materials varies in different localities of the United States. Commonly water pipe is either steel or hard drawn copper tubing. Thus, if one has a hard drawn copper tubing water supply system and one is connecting this copper tubing to a item that is of steel or iron, quite obviously one needs an isolating fitting which should be shown on the connection detail of these two dissimilar metals. Again it is quite common to find waste piping made of cast iron or extra heavy cast iron with either caulked or mechanical joints or sometimes hubless joints. As can be seen from our previous list of piping materials, this piping could have been constructed of other materials.

There is, however, another element that applies to all plumbing and piping. Frequently, in the installation of plumbing piping the work is done in a situation that requires a certain structural strength and rigidity. An example is waste and vent piping that is installed before the walls are built for a given toilet room. Hence the basic problem is not only that the piping must be installed correctly as proposed and sized correctly as required by the details but also that it has sufficient rigidity to stand alone since initially there will be no wall support. Further, there are no vertical hanger connections available. The waste and vent system is, effectively, self-supporting.

When that situation arises, the pipe material and the joints need to be rigid, and the system needs to be very firm and very solid in every aspect in order to stand alone and to take a certain amount of minor abuse during the total building construction process. The details or the specifications must clearly spell out that the connection are of such a type as will maintain a continuous rigid self-supporting structure. Generally, plumbing details comprise systems and types of piping which are for the most part installed above ground. Waste piping and water piping, to a smaller degree, may be installed below grade. The waste system, except for the main horizontal drain lines, and the main storm water lines are normally installed above grade and must be supported in a proper manner both horizontally or vertically. In effect, they are hung, braced, attached, fixed, connected in some fashion, and held in place by their rigidity since, at the moment of installation, they are not otherwise supported.

Finally, we arrive at the insulation of the piping. If, for example, your detail is simply one of a roof drain and the connection to it, your specification most likely will describe the insulation of its horizontal portion, especially if it is installed in a ceiling plenum where the temperature and humidity may cause condensation and dripping. Hot and cold water piping are normally insulated. As detailers we are not overly concerned with the type of pipe insulation but merely with the fact that the pipe is insulated and that the insulation requires additional space.

Joints. While we are not primarily concerned in our detail with type of joint made, we must take into account that the joint itself has a larger dimension that may be a problem. Typically a flange joint has a much larger diameter than the pipe itself. As noted in Chap. 1, we must be certain that our details can be successfully installed. Finally, the type of joint selected directly relates to rigidity. Frequently, as in the case of freestanding waste and vent piping, the joint connections will be either screwed or caulked, depending on the type of material, to ensure this structural rigidity. We either have to be certain that this is being specified or that we are noting it on our detail.

Caulking a cast-iron joint is not the only way to make a cast-iron joint connection. Certainly the mechanical joint and other joints have many areas of application. Whether or not the type of mechanical joint that is being specified is applicable in all instances is a subject that must be addressed by the specification or by the detail. Joints between plastic, copper, or screwed pipes are inherently rigid.

Finally, in our piping details there are two other subjects that are of considerable importance to our detailing. One of these is expansion and expansion joints, which should be carefully researched by the designer of the system. Another is the common pump detail. Here, there may or may not be a flexible connection or a vibration mounting of the base.

Hangers and Supports

The hanger or support for various types of plumbing fixtures and piping is frequently overlooked and sometimes casually treated. Common in specifications and on plans are words that say the piping shall be supported in a good and substantial manner in accordance with all local codes and ordinances as well as with the national plumbing code. This statement is very loose and misleading. The national plumbing code coverage of hangers and supports is on pages 103 and 104 of the 1981 *National Basic Plumbing Code* for example. On those two pages about the only major item covered is the horizontal and vertical spacing of hangers. While spacing is an important item, it is not a description of the type of hanger required.

Underground piping, for example, may be supported by earth when this is sufficiently firm or by concrete or masonry supports spaced as per the national plumbing code. For use above ground and in building piping there are a number of types of pipe hangers, some of which we illustrate in this portion of the book.

Pipe Hangers. We begin with a simple extended pipe clamp which is used for suspending piping runs when the exact distance from the structure to the pipe cannot be determined until the piping is in place (see Fig. 2.1). It is also used extensively in the marine field where the extended legs shown on the hanger can be welded to the structure. The clamp obviously is designed to tightly grip the pipe.

A second of our simplified hanger illustrations, shown in Fig. 2.1, is what is known as an adjustable split ring swivel hanger. This is primarily used for the suspension of noninsulated pipe lines. The hangers may be installed prior to the erection of the piping. The offset hinge provides enough support for piping during erection and allows removal of the pipe for further fabrication after the hanger is in place. This is frequently a very handy hanger to specify, particularly in difficult locations.

In a great many wood frame installations, especially those commonly seen in residential work and in some commercial work usually of a one- or two-story nature, the simple short clip is commonly used. We have provided a brief illustration of it in Fig. 2.1. Along with this simple short clip we have shown a different sort of pipe clamp. This is an offset type of pipe clamp to be used when you cannot rest the pipe on the surface (e.g., if you are running piping on a roof or on some flat surface). This sort of offset clamp will suppport pipe horizontally or vertically at a constant distance from the surface and provide a very firm and safe type of installation.

Pipe hooks. Figure 2.2 is another simple pipe hook that can come in very handy when pipes are run close to walls and hangers are difficult to install. This pipe hanger can be fastened from the wall and provides for a minimal 1-inch (in) clearance in all sizes.

Channel supports. The use of plastic pipe has created the need for another type of hanger which is normally supplied with a special channel and used to support flexible plastic pipe or rubber hose. The channel is the primary contact surface between the pipe and the hanger. The pipe is laid directly into the channel and wherever necessary the channel can be cut to allow for fittings and couplings, as can be seen in Fig. 2.2. The channel may extend from one hanger to the next since the steel channel is 8 feet (8 ft) long and is fairly rigid. Primarily this type of channel is used where a pipe must be supported continuously throuthout its length.

Clevis hangers. Frequently, designers and specification writers will refer to the fact that all pipes shall be supported from clevis hangers equal to figure so and so as made by such and such a pipe hanger company. There is nothing wrong with this statement except that the clevis hanger does change in size and thickness depending on the type of piping that is being supported. The designer is cautioned to recall that the difference in weight between a copper pipe and a cast-iron pipe is very considerable. The specification should reflect this fact. In addition there are clevis hangers with an extended U-shape spacing for piping where the hanger is expected to attach to the pipe itself and the insulation is expected to butt against the hanger. This requirement should be carefully noted by both detailer and specifier.

There are two normal ways to support the insulated pipe. One is to support the pipe directly and let the insulation butt up against the hanger. This is normal use

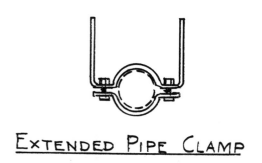

EXTENDED PIPE CLAMP

ADJUSTABLE SPLIT RING
SWIVEL HANGER

SHORT CLIP

OFFSET PIPE CLAMP

PIPE SUPPORTS
SHEET 1 OF 2

FIGURE 2-1

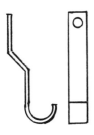

"J" HOOK

RISER CLAMP

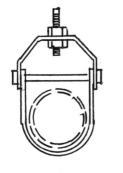

CLEVIS HANGER

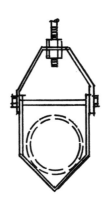

PLASTIC PIPE HANGER

PLASTIC PIPE CHANNEL

PIPE SUPPORTS

SHEET 2 OF 2

FIGURE 2-2

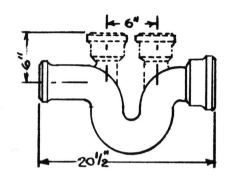

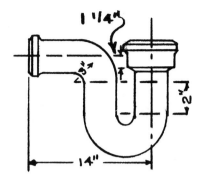

4×4 RUNNING TRAP WITH OR WITHOUT SINGLE OR DOUBLE VENT OR CLEANOUT

— NOT TO SCALE —

4×4 DEEP SEAL P TRAP

— NOT TO SCALE —

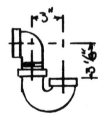

2" P TRAP

NOT TO SCALE

1½" BATH P TRAP

NOT TO SCALE

1½" ADJUSTABLE P TRAP

— NOT TO SCALE —

CAST IRON TRAPS

FIGURE 2-3

of the extended clevis hanger. The other way is to allow the insulated pipe to pass right through the hanger. When this is done, an additional supporting section of U-shaped galvanized pipe should be specified to be inserted inside the hanger to extend the contact surface area between the hanger and the insulating material and to provide rigidity at the point of hanging. This sort of U-shaped trough is used also for plastic pipe. The trough supports the insulated pipe for a length of approximately 8 to 12 in (longer for plastic) and is shown in Fig. 2.2.

Riser clamps. The simple pipe clamp can also serve as a normal or a heavy-duty riser clamp. The clamp usually locks around the pipe and extends beyond the cored or framed riser opening and rests on the floor to provide the support for the vertical pipe. The clamp, as illustrated in Fig. 2.2, is screwed and locked tight to the pipe to maintain the pipe in a fixed vertical position.

Traps and Cleanouts

In the plumbing piping system the engineer frequently specifies various types of traps and cleanouts. Commonly and frequently in every waste and vent piping detail a variety of traps and cleanouts may be shown or described. Regardless of the material, a trap described by any specific type such as P trap and the like is the same in terms of its physical shape. However, a great many engineers and especially a great many detailers are not aware of the physical space that these traps actually take.

Cast-Iron Traps. In order to illustrate the space problem we have deliberately chosen to illustrate in Fig. 2.3 five of the common traps, using cast-iron bell and spigot piping in the case of the larger traps and cast-iron screw piping in the case of the smaller traps, with a dual purpose in mind. The traps we are depicting in Fig. 2.3 clearly represent the types of traps that are common to most requirements. By using the cast-iron bell and spigot joint connection, we are depicting the trap that takes the most space. You may rest assured that any trap that you pick will not take more space for a given size than the traps we are showing.

In Fig. 2.3 we begin by showing a running trap commonly used in connection with a vent on a line leaving the building. We want to point out that this is not something unusual in terms of a trap but that it takes, in a 4-in size, a length of 20½ in to install. In the same figure we have depicted a deep seal trap. The difference between a standard trap and a deep seal trap is the height of the water seal depth in the trap. In the 4-in trap the water seal height instead of being the usual ½ in is noted as 2 in. The lower half of the figure depicts three common cast-iron screw type traps. The 2-in P trap might be commonly used in a bathtub installation. We have also shown for a single bathtub the 1½-in cast-iron screw trap and illustrated just how much space it takes. Finally, we have shown the common 1½-in adjustable P trap of

screw cast iron which frequently finds use in certain special equipment and kitchen installations.

Fittings. Frequently details do not completely or accurately represent the pipe fittings which are the means of connecting one section of piping to another or one piece of equipment to its piping. While not totally necessary in every instance, the depiction of the fitting should be accompanied by the understanding of the space required for its installation.

For the plumbing detailer the area of most extreme space constriction is usually the chase between back-to-back toilet rooms serving office buildings or similar situations. In addition, within that given area the waste piping installation is usually the most constricted of any part of the system. This is where the all-copper or all-plastic installation has somewhat of an advantage. The problem did not go away merely because these materials were selected, but it became, because of the nature of the connections of these materials, a little easier to install.

Certain types of combined fittings are commonly used in the toilet fixture installation of waste piping. Figure 2.4 illustrates some of the more common fittings used. Again, as in our trap installation discussion, we have picked cast-iron bell and spigot piping which normally uses caulked joints as the material to be detailed. Obviously cast-iron piping comes in a variety of sizes and the dimensions as shown would vary with the size. Since the most commonly used sizes are 3 in and 4 in, we have shown the 4-in size.

Our purpose here is the same as it was in our trap discussion. These fittings require the exact space dimensioned. It does not get any smaller if you use this type of material. If your space will not allow for the dimensions shown, you are going to have to conduct a literature search to find some other material that takes less space or the space you have must be enlarged.

Luck, fortunately, is frequently on the side of the designer and detailer. The space turns out to be adequate although sometimes just barely. Perhaps the most logical use of this detail is to check its dimensions against the space allocated for your waste and vent piping before starting your detail. You can quickly ascertain an approximately accurate opinion on the space available and required with this simple step in your detail design procedure.

Plastic Pipe

To summarize we would like to call your attention to certain areas in the BOCA basic plumbing code which have become the standard for most plumbing codes in the United States. As noted in our previous discussion, plumbing materials have in the past primarily been steel and copper. However, there has been a growth in the use of the plastic piping, and the manufacturers of plastic piping have vastly improved the quality, type, and specific use of plastic pipe. Since this subject of plastic piping is rather new, we would like to devote a small amount of space to a discussion of plastic piping.

There are a number of advantages to plastic pipe and seemingly everyone feels that it is the ultimate in modern technology insofar as plumbing piping is concerned. This is not totally true, but it is sufficiently true for the plastic pipe manufacturers to have considerably enlarged their

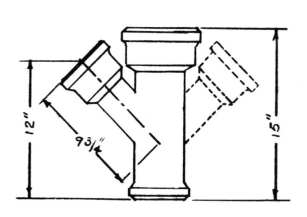

4" WYE, SINGLE OR DOUBLE
— NOT TO SCALE —

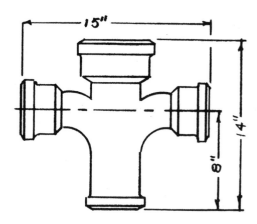

4" SANITARY T, DOUBLE
— NOT TO SCALE —

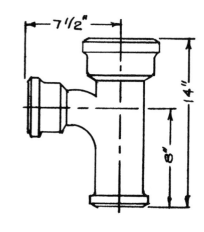

4" SANITARY T
— NOT TO SCALE —

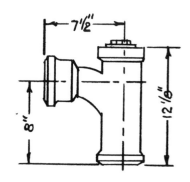

4" SANITARY T
CLEANOUT PLUG ON MAIN
— NOT TO SCALE —

CAST IRON PIPE FITTINGS

FIGURE 2-4

share of the plumbing and piping market. Primarily, the advantage of plastic piping is the relative ease of making joint connections and the fact that the plastic surface is immune to a great many types of corrosive liquids which normally cause damage to copper, steel, and cast-iron pipe. The newer, more rigid types of plastic piping have become an alternative way to install vent and waste piping when, in construction, the vent and waste piping in a typical situation is installed first and is freestanding.

The BOCA *Basic Plumbing Code* permits the use of polyethylene and polyvinyl chloride plastic pipe for water service pipe as well as for water distribution pipe. The code also permits both styrene plastic pipe and polyvinyl chloride pipe for above ground drainage pipe. Below ground drainage and vent pipe may be either styrene or polyvinyl chloride. An advantage of plastic piping is that it is immune to galvanic underground corrosion that is created by stray electrical currents which set up a positive-to-negative field around a steel, iron, or copper pipe and cause rapid deterioration.

With the advantages of plastic piping the question naturally is, "What are the disadvantages?" The first hint of a possible disadvantage of the plastic pipe is contained in the plumbing code which very carefully describes the distance between hangers for various types of pipes. For most of the ordinary cast-iron or steel pipe, hanger space varies between 5 and 10 ft depending on type size. For plastic pipe the hanger spacing varies between 4 ft and 2 ft 8 in. Thus there are more hangers required for the installation of plastic piping.

However, hanger spacing is not the major area of concern to code officials and fire protection officials who are both becoming sensitive to the fire characteristics of all kinds of plastic materials used in modern building construction. Basically, all plastics used in plastic piping will burn including the flame retardant materials and those which cease to burn when the initiating flame is removed. And, when they burn, all plastic materials release toxic and nontoxic gases. The nontoxic gases are carbon dioxide and water vapor. The toxic gases are carbon monoxide, hydrogen chloride, and hydrogen cyanide. In addition, most burning plastics release large quantities of dense smoke. Other chemical products created by the burning of plastics include small quantities of aldehydes and ketones as well as other minor chemical compounds. Of all of these, the most dangerous has been proven to be carbon monoxide which represents by far the majority of the fire load. In fairness to the plastic pipe industry the fact remains that the greatest danger from plastic materials does not come from the relatively small quantity of plastic piping but rather from the large quantity of plastic materials contained in the spaces in which the fire occurs. Thus, the real danger is primarily from the room contents and not the piping system.

Valves

The word "valve" has its origin in the Latin word "valvae" which means folding doors. Valves have been used since at least the year 2700 B.C. Some that have been excavated and dated were made of leather and wood. In modern detailing the valves, by type, are usually described in the project specifications. But it is the detailer who is, or should be, noting what type goes where in each detail being designed. There are four basic areas of valve classification applicable to all details.

The first area is on-off service. This surely fits the Latin origin of the folding door description. On-off valves are gate valves which get their name from the gate-like closing member called the wedge or disk. They are designed for unobstructed flow when open and no flow when closed. They are not suitable for throttling flow because of the very large amount of flow change that occurs at high velocity when the valve is almost closed. If they are barely opened to provide some form of throttling control, both the disk and seat of the valve are subject to severe wire drawing and erosion that eventually will preclude tight shutoff of flow.

The second area is throttling service. This requirement of flow control, or throttling, is provided by a globe valve. The valve is designed for bubble-tight shutoff and for intermediate control of flow between the full-open and full-closed positions. The globe valve should be used when there is a rapid cycle of operations and when partial as well as full flow is required. The globe valve is not a straight-through flow valve as is the gate valve. Its flow pattern creates turbulence as it is closed, and it is more costly than a gate valve.

The third area is the prevention of backflow. Check valves are designed to "check" or preclude flow reversal. They are completely automatic in their operation and are important health and safety devices. There are several types of check valves to consider. The swing check valve has a swinging disk or clapper which is installed in the horizontal position and swings open in the direction of flow. The lift check has a disk that lifts in the direction of flow and has a body similar to a globe valve.

The fourth area is the special purpose valve. Among the special purpose valves is the plug valve which is a modern version of the simple cock valve. A quarter turn fully opens or closes the valve. Like the gate valve it is used for on-off service and is commonly seen on gas lines. The ball valve is another version of the gate valve. The ball usually sits on a plastic or rubber seat for positive closure. The butterfly valve got its name from the hand-operated damper in the old-time coal stove smoke pipe. The modern butterfly valve, when properly specified and selected, can be used as a combination gate and globe valve and can provide both throttling and on-off functions.

THREE

Water Supply Systems

One of the major engineering endeavors of mankind has been the construction of a central water supply system. This sort of public works program has many examples of which the ancient Roman aqueduct system, still partially in use, is particularly well known. But the supplying of water to a building is a fairly new situation. The best known early examples are buildings in New York in 1842 which were the first ones to be supplied with water under pressure from a central source.

Our concentration in this chapter is on sources of both hot and cold water and how they should be detailed. Water supply to a building begins our detailing. Commonly the source of water is a public or private utility company. Thus the first plumbing details would rightly be concerned with the proper service entrance and metering details for this source of water.

However, there is another large and fairly complex area of building water supply. This source is a well of some type. Both the civil engineer and the mechanical engineer in the plumbing field feel that this well and its related pumping and storage systems are a part of their work. In our judgment we tend to concur with the mechanical engineer's opinion. There are always going to be large water and fire protection requirements for very large buildings that will require special expertise. Usually certain civil engineering firms who commonly design municipal waterworks systems are the ones who should be retained when such a large scale design requirement occurs. In many cases the system design is readily within the area of the mechanical engineer's expertise.

Water supply systems also cover systems within systems. For most buildings over 12 stories high the water supply system requires a booster pump and a storage system that takes utility- or well-supplied water and raises its pressure sufficiently to serve all floors of the building regardless of height.

Hot water supply is very definitely a separate system of varying complexity, application, and use. There are a number of different ways to heat hot water, and there are also applications that are similar but vary to some degree depending on the type of fuel used as the water-heating energy source. This chapter will cover these variations also.

Well Investigation

In the detailing of the plumbing water supply system the firm and its engineers, designers, and detailers usually begin with the appropriate details related to the incoming supply of water. This water source is frequently, but not always, a public or private utility company. Here we will deal with the other source—the driven well.

There is only one reliable way to find out what quality and quantity of water is available. Engage a competent experienced well driller who normally operates at a base set-up cost plus a charge per foot of test well drilled! This decision should be based on some elementary logic. First, it is unlikely that water is going to be found in an area in which none has ever been found before. Second, if most wells in the area of the planned test are only successful at a depth of 900 feet (ft), good judgment would require testing to that depth unless an unusually high and unusually large quantity is found at a lesser depth.

Most experienced reliable well drillers are exactly that—experienced and reliable. Trying to second guess their opinions is not a wise choice. You simply may have to face the fact that there is no water in the area drilled. A second test, sometimes a third, may be required at alternative locations on the same site. On occasion there may be no water available within the site boundaries of your project and either some off-site arrangement must be made for the water supply or the project will have to be abandoned.

Not many years ago farmland strategically located next to newly built interstate limited-access highways was considered ideal commercial and industrial as well as residential subdivision real estate property. It was better if the property was at, or near, the junction of two separate interstate highways. Wells were dug and perfunctorily tested, and all sorts of structures were erected on these sites. Today, some of this lovely farmland, now changed into building lots away from noise and pollution, has been found to have a water pollution problem far greater than any imagined. Certain types of pesticides and herbicides once thought to be rendered harmless by earth filtration have been discovered to be actually life threatening. Our book on plumbing details is not an environmental treatise. If your firm plans to start the design of a project utilizing well water, we feel you must not only have the water from your new well tested by a qualified testing laboratory but you must also thoroughly check the overall area surrounding your project to be certain that no problem exists. An area problem can very easily become your problem.

There are certain criteria in judging the output of a well. These are usually expressed in the language of the well driller. For example, the capacity of the well is the discharge per foot of drawdown at the well. For the yield-test rate determination, the potential in the well is presumed to be the same from the water surface to the bottom. The maximum safe yield of a well is limited by the capacity of the underground source to supply water without constantly lowering the water level in the well.

In basic terms the extraction of water from the well lowers the water's elevation in the well. Too rapid pumping will cause the well to run dry. Thus your pump selection must be related to the maximum safe yield. At this tested gallons-per-minute output the water level will drop a certain fixed value which is normal for all wells and then it will feed water in as fast as it is being pumped out.

Wells

Depending on the construction method, wells are classified as dug, driven, drilled, or jetted. The choice depends on project requirements, water table elevation, and other factors. Dug wells are usually shallow wells constructed in a similar manner to that of any open shaft. Driven wells are constructed by driving into the ground a small pipe, usually having a diameter of 2½ inches (in) or less and a well point at the tip. In essential detail a jetted well is constructed by the washing action of a pressurized stream of water in a jetting pipe that is lowered into the ground. Generally a drilled well will serve the plumbing detailer's project.

In the usual course of events the plumbing detailer's work begins after the well has been drilled, tested, and found acceptable as a water source. Once the quality and quantity of well water has been determined, there are a number of decisions that must be made. Primarily these relate to the type of pump to be applied to the well and the type and size of storage and pressure tanks. If final pressure at the tank source is a problem, a booster pump system may also be part of this overall design decision.

While wells are really not part of plumbing detailing, some firms prefer well details to be part of their details. In ensuing details we will present the typical well and pump detail. This commonly involves a single-cased well. However, there are double-cased wells and wells in rock which may end up with the same pumping arrangement as for the single-cased well. The wells themselves are depicted differently. For this reason in Fig. 3.1 we present the double-cased gravel wall well and the typical well in a rock stratum situation.

Some experts prefer the description of a gravel wall well to be properly entitled a gravel packed well. Regardless of what it is called, this type of well is a common and popular way of obtaining subsurface water from an unconfined source. The well has an inside and an outside casing. The larger outside casing has a diameter of 10 to 12 in and is first sunk to the full depth of the well. The inner casing, with a 4- to 6-in diameter, is placed inside the outer casing, and the space between the two casings is filled with gravel. The size of the gravel pebbles depends on the material in the subsoil. As the depth of gravel increases, the outer casing is pulled up to provide a gravel surface between the subsurface material and the gravel. We show grout for the final few feet and a pipe to permit the addition of gravel created by settlement problems. The well screen is positioned in the inner casing, and the inner casing is withdrawn to expose the screen.

In rock situations the double-wall installation may have a separate, different variation. The drilled well may be in two diameters. The outside, larger casing may be drilled as we show in the right side of the detail to a certain depth and the final, deeper hole dug by a smaller drill. Thus our inside casing is some 5 ft 0 in deeper than the outside casing. Where we normally had our perforated screen, we now have simply an unlined rock wall into the bearing strata. Since the rise of water is from a vein in the rock, there really is no need for a screen, and the drilled rock is a permanent fixed passage for the vein of water.

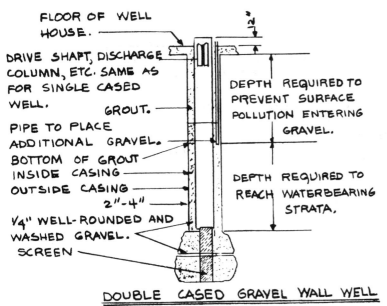

DOUBLE CASED GRAVEL WALL WELL

NOTE:
FOR USE WHERE WATER-BEARING STRATUM
IS OF UNIFORM FINE SAND WHICH IS SUBJECT
TO FLOWING AT THE VELOCITY THE WATER
ENTERS THE SCREEN. THE GRAVEL PROVIDES
A MUCH LARGER AREA THAN THE SCREEN AND
THE VELOCITY OF WATER AT THE OUTER LINE
OF GRAVEL MAY BE REDUCED TO SUCH VELOCITY
THAT THE SAND WILL NOT FLOW.

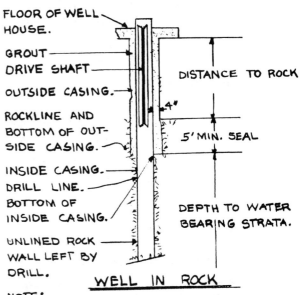

WELL IN ROCK

NOTE:
WELL HOUSE, PUMPING LEVEL, SCREEN AND
OTHER DETAILS ARE SAME AS FOR SINGLE
CASED WELL.

GENERAL NOTES:

1. WELL SITE SHOULD BE 200 TO 500 FEET FROM POSSIBLE SOURCE OF POLLUTION.
   LOCATE AT ELEVATED POINT IF POSSIBLE TO PREVENT FLOODING. FENCE IN
   WELL SITE.

2. WELL BUILDING SHOULD BE FIREPROOF, VENTILATED AND IN COLD CLIMATES,
   INSULATE THOROUGHLY AND PROVIDE HEAT. ELEVATE ABOVE GRADE TO PROVIDE
   DRAINAGE AWAY FROM BUILDING.

3. PUMPING LEVEL IS DETERMINED BY CONTINUOUS FLOW TEST AT REQUIRED
   WELL CAPACITY, AND THE MEASUREMENT OF THE RESULTING DRAWDOWN
   BELOW THE STATIC WATER TABLE.

4. THE TOP OF THE SCREEN SHOULD BE 50 FEET MINIMUM BELOW GRADE
   TO AVOID SURFACE POLLUTION UNLESS UNUSUALLY IMPERVIOUS EARTH
   (10 FEET OF COMPACT CLAY) OCCURS AT THE SURFACE, OR WELL IS
   FAR REMOVED FROM POSSIBLE SOURCES OF POLLUTION.

FIGURE 3-1

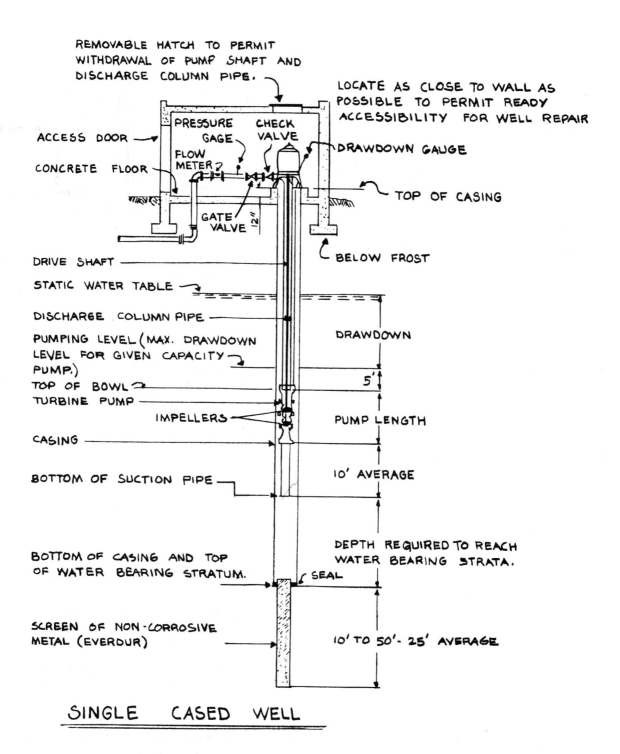

REMOVABLE HATCH TO PERMIT
WITHDRAWAL OF PUMP SHAFT AND
DISCHARGE COLUMN PIPE.

LOCATE AS CLOSE TO WALL AS
POSSIBLE TO PERMIT READY
ACCESSIBILITY FOR WELL REPAIR

ACCESS DOOR

PRESSURE GAGE

CHECK VALVE

DRAWDOWN GAUGE

FLOW METER

CONCRETE FLOOR

TOP OF CASING

GATE VALVE

12"

BELOW FROST

DRIVE SHAFT

STATIC WATER TABLE

DISCHARGE COLUMN PIPE

DRAWDOWN

PUMPING LEVEL (MAX. DRAWDOWN
LEVEL FOR GIVEN CAPACITY
PUMP.)

5'

TOP OF BOWL

TURBINE PUMP

IMPELLERS

PUMP LENGTH

CASING

BOTTOM OF SUCTION PIPE

10' AVERAGE

DEPTH REQUIRED TO REACH
WATER BEARING STRATA.

BOTTOM OF CASING AND TOP
OF WATER BEARING STRATUM.

SEAL

SCREEN OF NON-CORROSIVE
METAL (EVERDUR)

10' TO 50' - 25' AVERAGE

SINGLE CASED WELL

FOR USE IN ORDINARY SANDY &
GRAVELLY SOIL

FIGURE 3-2

Single-Cased Well. In Fig. 3.2 we depict a single-cased well that represents a fairly common well and pump installation. The well casing is supplied by the well driller, as is the screen. The basic equipment in the well contract consists of the pump, drive shaft, discharge pipe, and motor. Well casings are usually some type of ingot iron, wrought iron, or alloyed or unalloyed steel. Our detail shows a well screen of noncorrosive metal called Everdur which is a trade name for an alloy of 96 percent copper, 3 percent silicon, and 1 percent manganese. This is only one of a variety of alloys that are available.

Since the detail is a typical composite, there are a number of notes that should be refined to contain data from the drilling and testing log reports on the well. These refinements include actual elevations of the top of the casing, the static water table, the pumping level or maximum drawdown level, the bottom of the casing level, and the length of the intake screen. Originally the casing bottom is at the same depth as the bottom of the screen and the screen sits inside the casing. The casing is pulled up and slides around the screen leaving it exposed and in place as our detail shows.

The pump depicted is a typical water-lubricated turbine pump which commonly provides a water system for 4-in and 5⅝-in inside diameter and larger wells with lifts up to 250 ft and outputs up to 100 gallons per minute (gpm).

Submersible Pumps

As we noted in the description of the turbine pump sitting in the well house, the depth limit figure for this type of pump is 250 ft and the gallon limit rating is about 100 gpm. However, suppose your well-drilling consultant advises that water has been found at 380 ft below grade. While the submersible pump can work at shallow depths, it also can, to a limited capacity, work to depths up to 1,000 ft.

In Fig. 3.3 we show a typical submersible pump in a typical single-cased well. To further clarify your detail we have indicated in simplified fashion the control wiring which is actuated by a storage tank pressure switch. Here, as in all other pumping details, is a key issue made a little more obvious. The well pump cannot be directly tied to your water-using equipment. The constant starting and stopping would quickly wear out the pump. Commonly the well pump can deliver water at sufficient force to fill an air pressurized storage tank. It is the water under pressure in the storage tank, acting like water under pressure in a utility main source of supply, that feeds your plumbing system.

Our completely submerged pump at the bottom of our well accepts water from the strainer to the pump. Since everything is below water—pump, strainer, and motor—the system is no different from a submerged sump pump. Since they are designed to operate in this environment, your selection process is based on the pump's ability to overcome static head plus storage tank pressure. Selection of both head and pressure is made simple in the manufacturer's catalogs. Depending on depth, the submersible turbine pump in a 12½-in well can have a capacity as high as 1,600 gpm.

Pneumatic Tanks

In our previous discussion of wells and especially in our discussion of a well pump installation described in Fig. 3.3, the inevitable storage tank which is a common part of any well system made its appearance. The tank is used in conjunction with a water pump and acts as an equalizer of the system water demand. The pump shown in Fig. 3.3 is regulated to stop at a certain maximum pressure which, when reached, provides a supply of water at the proper pressure to meet the intermittent demands of the system. Fig. 3.4 is a simplified diagram of a pneumatic water storage tank which serves the purpose of explaining the total pneumatic tank detail in Fig. 3.5.

The essence of the storage system is the proper selection and arrangement of piping and controls for the pneumatic tank. It must be clearly understood that every gallon of water used in the system comes from the pump. The air in the tank shown as V_a in Fig. 3.4 is a cushion created by the force of the pump delivering water into the tank. Thus the pressure in the tank cannot exceed the discharge pressure capability of the pump. In the normal sequence of events a pressure controller on the tank stops the water pump when a certain preselected pressure is reached. This is the high level shown in Fig. 3.4. Intermittent or constant fixture water demand reduces the volume of water in the tank, lowers the air pressure, and, at a preselected minimum pressure related to the low level, the pump starts.

In Fig. 3.4 the air space, high and low water levels, water intake and discharge, and the water seal are based on an arbitrary depth of water in the tank selected to assure that water, not air, is always delivered to the plumbing system.

In the perfectly ideal situation, which almost never occurs in any actual installation, the water demand would be constant and the pump would operate constantly. While unusual, there are situations in certain industrial applications in which the demand is sufficiently constant to operate a supply system without a tank. There is no rule governing the location of the tank. It can be at any point in the system. Tanks placed at the highest point in the system can be made into a combination equalizer and storage tank. In this instance the capacity must be increased to provide for emergency use by gravity.

Tank Sizing. There are three volumes to consider in Fig. 3.4. As we noted, the water seal volume V_s is an arbitrary selection. The variable volume V_{uw} is based on how frequently you desire the pump to operate. If you desire a rest period that we can call T minutes and you have selected a pump rated at G cubic feet per minute (cfm), the usable water space V_{uw} must equate to GT cubic feet (cu ft) of water.

Normally the pump is selected to meet maximum demand conditions, ideally at the least acceptable pressure discharge head. Having decided the time interval and having the system requirement set our gallons-per-minute demand, all that is remaining is to solve the volume of air space. Our system should try to maintain a steady system pressure. As we'll see in the formula below, the smaller the variation in pressure, the larger the tank. A general design criteria is 5- to 10-pounds per square inch

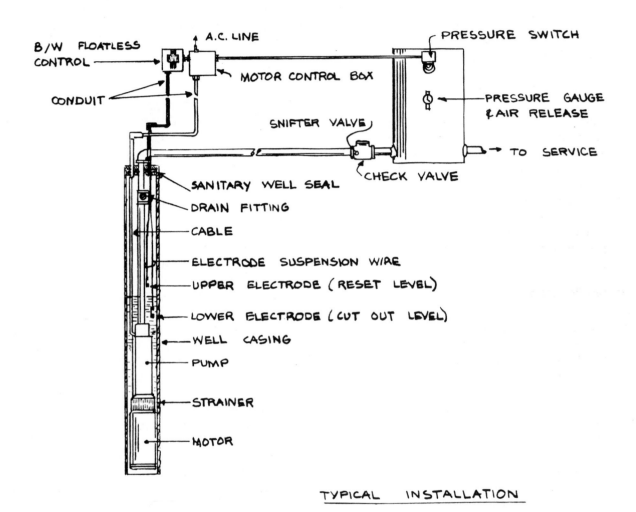

TYPICAL INSTALLATION

SUBMERSIBLE PUMP CONTROL

FIGURE 3-3

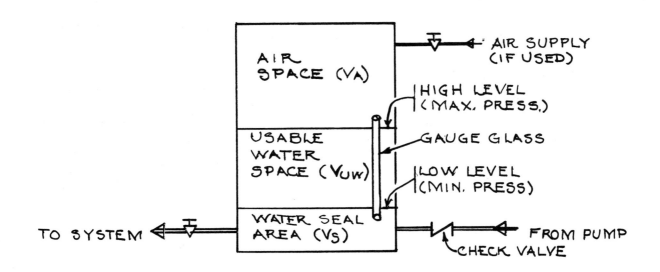

ELEMENTARY DIAGRAM
PNEUMATIC STORAGE TANK
—— NO SCALE ——

FIGURE 3-4

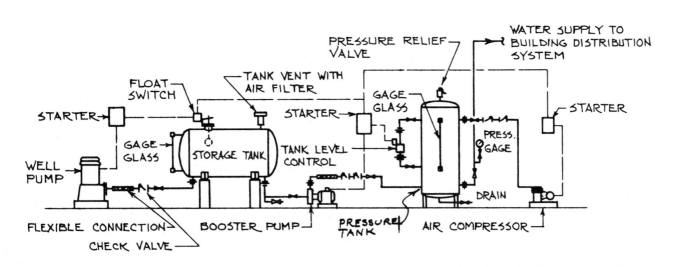

COMBINED STORAGE & HYDRO-PNEUMATIC WELL WATER SYSTEM

NO SCALE

FIGURE 3-5

gravity (psig) variation in system pressure. The basic formulas are as follows:

1. Usable water space: $V_{uw} = G \times T$
2. Air space required: $V_a = \dfrac{V_{uw} \times P_2}{P_1 - P_2}$

where P_1 = maximum air pressure (psi absolute)
P_2 = minimum air pressure (psi absolute)
V_{uw} = variation in volume, $G \times T$
G = maximum water demand, cubic feet per minute
T = time factor, minutes
V_a = minimum air space

As an example let us select a typical problem of 150-gpm demand at 60 psig at a 10-psi variation. Thus we have the following list of data based on the value of 1 cu ft of water equals 7.5 gallons (gal).

Pump capacity G = 150 gpm/7.5 = 20 cfm
Minimum pressure = 60 psig or 74.7 psi absolute (psia)
Maximum pressure = 70 psig
Difference = 10 psig
Time = 0.5 minutes (min) (our selection)

Therefore,
space V_{uw} = 0.5 × 20 or 10 cu ft
space V_a = 10 cu ft × 74.7 psi/10 psi = 74.7 cu ft
$V_a + V_{uw}$ = 10 + 74.7 or 84.7 cu ft

Based on the above we now make a decision on space height. Suppose our room available is 8 ft 0 in and we decide the tank should have an overall height of 7 ft 0 in. Further our arbitrary decision on water seal is 1 ft 0 in. This leaves us with a tank whose height available for $V_a + V_{uw}$ is 6 ft 0 in. Dividing 84.7 cu ft by 6 ft gives us 14.12 square feet (sq ft). A round tank with a volume of 14.12 sq ft has a diameter based on area equal to $3.14D^2/4$ gives us a diameter of 4.25 ft. Thus our tank is 4.25 ft in diameter by 7 ft 0 in high.

Residential Systems

Before going to Fig. 3.5 we can view Fig. 3.4 again as a common residential system in which the well pump is the only pump and the pneumatic tank does not have an air compressor. This is the usual case. In this instance the seal volume is what remains after we deduct volume V_aD and V_{uw} from the total tank volume. The total tank volume is the required air volume under atmospheric pressure and equals $V_a \times P_2/14.7$. The volume V_{uw} is selected on the basis of any maximum load expected and should be carefully selected to avoid unduly large tanks. The air volume $V_a = V_{uw} \times P_2/(P_1 - P_2)$.

Usually a maximum discharge rate for residential use

is 5 gpm maximum or about 0.75 cfm. The normal values are usually as follows:

Maximum discharge rate = 0.75 cfm
Maximum gauge pressure = 40 psig or 54.7 psia
Minimum gauge pressure = 30 psig
Variation = 10 psig
Space V_{uw} = 0.75 cu ft
Space V_a = 0.75 × 54.7/10 = 4.10 cu ft
Total theoretical tank volume = 4.10 × (54.7/14.7)
= 15.25 cu ft

Thus a 2-ft 0-in diameter tank which is 3.14 cu ft high would be 5 ft 0 in (rounded) in height to serve the purpose.

Well Water System

As we noted in the discussion of Fig. 3.4, the real detail of the normal well system would be shown in Fig. 3.5. Figure 3.5 shows what is described as a combined storage and hydropneumatic well water system. In the beginning of this chapter we depicted wells and their related details. The pumps and other related equipment were only shown in limited diagrammatic form. We also noted that the well is usually presented to the plumbing designer as a completed and installed ready-to-use source similar to what a utility would provide in the way of a service tap to the building property line.

Commonly in the well source situation the plumbing engineers have had considerable input into the ultimate approval of the well source and frequently have specified or been involved in the specification of the well pump.

Figure 3.5 depicts two pumps and two tanks because we have two different problems to resolve. In our discussion of pneumatic tanks and pumps in Fig. 3.4 we described at length the sizing of the pneumatic tank using assumed values of demand for our building. In the data for Fig. 3.4 we assumed a demand of 150 gpm for a very short duration of time. Now let us look at that figure of 150 gpm from another standpoint. Assume that although the 150 gpm value is only for a short period, there is a sustained demand that lasts for 2 hours (hr) and equals 100 gpm. And further assume that our well has a safe pumping rate of 30 gpm. Finally assume that the rest of the time our building drops off to 15 gpm. All of this is not uncommon.

For the 2 hr of sustained demand we will need 18,000 gal of water (2 hr × 60 min/hr × 150 gpm). During this period our well can only produce 3,600 gal (2 hr × 60 min/hr × 30 gpm). Obviously we need storage. Our storage tank shown in Fig. 3.5 must hold 14,400 gal of water (18,000 − 3,600). In addition we said that for the remaining 22 hr we require 15 gpm. Thus for the entire 24 hr we need a total of 37,800 gal (18,000 gal + 22 hr × 60 min/hr × 15 gpm = 19,800 gal). Thus, if our well operates 6 hr a day (a normal value), it can produce 10,800 gal (6 hr × 60 min/hr × 30 gpm). To produce 37,800 gal it would have to run 21 hr (37,800 gal/30 gpm × 60 min/hr). This is unrealistic; the well will run dry.

Putting everything in proper order, our design of 150 gpm has got to be reduced or we need more wells or a well of much greater capacity. In every case the detail shown in Fig. 3.5 must be based on the well capacity. Our 30-gpm well might reasonably be expected to operate at capacity for 8 hr and, depending on test data, possibly 10 hr. Using 8 hr as a reasonable figure, we have available 14,400 gal (8 hr × 60 min/hr × 30 gpm) of water. We can store all of it in a 15,000-gal storage tank. No matter how slowly or rapidly it is used, that is all that is available for one 24-hr period. What is required of the plumbing system designer is not merely the peak demand, which can readily be resolved, but the total demand calculation. If this building were an office, school, etc., perhaps a usage figure of 10 gal per person per day could be presumed, and a 30-gpm well could easily handle an occupancy of 1,000 persons.

The detail in Fig. 3.5 is fairly straightforward. Once the demand has been determined, the storage tank is sized to handle the building's peak and sustained demands and is fed by the booster pump. Since the booster pump is larger than the well pump, the storage tank provides the available high-usage water. The well pump operates independently of the building system. Its job is to maintain a fixed level of water in the storage tank. During periods of high demand both the well pump and the booster pump are operating, as is the air compressor.

By a dashed line we indicate the fact that the controls are linked together. Usually this is a one-way arrangement that precludes the system from operating if the well pump fails. In some systems the designer may include bypass arrangements in the electrical controls to allow for limited system operation if the well is out of service. This should be carefully investigated as it is possible to create air-bound systems and burned-out pumps if all the water is used up.

All of the items depicted are readily available in a variety of configurations from a number of different manufacturers. Ideally both the storage tank and the pneumatic tank would be installed below grade. In multistory buildings (covered in Chap. 4, Water Distribution System) the storage and pneumatic tanks may very well be installed in the upper floors of the building or on the roof.

The detail uses standard symbols for gate and check valves and flexible connections. We only noted a check valve and a flexible connection in one instance as a sample of others. The flexible connection at the booster

pump may not be necessary depending on pump selection and installation arrangements.

Large Well Systems. As was noted in our opening discussion on wells and well supplies, large systems involving deep wells of high volume with elevated storage tanks and the capability of supplying domestic water, process water, and fire protection systems are rightly the province of the water supply specialist. These systems should not be designed by people whose expertise is in the plumbing field. This is not to say that the basic premises are any different than for a smaller system. But there are tank and well systems and treatments, special storage tank design, tank heating, and a host of items that are not common to plumbing design nor commonly part of the plumbing designer's expertise.

Figure 3.6 is an attempt to put into perspective in a composite form some of the uses and arrangements of large well water supply systems. If the plumbing designer is part of a design team responsible for a large new project which uses a well supply source, all of the piping shown on Fig. 3.6 plus the well pump and its piping (not shown) will be designed by the water supply specialist. However, the building system most likely will have supply pumps and possibly a pneumatic tank as we depicted in Fig. 3.4.

Since the storage tank and piping may already exist in a large modernization project, the system changes may require some piping connection, control, or valve changes. In this event the arrangements shown in Fig. 3.6 with, most likely, some other refinements could logically be part of the plumbing design. The altitude valve, for example, is somewhat different in design but similar in function to the float control valve in our detail shown on Fig. 3.5.

Our chemical treatment is fairly simplified and schematic. This is deliberately so because chemical treatment is a very special area about which almost no plumbing engineer is knowledgeable, and it should definitely not be planned or detailed except as per the advice of a chemical and/or water treatment expert.

Booster System

One of the problems that must be resolved by the plumbing system designer is the increasingly common occurrence of sufficient water at too low a pressure. This is frequently caused by the constant addition of users to a fixed-supply distribution network. If the network had all of its users on the same level as originally contemplated, the problem would not have occurred. But, as is fairly common, the original town or city was at a lower level than the newly created suburbs, and the pressure available to the newer higher-elevation suburb is therefore decreased.

Figure 3.7 details a pneumatic water system. If you will refer to Fig. 3.5, you will quickly note some striking

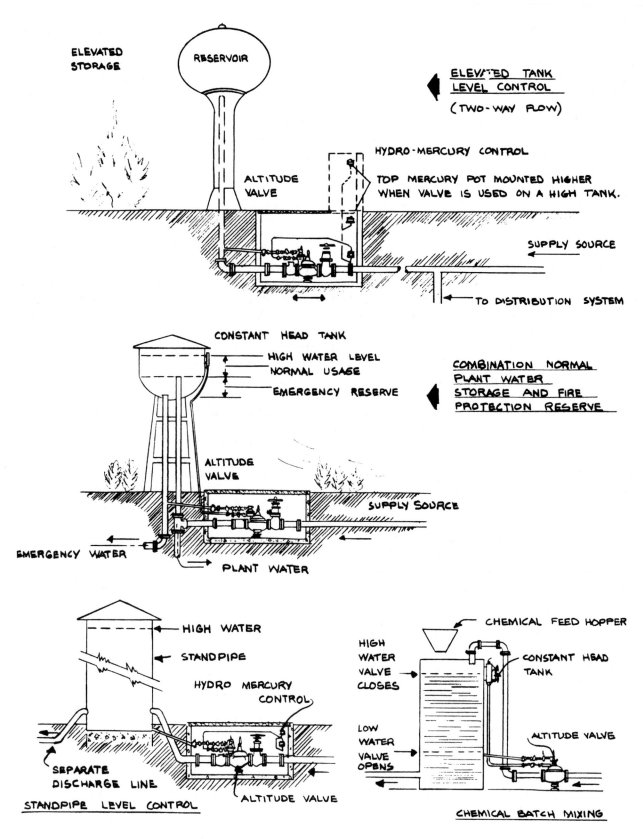

ELEVATED STORAGE

RESERVOIR

ELEVATED TANK LEVEL CONTROL
(TWO-WAY FLOW)

HYDRO-MERCURY CONTROL

TOP MERCURY POT MOUNTED HIGHER WHEN VALVE IS USED ON A HIGH TANK.

ALTITUDE VALVE

SUPPLY SOURCE

TO DISTRIBUTION SYSTEM

CONSTANT HEAD TANK

HIGH WATER LEVEL
NORMAL USAGE
EMERGENCY RESERVE

COMBINATION NORMAL PLANT WATER STORAGE AND FIRE PROTECTION RESERVE

ALTITUDE VALVE

SUPPLY SOURCE

EMERGENCY WATER

PLANT WATER

HIGH WATER

STANDPIPE

HYDRO MERCURY CONTROL

SEPARATE DISCHARGE LINE

ALTITUDE VALVE

STANDPIPE LEVEL CONTROL

CHEMICAL FEED HOPPER

HIGH WATER VALVE CLOSES

CONSTANT HEAD TANK

LOW WATER VALVE OPENS

ALTITUDE VALVE

CHEMICAL BATCH MIXING

TYPICAL APPLICATIONS

FIGURE 3-6

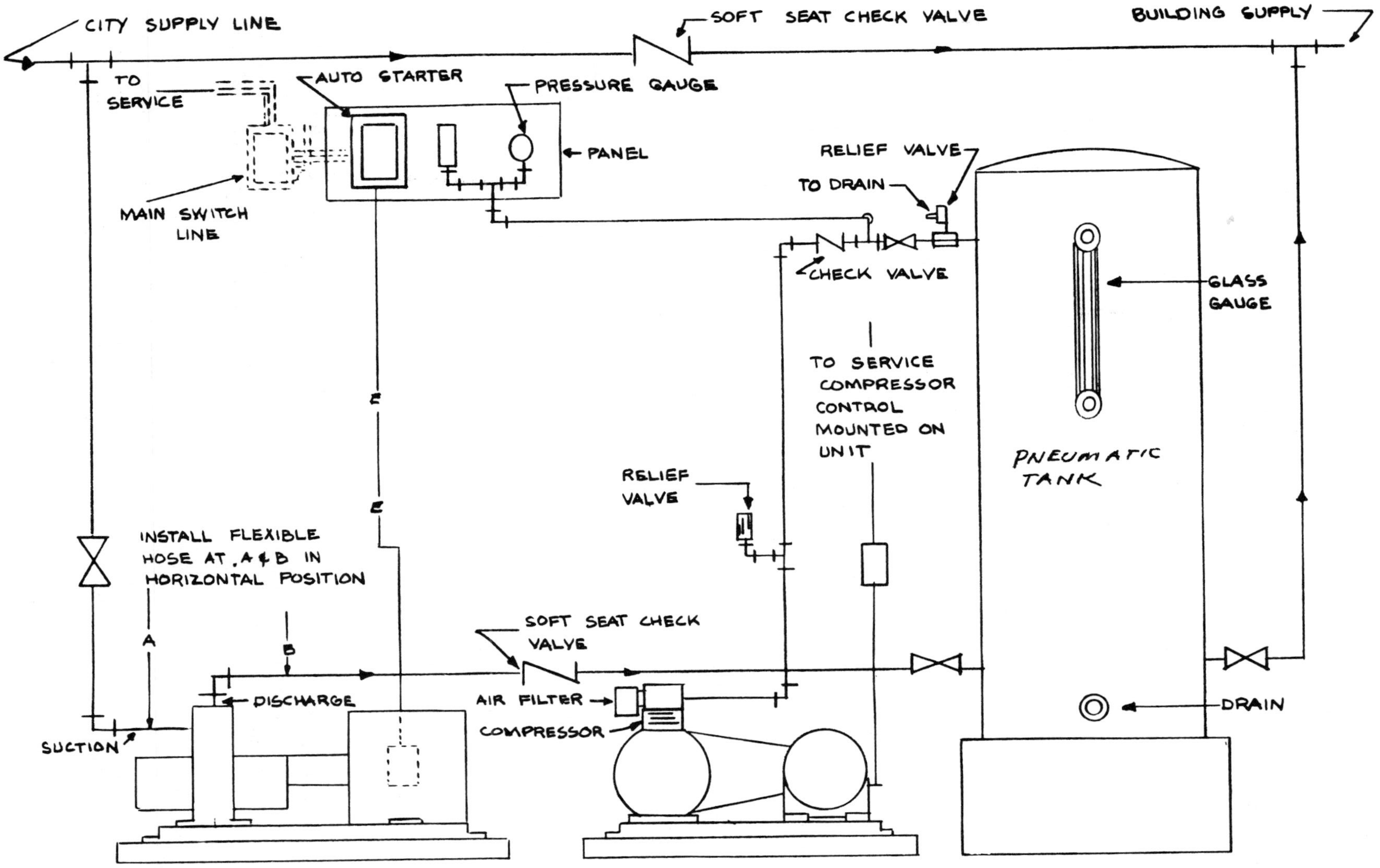

PNEUMATIC WATER SYSTEM

FIGURE 3-7

similarities between a portion of Fig. 3.5 and Fig. 3.7. While there are similarities, there are also differences. In Fig. 3.7 there is no concern over the quantity of water. As was noted above, the city water supply is sufficient for our needs. The problem is pressure.

The water supply pump shown in the detail is sized to handle our peak demand and to discharge at a pressure above that in the pneumatic tank. On occasion one of the mistakes that is made with this system, as shown in our detail, is the selection of a pump with too low a discharge head pressure. The controlling factor in this situation is the system head pressure as required by the equipment served. This pressure is maintained by the pneumatic tank air compressor.

The pneumatic tank has to be designed using design criteria similar to that which we outlined in our discussion of Fig. 3.4. The purpose is very similar, to preclude rapid, short on-off operation of the main supply pump when the use of water is in small limited amounts.

What is different about Fig. 3.7 is the fact that pressure is also imposed against the city supply line. This pressure accomplishes two goals at one time. It supplies the plumbing system, and it holds a check valve closed, ensuring that the city water goes to the pump. But this is not the total resolution.

The city water supply does have some pressure and can at least partially satisfy some of the building's plumbing system requirements. The pneumatic system can always fail at an unexpected time and for a variety of reasons, including loss of electricity. At that point the drawing of water from the system will quickly reduce the pressure. When the pressure drops below the city supply pressure, it will cause the check valve to open and city water will be available at reduced pressure.

The above sequence of events will also create a backward flow through the pressure tank and toward the supply. The check valve on the supply pump will stop the backward flow. Thus, with the insertion of these two simple, strategically located check valves the system has an automatic change-over capability that is about as trouble free as can be found anywhere.

Engineering Data

Head is measured in feet, pounds per square inch, or in inches of mercury. However, so that a common means of head measurement is understood, it is recommended that all heads be expressed in feet of water. Measurement of liquid should be expressed in U.S. gallons.

At sea level atmospheric pressure is 14.7 psi. This will maintain a 29.9-in column of mercury or a 33.9-ft column of water. This is the theoretical height to which water may be lifted by suction. The practical limit for cold water (60°F) is 25 ft.

It is the pressure or weight of the air that pushes water up the suction line. In addition, this air pressure must impart velocity to the water to get it into the pump and must overcome friction resulting from the flow of water in the suction line. Therefore, the lower the suction lift, the greater will be the percentage of air pressure that is available. For this reason, the lower the suction lift, the more water the pump will get.

Suction Lift. This is the vertical distance from the center line of the pump's suction inlet to the constant level of the water.

Positive suction head: The vertical distance *above* the center line of the pump's suction to the constant level of the water. This is *subtracted* from the total head.

Dynamic suction head: The suction lift plus suction line friction loss.

Static discharge elevation: The vertical distance from the pump's discharge to the highest point in the discharge line.

Total dynamic head (*TDH*): The total head and the summation of static suction lift, friction loss in suction line, static discharge elevation, friction loss in discharge line and fittings, plus discharge pressure, if any.

To be hydraulically correct, we should not include static head in total dynamic head. Dynamic means "moving" and dynamic head only includes velocity head and friction loss. Many pump people use TDH interchangeably with total head (TH).

Net Positive Suction Head.
Net positive suction head (NPSH) is defined as head that causes liquid to flow through the suction line and enter the impeller eye. This head comes from either atmospheric pressure or from a static suction head plus atmospheric pressure. Two types of NPSH will be considered.

Required NPSH: A function of pump design. It varies for different makes and models and with capacity of any one pump. This value (shown on the pump performance curve) is supplied by the manufacturer.

Available NPSH: A function of the system in which pumps operate. It can be calculated for any installation. For a pump to operate properly, available NPSH should be greater than the required NPSH plus 2 ft for safety factor at a desired head and capacity.

NPSH *does not* indicate the priming capabilities of self-priming centrifugal pumps. This capability is shown, generally on engine driven pumps, to be respective "break-off" lines representing static suction lifts of 10, 15, 20, or 25 ft.

Useful Factors

Useful Abbreviations

Abbreviation	Meaning
bhp	brake horsepower
eff	pump efficiency
gpm	gallons per minute
psi	pounds per square inch
rpm	revolutions per minute
sp gr	specific gravity of water
TDH	total head or total dynamic head
whp	water horsepower

Useful Equivalencies

psi	feet head $\times$ .432
feet head	psi (water) $\times$ 2.31
sp gr water	1.0
1 U.S. gal	8.33 pounds (lb)
1.0 imperial gal	1.2 U.S. gal
1 cu ft water	7.48 gal
whp	feet head $\times$ gpm/3960
bhp	whp/eff or feet head $\times$ gpm/(3960 $\times$ eff)
eff	whp/bhp $\times$ 100

Pump Data

Explanation	Equations
Capacity varies *directly* as the speed or the impeller diameter	$gpm(1)/gpm(2) = rpm(1)/rpm(2)$ $gpm(1)/gpm(2) = diameter(1)/diameter(2)$
Head varies as the *square* of the speed or the impeller diameter	$head(1)/head(2) = rpm(1)^2/rpm(2)^2$ $head(1)/head(2) = diameter(1)^2/diameter(2)^2$
Power (bhp) varies as the *cube* of the speed or impeller diameter	$bph(1)/bph(2) = rpm(1)^3/rpm(2)^3$ $bph(1)/bph(2) = diameter(1)^3/diameter(2)^3$

Individual Water Systems

The complete individual water system consists of a pressure tank and the pump. The pump has controls which have a definite function for the proper operation of a complete system. The controls are the pressure switch, the air volume control, the pressure gauge, and throttle valve. One other item necessary to the proper functioning of the jet system is the ejector, which may be either a shallow well, a deep well, or the convertible type which may be used either as a shallow well or a deep well. There are limitations to the use of the convertible unit in that it is limited to a minimum of 4-in well casing to convert to a deep well operation.

The Pressure Tank. The purpose of the pressure tank is to provide a constant supply of water in a limited pressure range. The pump supplies water to the tank and compresses the trapped air to 40 lb. At this point, the pressure switch turns off the pump. Then, when water is drawn off, the pressure drops, and when it reaches 20 lb, the pressure switch turns on the pump.

The Pressure Switch. The pressure switch senses the pressure in the system at all times. The normal limits (factory preset) are 20 to 40 lb. In other words, the switches are preset to cut in at 20 lb and cut out at 40 lb. If higher pressure is needed such as 30 to 50 lb, the installer may increase to these limits simultaneously by following the directions enclosed in the switch.

Pressure Gauge. The purpose of the pressure gauge is to provide a means of setting the proper operating pressure to the ejector and resetting the limits on the pressure switch.

Throttle Valve. The throttle valve is used to set the operating pressure of the ejector only, and it has nothing to do with the pressure switch or tank pressure.

Air Volume Control. The air volume control is incorporated into the individual system when the air-seal-type tank is not used. The air seal, as you know, can only be incorporated into the vertical-type tanks and separates the air from the water by a float that floats on the water in the tank. The addition of air on each pumping cycle is not necessary in this type unit; however, in the horizontal type, an air volume control is necessary and adds air to the tank on each pumping cycle to maintain an even pressure range between the air and water.

Meters. The use of water is commonly billed according to the quantity used, which is measured by a meter which continuously records the amount taken from the utility system. In our well details no meter was detailed or described since the user owns the source of water. If there were any cost concern at all, it would undoubtedly be concerned with the amount of electricity used by the various system pumps.

There are certain basic concerns of the water utility company which involve and affect the plumbing system designer. Metering installations involve not only the meters themselves but the various shut-off valves and the piping arrangement of the meter and valve installation.

Figures 3.8 through 3.10 illustrate typical valve and meter installation details. In residential, very small commercial, and fire hydrant installations the utility company commonly has an exterior shut-off valve such as we depict in Fig. 3.8. The actual meter may be in a basement or utility space. In the small installation the meter commonly has shut-off valves on both sides of the meter supply and discharge. These are not controlling shut-off devices but are primarily meter service devices. To replace or repair the meter simply close both valves and thus isolate the meter from both the building plumbing system and the utility supply system.

There are various types of housing and valve arrangements for an exterior shutoff. In the usual plumbing system design the utility has, or will provide, a tap-off of the utility system distribution main. This may or may not be a valved tap-off. Usually it is not. The plumbing design will include an exterior shutoff shown in Fig. 3.8 with the water service line continuing to the utility property water service tap. The final connection to the utility main is almost always made by utility employees or their authorized contractor.

The shutoff shown in Fig. 3.8 may vary to some degree, but our detail is a reasonable representation of the common type of valve used. The long handle depicted has an inverted U-shape that fits into two grooves on the shut-off valve body. The user removes the screwed cover at grade, inserts the rod until its end slides into the valve grooves, and turns counterclockwise to close or clockwise to open.

Engineering consultants are normally not retained for very small installations. As such they seldom are involved in the work depicted in Fig. 3.8 except when their fire protection or sprinkle system design involves

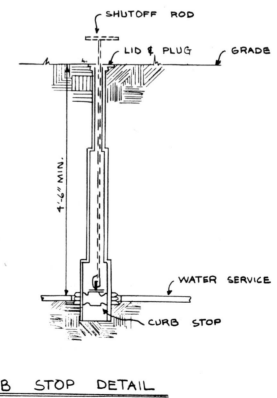

CURB STOP DETAIL

NOT TO SCALE

FIGURE 3-8

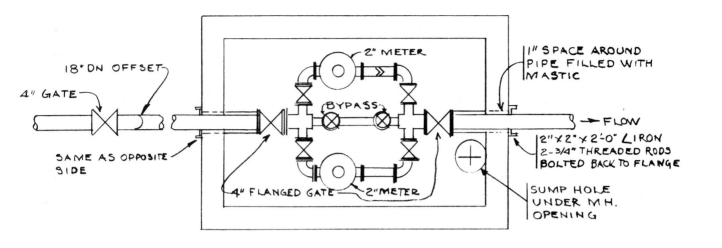

4" SERVICE – DUAL METERS

——— NOT TO SCALE ———

FIGURE 3-9

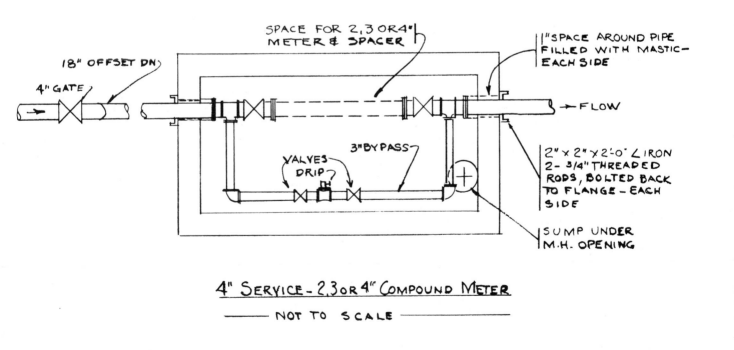

4" SERVICE – 2,3 OR 4" COMPOUND METER

——— NOT TO SCALE ———

FIGURE 3-10

fire hydrants. In Chap. 9 we detail the various items involved in sprinkler and fire protection systems. There you will note a very similar type shutoff for a fire hydrant.

The vast majority of the plumbing engineering design work is in the area of 2-in to 4-in water service supply systems. Figure 3.9 depicts a dual meter installation and Fig. 3.10 depicts a single meter installation. These two details are typical installations in an exterior meter pit. They are not necessarily the exact arrangement required by your particular local utility company. Thus the very first step taken by your firm must be to ascertain exactly the metering, piping, and arrangement requirements of your local utility company.

Second, to make our details more all-inclusive, we also show the pit itself in details depicted in subsequent figures. In your case the meters may well be in the building equipment room area; therefore no pit is required or depicted in your detailing.

In both Figs. 3.9 and 3.10 we depict a shut-off valve away from the pit itself. If this piping as depicted were in the building equipment room, this valve would also be away from the room and outside the building. Whether or not this valve is simply buried or is in its own separate valve pit is another matter to be cleared with your utility company.

Water service piping 3 in and larger may be steel, iron, or copper. Commonly, as depicted in our meter details, it is flanged cast-iron water service pipe conforming to the American Water Works Association (AWWA) standards. This pipe material is also an item to be cleared with your utility company.

Of the common sizes we chose the 4-in cast-iron flanged service pipe and have carefully depicted the meters, shut-off valves, and piping. Our details, reduced from larger original sizes, are no longer to scale but yours very definitely must be to scale. Ultimately there will be a flanged space of the correct size and spacing to permit the utility-furnished meter with its flanges to slip into place and be bolted tight.

The piping at the meter should be sized to fit the meter openings. This is especially critical in Fig. 3.10 in which a single meter is involved. While your service pipe is 4 in, the utility company may decide, based on their use analysis, to provide a 3-in meter. Consequently in Fig. 3.10 the meter space allocated is much longer to permit, if required, flanged reducing connections.

Within the confines of the pipe the size and rigidity of the pipe eliminate the need for pipe hangers and floor supports. If this same installation were in a large equipment room, it is quite likely that meter inlet and outlet horizontal piping would be much longer and hangers or floor supports would be required.

In a pit installation surface water entering the pit is always a problem to some degree. If the groundwater table level permits, a simple sump and drain that allows water to be carried to a site storm drainage system is sufficient.

Usually the water line is some 4 ft below grade and the exterior pit depth is such that the meters are about 5 to 5½ ft below grade. This is effectively what occurs in our meter details. If the groundwater level is above that elevation, you may well have to consider specialized waterproofing of the pit or the installation of a sump pump.

Figure 3.9 may appear to have redundant valving in that there are shutoffs on both the 4-in line and the individual meter lines. However, this is where checking with the utility company is important. In the actual project installation which our detail depicts, the utility company viewed the dual meter installation as a distinct unit and requested the additional 4-in valves.

Meter Pit. Figures 3.11 and 3.12 are one answer to the detail requirement of an exterior enclosure for the incoming water service metering installation. While many service installations are located in basement boiler or mechanical equipment rooms, a larger building constructed as a slab-on-grade with no basement installation may very well have a limited amount of space for mechanical equipment. Further, the water utility company may have specific rules which require an exterior meter installation.

These two figures are not completely dimensional since they will vary in overall size depending on the particular meter application and installation clearance requirements. A common size which approximately corresponds to our detail is 10 ft by 6 ft by 6 ft deep. In Fig. 3.11 the two means of access to the pit are depicted. The standard 24-in manhole with its cast-iron frame and cover extends to the surface. The cover may be locking or nonlocking as required by your particular project. The second means of access is covered by the paving or earth depending on the particular situation of the installation. The second cover is a usable, but obviously seldom used, means of access for major replacement or repair of the meters and valves.

The entry ladder, a mandatory requirement, is depicted in Fig. 3.11 and further detailed in Fig. 3.12. The walls and roof are poured separately and the walls are keyed. Reinforcing is depicted in both directions for the pit roof. The walls, roof, and floor thicknesses are all dimensioned. For many installations of normal wheel loading, the pit, if built of 3,000-lb concrete, may be used as is provided you add the proper dimensions of length, width, and height. However, when there is any question of wheel-loading weight on the pit, or when your dimensions vary considerably from the basic 10 by 6 by 6 size, the structure should be checked by a competent licensed professional structural engineer.

Water in the pit may be a problem as it frequently is in any underground chamber. The design is relatively watertight, but there is no guarantee that it is absolutely watertight. One solution, if an underground storm drain at the proper elevation is available, is to utilize the pit drainage system depicted in Fig. 3.12. In addition we have added some possible water seepage resolution notes to Fig. 3.12. If the water table is high, it may be necessary to install a pit sump pump in lieu of the pit drain.

Water Treatment. Water treatment is a specialized subject that is best left to experts in the field. Normally

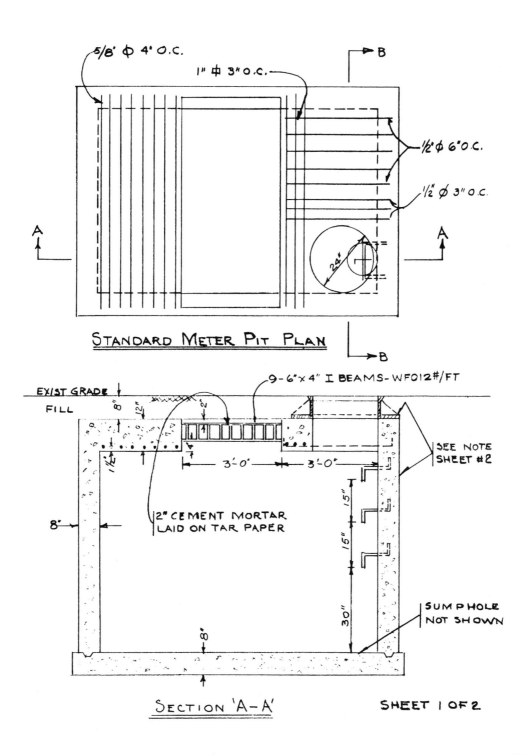

FIGURE 3-11

no treatment is required for utility-supplied water that is acceptable to local health authorities. There is a wide variety of acceptability standards and what might be acceptable is not necessarily what is desirable to the water user. When such a condition occurs, the product or device added to the system is one that removes dissolved iron and other materials that cause water to be hard or one that improves color or odor or taste of the water.

None of the above-noted devices which improve water quality have particular detail requirements. In all cases the water is passed directly through the device, usually

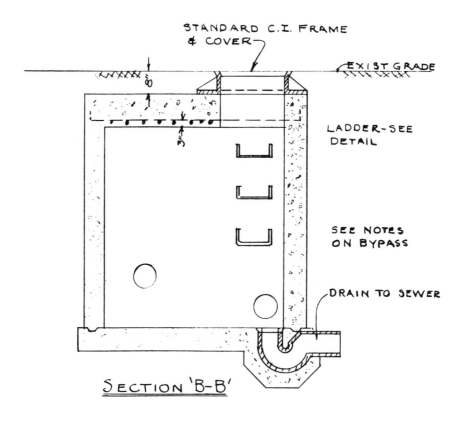

SECTION 'B-B'

NOTES
1. PIT NOT TO BE USED UNDER ROADS OR WHERE HEAVY LOADS ARE IMPOSED.
2. WATER PROOF TOP & SIDES OF PIT WITH 2 PLY TAR PAPER & TAR OR CONSTRUCT OF WATER PROOF CEMENT.
3. BYPASS MAY BE CONSTRUCTED OUTSIDE & AROUND PIT WITH DRIP EXTENDED THRU WALL. BYPASS MUST ALWAYS BRANCH TO RIGHT LOOKING ALONG DIRECTION OF FLOW.

SHEET 2 OF 2

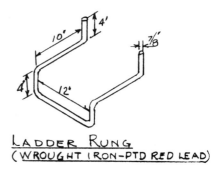

LADDER RUNG
(WROUGHT IRON-PTD RED LEAD)

FIGURE 3-12

with a valved bypass arrangement, so that service may continue when the device is being repaired, serviced, or replaced.

Figure 3.13 is a somewhat special situation that occurs in a well installation on certain occasions. All water supplies, regardless of origin, must be tested by a properly licensed laboratory and the test results approved by the proper local and state health organizations before that water is approved for domestic use. In our earlier discussion in this chapter we noted that commonly the well and its flow capability is presented to the plumbing system designer as an accomplished fact. This being the case, it is very important that the plumbing system designer have in his or her possession not only the test results of the well but also the health and laboratory data attesting that the water is approved for domestic use. The plumbing designer could be held liable if the designed system is connected to an unsafe source.

The detail depicts the proper installation of a chlorinator to a well source. This chlorination of water is a common requirement in a utility-supplied water system and is the usual method of bacteria control. Every well that is drilled does not automatically need a chlorinator. The laboratory test would determine this requirement. Our purpose in presenting this detail is to provide the detailer with information to use when this special treatment is mandated. Chlorine in a specially controlled dosage is delivered.

Figure 3.13 illustrates a gas chlorinator on a typical deep well pump installation. The design premise is to inject the proper metered amount of chlorine into a pressure main varying from 30 to 120 psi or higher. Piping of 1 in is connected from the supply main to the booster pump and then to the ejector. A piece of flexible hose is used where indicated. The chlorinator also comes with a vent connector to be extended to the outside atmosphere in plastic tubing. The electrical wiring to the booster pump is interconnected to the well pump starting contacts so that both pumps operate at the same time. Water flows from the main through the pump where its pressure is increased and then to the ejector. The high velocity of water passing through the ejector

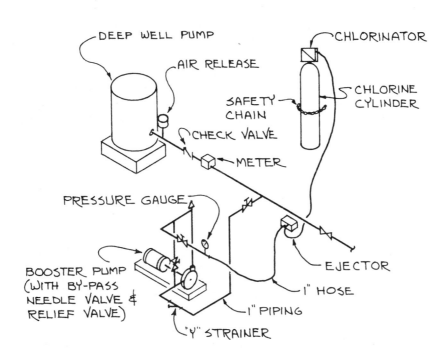

CHLORINATOR INSTALLATION USING
TURBINE TYPE BOOSTER PUMP

NO SCALE

FIGURE 3-13

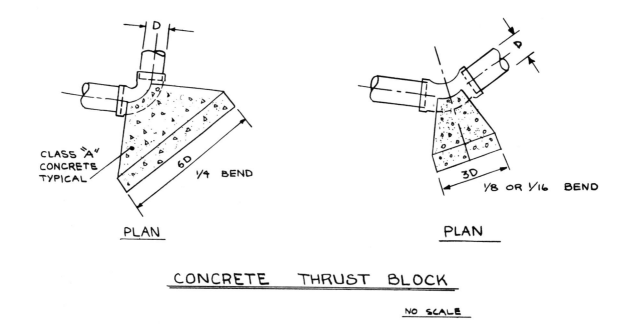

CONCRETE THRUST BLOCK

NO SCALE

FIGURE 3-14

creates a vacuum at that point. This vacuum feeds back through the tubing to the chlorinator valve to admit a preset correct amount of chlorine as determined by the lab test and the chlorine residual test. Stopping the pump breaks the vacuum and stops the flow of chlorine.

Sizing of the chlorinator can be very tricky. A standard treatment of 0.5 parts per million (ppm) may or may not be the correct dose. Some types of water, especially those containing sulfur compounds, may need as much as 10 times normal amounts. An expert should be employed to determine tank and meter sizes.

Thrust Block. The frequent bang of the water supply piping in your house or apartment that occurs when someone rapidly closes a faucet or some appliance rapidly stops the flow of water at any device makes us all aware of the fact that halting the flow of moving water can create pressure.

In the street or the site main connection any requirement to support your incoming service main against back pressure caused by project building operations is the responsibility of the utility company or the site engineers. However, there are, and will be, occasions when the service shown on the plumbing plans is not a straight line from the main to the building service or meter location. One of the common mistakes in depicting this incoming line is to create on the plans an angle that is required to make the piping enter the building at the desired location. While it may be drawn that way, the

pipe installer cannot put it in as drawn. The installer must deal with standard fittings.

If the pipe is copper, a slight offset or bending can usually be accommodated by gently adjusting the angle slightly at each connection point, creating a deviation over a reasonable distance from a straight line. But do not count too heavily on this accommodation. Usually the installing contractor has to make all the changes in direction using $\frac{1}{16}$ (22.5°), $\frac{1}{8}$ (45°), or $\frac{1}{4}$ (90°) standard fittings.

Abrupt pressure changes in the water supply system can adversely affect these fittings causing them, over time, to loosen, break, crack, and leak. To guard against this possibility a bracing system known as thrust block is installed at all offsets of 45° angles and larger.

Figure 3.14 depicts a standard thrust block. The depth of concrete varies from 12 to 24 in. In addition the square outer end of the base is 12 in long to create the rectangle as shown. Notice that in both the 45° and the 90° block, the concrete thrust block sides are angled so that the sides are at right angles to the pipe. The 6D or 3D simply means that the distance is 3 or 6 times the pipe diameter that is being supported. This is a standard design. For the unusual pressure condition a force calculation should be made using, for example, Crocker's *Piping Handbook* published by McGraw-Hill or a similar source. A special base designed to meet these calculated results should then be installed.

FOUR

Water Distribution System

For water system design there are, in addition to this text, many sources of system design information. Two of the best known publications of these data are the ASHRAE *Fundamentals Handbook* and the McGraw-Hill *Piping Handbook*. In addition an authoritative source for both publications is the National Bureau of Standards (NBS) Report BMS79 entitled *Water Distribution Systems for Buildings*. Finally there is the Building Officials and Code Administrators International, Inc. (BOCA) *Basic Plumbing Code*.

Our purpose in this chapter is to present solutions to typical design and detailing problems using standard information primarily from the above named sources as well as to present field-tested, practical installation details. In the ASHRAE handbook, for example, a number of pipe sizing graphs used are credited to the original NBS data and are used in their calculations. Since data in other texts on various types of pipe are given in detail, we will limit our sizing information to a few common types of pipe material. Nothing in sizing changes when different types of pipe are used except that the pipe capacity varies at a given size and pressure drop per hundred feet. This is due to the variation in surface roughness and, consequently, the resistance and available capacity at a given pressure drop. The approach to the pipe-size problem remains the same.

In the water distribution system the resolution of the design and detailing begins with the service detailing shown in Chap. 3 and ends at the fixture or equipment connection. There is one caveat. The water pressure coming into the building must be sufficient to properly serve the highest fixture in the system. If it is not, something must be done immediately. Thus our chapter begins with the incoming service, the usual metering, and the detail solution if the pressure is insufficient for the use intended.

Water Pressure

Commonly the pressure of the domestic water supply systems available to the project being designed by the plumbing engineer is in the range of 40 psi [275.76 kilopascals (kPa)] to 60 pounds per square inch (psi) (413.6 kPa). But this is not always the case. The BOCA plumbing code, which is the basis of most codes, states the requirements very clearly. "When street main pressure exceeds 80 psi (551.52 kPa), an approved pressure-reducing valve shall be installed in the water service pipe near its entrance to the building to reduce the water supply pressure to 80 psi (551.52 kPa) or lower."

And conversely "whenever water pressure from the street main or other source of supply is insufficient to provide flow pressures as required in the table below, a booster pump and pressure tank, or other approved means, shall be installed on the building supply system."

Finally the code has one more requirement that can easily be overlooked. "Water supply to shower heads, lavatory faucets, kitchen faucets, and similar equipment shall be equipped with approved water saving devices, restricting and controlling flow to not more than 3 gallons per minute (gpm) [11.34 liters per minute (lpm)] at 80 psi (551.52 kPa) flowing."

In looking at Table 4.1, the chart of minimum pressures, one can quickly conclude that the final pressure to the farthest and highest fixture from the service entry could be as low as 8 psi and in most cases a safe all-

Table 4.1 Water Distribution System Design Criteria Required Capacities at Fixture Supply Pipe Outlets

Fixture supply outlet serving	Flow rate (gpm)	Flow pressure (psi)
Bathtub	4.00	8
Bidet	2.00	4
Combination fixture	4.00	8
Dishwasher, residential	2.75	8
Drinking fountain	0.75	8
Laundry tray	4.00	8
Lavatory	2.00	8
Shower	3.00	8
Shower, temperature controlled	3.00	20
Sillcock, hose bib	5.00	8
Sink, residential	2.50	8
Sink, service	3.00	8
Urinal, valve	15.00	15
Water closet, blowout, flushometer valve	35.00	25
Water closet, tank close coupled	3.00	8
Water closet, tank one piece	6.00	20
Water closet, siphonic, flushometer valve	25.00	15

around design value would be 15 psi. The value of 15 psi is also a code requirement.

Interior Meters. Most of the time the plumbing system under design in the detailer's office has a water meter installation in a boiler or equipment room. In all of these situations the meter installation is simply a building interior version of the meter installation depicted in Figs. 3.9 and 3.10 in Chap. 3. In addition there is an exterior shut-off valve or curb stop such as depicted in Fig. 3.8 in Chap. 3.

Figure 4.1 depicts two typical versions of the standard interior water meter installation. There are several plumbing code requirements that are included in our detail. The first is that there must be a gate valve or other full-way valve on the inside of the building where the service enters the building. In addition there must be a gate valve or other full-way valves on the discharge side of the meter. Both of these valves shall be full pipe size.

While not specifically required, the meter bypass depicted in our detail seems simply to be a common-sense requirement. Anytime you have any sort of metered service, gas, water, or electric, with a means of securing service without passing through the meter, there will always be a temptation to cheat. These bypasses do not have to be full size and they can be locked. If one wanted to cheat badly enough, one could, without a bypass, shut off both sides of the meter, remove the meter and install a piece of pipe in its place to get free service for as long as desired. Detection would be the utility company's problem.

On one of the two details we've added a couple of items that may easily be overlooked. These are a pressure-reducing valve (PRV) and valved system feeders. As noted in the beginning of this chapter, the allowable maximum water pressure is 80 psi. There are communities in many locations where the street or incoming building pressure exceeds 80 psi. This PRV is a code requirement when the excess pressure situations occur.

Not shown in the water meter detail is the clearance from the sewer line requirement. This is an item frequently and easily overlooked. The general code rule is that sewer and water lines to the building must be 10 feet (ft) apart horizontally. The plumbing code does give the local inspector some leeway in allowing sewer and water lines in the same trench provided the water line is on a shelf on one side of the trench and is at least 12 inches (in) above the sewer line based on comparative invert elevations. This rule is complex and, when a water line installation clearance problems exists, it should be checked with the appropriate code authorities.

High-Rise Supply Systems

As in a low-rise structure, the basic problem of the plumbing designer of a high-rise structure is to provide a potable supply of water to all fixtures. Added to that basic requirement is the need to supply a proper quantity of water for sprinkler and standpipe fire protection systems. In the discussion that follows the various solutions to this problem are covered under the heading "Supply Zoning."

The building's use of water plays a very large role in the type of zoning that is applied. The city or town zoning requirements may also limit the height of a building to 25 stories or less. As will be noted in the details presented in the supply zoning discussion, the water supply can be delivered with or without storage tanks. It is generally recommended as good practice to limit the size of storage tanks to 30,000 gallons (gal) [113,000 liters (L)].

Supply tanks have a number of requirements. They

should obviously be watertight and capable of containing the pressures under which they operate. In addition they should be rodent proof, vermin proof, and corrosion resistant. Tank supports should be noncombustible. Piping to and from the tanks must pass through floors using watertight sleeves.

The supply lines to the roof storage tanks should enter the tank above the tank overflow pipe to create an effective air gap. Overflow pipes should discharge within 6 in of the roof and onto a surface that effectively protects the roof or into a specifically designed overflow drain. The tank should also have a drain line to completely empty the tank, and this drain outlet should empty onto a splash block on the roof or into some other properly designed indirect waste.

Since the system must be safe from contamination, certain other factors must be considered. All overflow lines should be provided with size 100 mesh or greater

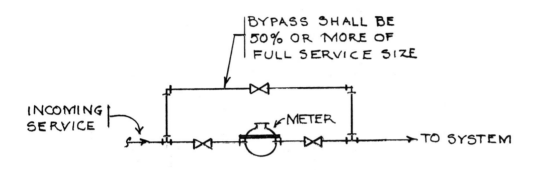

STANDARD METER INSTALLATION

— NO SCALE —

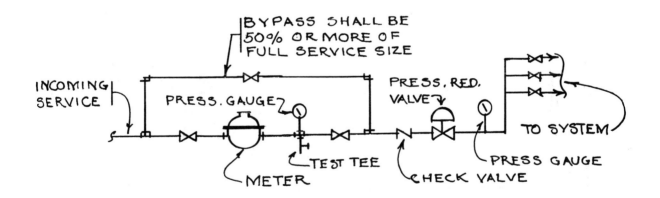

PRESSURE REDUCING METER INSTALLATION

— NO SCALE —

TYPICAL WATER METER INSTALLATIONS

FIGURE 4-1

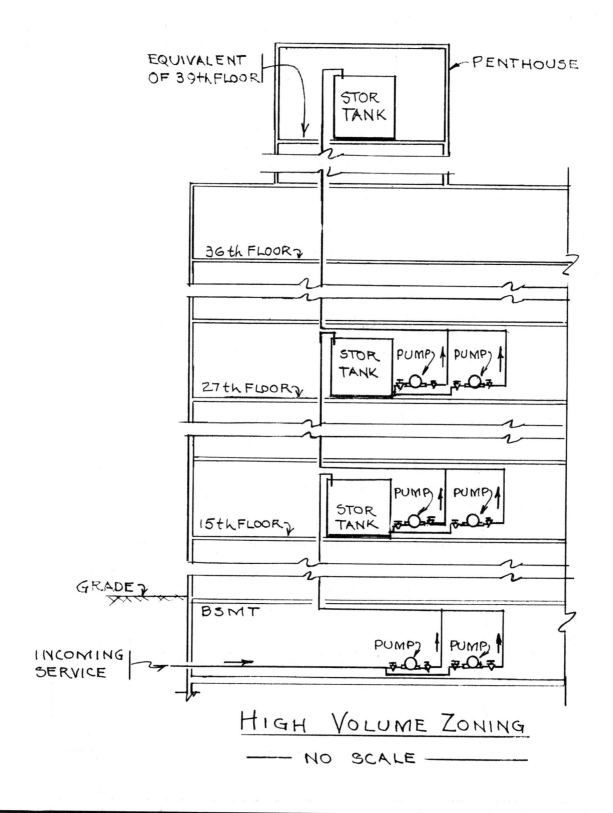

EQUIVALENT OF 39th FLOOR

PENTHOUSE

STOR TANK

36th FLOOR

STOR TANK   PUMP   PUMP

27th FLOOR

STOR TANK   PUMP   PUMP

15th FLOOR

GRADE

BSMT

INCOMING SERVICE

PUMP   PUMP

HIGH VOLUME ZONING

— NO SCALE —

FIGURE 4-2

screens. The tank and its service manholes should be located directly below soil or waste piping lines. Tanks under repair of any sort should be shut off from the system and repairs must not include material that alters the taste or quality of the water supplied.

The storage tanks illustrated in the details that follow must meet not only the domestic water supply need but also the fire supply need. The tanks depicted do have that capability, but the fire service piping is not shown to avoid cluttering the detail. The fire service piping is depicted in Chap. 9. For the same reason we have shown hot water supply piping in only one nonstorage-tank detail.

Supply Zoning

There are two basic requirements of any water supply system. The first, and most obvious, is to provide a service large enough to handle the peak demand of the building. Selecting, calculating, and sizing for this demand will be covered later in this chapter. The second, and what should be equally obvious but sometimes is overlooked, is the need to consider the static head.

We have made the point that a proper pressure at the highest fixture level is 15 psi. The plumbing code very clearly makes the point that the highest acceptable pressure is 80 psi. The difference of 65 psi (80 − 15) is what you have available under design conditions to overcome the static head created by the requirement to meet the 15 psi condition at your highest fixture. Adding a reasonable loss for pipe friction and a meter of 10 psi, the total loss is probably 25 psi.

A pressure of 80 psi is neither standard nor common for many utility-furnished water supply systems. A much more common value is 40 to 60 psi. In terms of pressure a water column of 1 ft in height exerts a pressure of 0.433 psi. In our ideal case of 80 − 15 − 10 the 55 psi available is a force to offset static head equaling 127 ft (55/0.433). Thus, if your highest fixture is no more than 127 ft above your service entry elevation, your system should function properly.

If we presumed a more normally encountered system pressure of 50 psi, the height of your highest fixture would be limited to very nearly 58 ft [(50 − 25)/0.433]. A normal five-story building with a floor-to-floor height of 12 ft is 60 ft high (5 × 12). Since your plumbing service is normally above the basement floor and your top floor plumbing fixture is relatively near the fourth floor level, the vertical difference is usually 58 to 60 ft. Before you quickly conclude that this very minor possible inadequacy of pressure is not important, we should point out that our simplistic assumption did not include other pipe pressure losses of probably 3 to 6 psi.

As we are going to cover pipe sizing in more detail later in this chapter, we shall continue to focus on the elevation and static pressure problem. Our representative sample shows that street pressures of 50 psi can certainly serve a four-story building and possibly serve a five-story building. Beyond five stories, we definitely need a pump to boost the pressure.

To look at the problem in the overall, Figs. 4.2 through 4.6 depict five ways of resolving the problem in multistory tall buildings. These five details are five ways of handling water supply pumps with storage tanks. As we noted in Chap. 3, energy use is conserved and pressure variations reduced or virtually eliminated in our depicted details. The pump only runs when the tank volume falls to a predetermined level. Its operation then refills the tank to a predetermined high level.

Easily overlooked in all of these cases is that the building water pressure, except where pneumatically supplied for the lower level as depicted in Fig. 3.6, is a function of the difference in elevation between the level of the water in the tank. The BOCA *Basic Plumbing Code,* a common code source for many cities and towns, requires a minimum of 8 psi at most fixtures with 15 psi for flush valves and similar type fixtures. Using our basic formula, the elevation of tank water surface must be 18 ft (8/0.433) or 35 ft (15/0.433) *above* the fixtures it serves.

Figure 4.2 depicts a system of two or more zones of supply. Since the maximum allowable pressure is 80 psi, the distance of the first tank to its lowest fixture served is 180 ft (80/0.433). Thus the effective usable height of structure served is 145 ft (180 − 35). At a building floor-to-floor height of 12 ft the usable 145 ft will cover 12 stories. Each zone height is 12 stories with the tank 35 ft above, which puts a tank on the fifteenth, twenty-seventh, thirty-ninth, etc. floors.

There are a number of advantages in the system depicted in Fig. 4.2. Each zone has an operating pump as well as a spare one. Since the pressure in each zone is nearly the same and is relatively low, higher pressure pipe and fittings are eliminated. The piping and control systems are both simple and repetitive, creating savings in installation cost. The pump efficiency is higher because the discharge head requirements for each pump are well within the limits of efficient operation for centrifugal pumps.

Figure 4.3 is a way to save money on the first cost of pumps and piping. The system, as depicted, has only one spare pump, which can serve any zone as required. There is nothing wrong with this design approach so long as it is clearly understood that certain limitations do exist. Realistic sizing of the one spare pump is based on the capacity to serve any one zone. Thus the pump must have the capacity to pump the required gallons per minute to the tank in the highest zone and at the same time must not overload its motor under a variety of operating conditions. Obviously when two zone pumps fail at the same time, there will be a shortage of capacity. Again the piping and controls are quite simple and repetitive. However, the pressure on the pipe of the spare pump requires special study and usually will result in extra heavy pipe and fittings being specified for this pump. In normal practice this sort of detail is used in

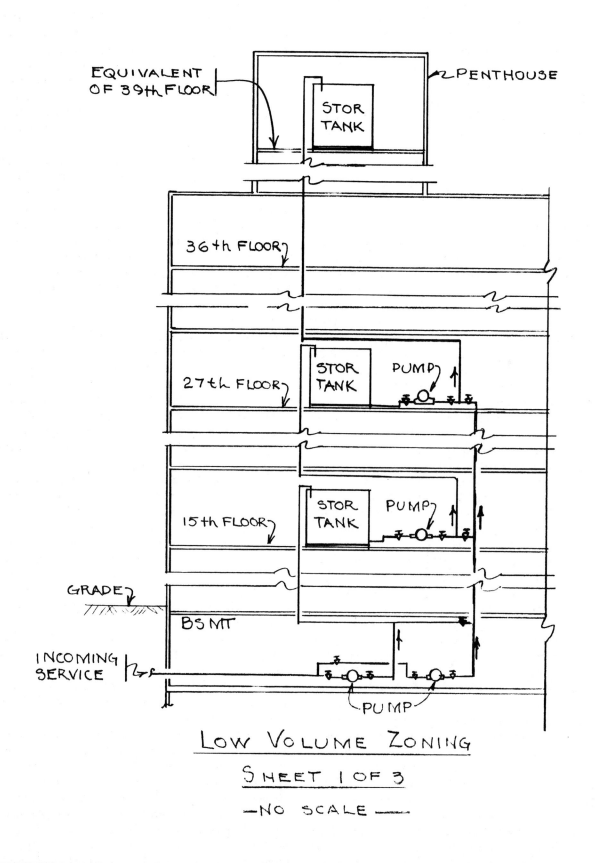

EQUIVALENT OF 39th FLOOR

PENTHOUSE

STOR TANK

36th FLOOR

27th FLOOR

STOR TANK

PUMP

15th FLOOR

STOR TANK

PUMP

GRADE

BSMT

INCOMING SERVICE

PUMP

LOW VOLUME ZONING

SHEET 1 OF 3

—NO SCALE—

FIGURE 4-3

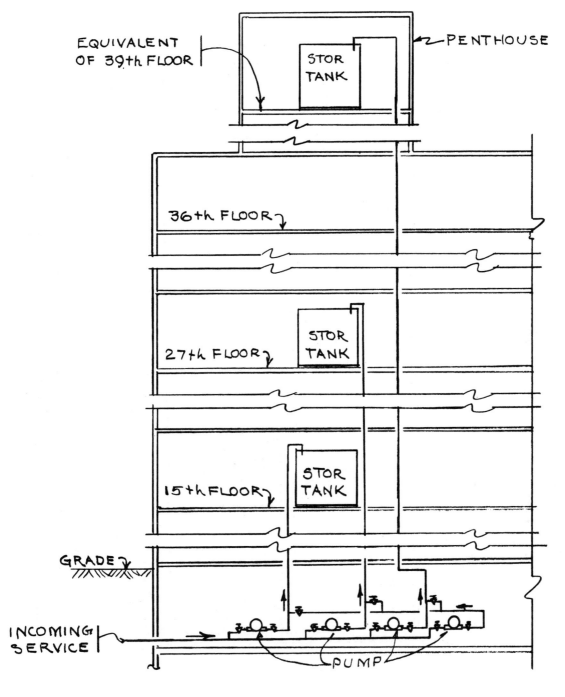

EQUIVALENT OF 39th FLOOR

PENTHOUSE

STOR TANK

36th FLOOR

27th FLOOR

STOR TANK

15th FLOOR

STOR TANK

GRADE

INCOMING SERVICE

PUMP

LOW VOLUME ZONING

SHEET 2 OF 3

— NO SCALE —

FIGURE 4-4

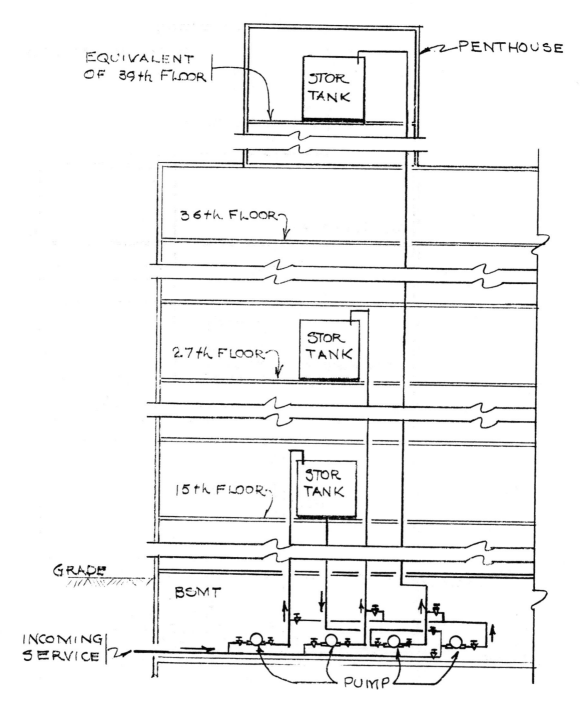

EQUIVALENT OF 39th FLOOR

PENTHOUSE

STOR TANK

36th FLOOR

27th FLOOR

STOR TANK

15th FLOOR

STOR TANK

GRADE

BSMT

INCOMING SERVICE

PUMP

LOW VOLUME ZONING

SHEET 3 OF 3

—NO SCALE—

FIGURE 4-5

situations in which there is a low flow, low consumption requirement. Figure 4.4 is another possible solution to a low flow, low consumption situation. All of the pumps are located at the bottom of the riser stack, which is usually a basement equipment room. The lower-level pump suction pressure is the utility supply pressure. The higher-level pumps take their suction from the first-level house tank. While this results in more frequent operation of the lower-level pumps, it also reduces the high discharge head requirement of the higher-level pumps. As in Fig. 4.3, all pumps are still in the basement.

Figure 4.5 is very similar to Fig. 4.4 with one basic difference. The supply for each pump is from the utility source. This reduces the operation of the first pump and, as an energy conservation measure, reduces pump operating cost. Offsetting this fact is the extra heavy piping requirement of the upper-level pumps as well as their lower efficiency at higher discharge pressure requirements.

Figure 4.6 is a combination of the pressure tank described in Chap. 3 and a gravity tank system. The first level is supplied by a pneumatic tank located in the

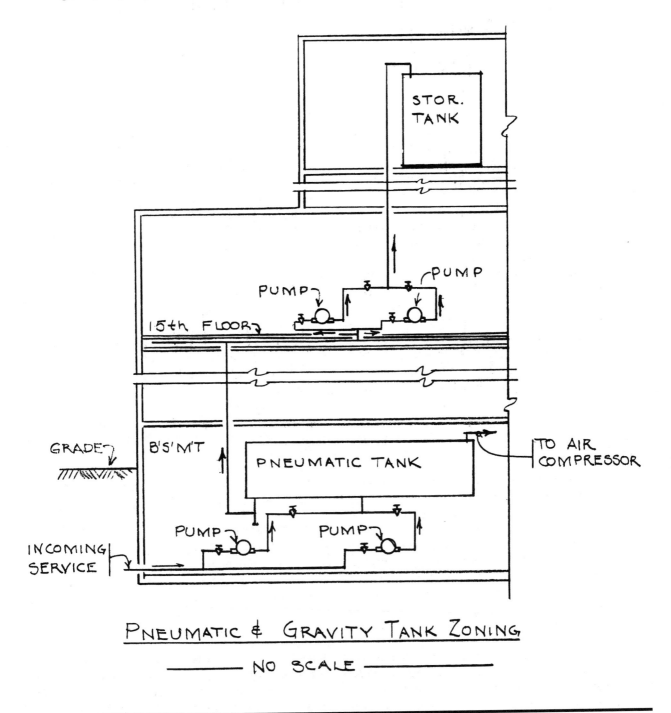

PNEUMATIC & GRAVITY TANK ZONING

—— NO SCALE ——

FIGURE 4-6

Table 4.2 Sizes for Overflow Pipes for Water Supply Tanks

Maximum capacity of water supply line to tank (gal/min)	Diameter of overflow pipe (in)	Maximum capacity of water supply line to tank (gal/min)	Diameter of overflow pipe (in)
0–50	2	400–700	5
50–159	2½	700–1000	6
150–200	3	Over 1000	8

Size of Drain Pipes for Water Tanks

Tank capacity (gal)	Drain pipe (in)	Tank capacity (gal)	Drain pipe (in)
Up to 750	1	3,001–5,000	2
751–1,500	1½	5,005–7,500	3
1,501–3,000	2	Over 7,500	4

basement equipment room. This is, as depicted, usually a duplex pump arrangement to resolve standby requirements. The upper zone of the building (and this may be more than the one zone depicted, although not usually) is supplied by operating and standby pumps which take as their source of supply some of the output of the pneumatic system. This installation, properly selected and designed, can provide good overall efficiency. The system also requires very little building area as the tanks, for the installation as depicted, are located in the building and in a rooftop penthouse. Within the floor levels depicted, the pump discharge heads are within the limit of good pumping efficiency and very little, if any, higher pressure pipe and fittings are required. One of the real secrets of this detail is the use of the first-level pressure loop as the suction pressure for second-level pumping. When space for equipment is at a premium, this is an ideal solution. Since both systems depicted have spare pumps, it will serve either light or heavy water use.

Tank System Notes.   One of the design criteria applicable to all of these five details is that all pumps or pairs of pumps shall be calculated to deliver a capacity as required by their respective zones served plus an added capacity of the pumps which receive their supply from the tank serving that zone. Thus, the capacities of the pumps will diminish with each succeeding zone. Another requirement is that hot water supply tanks, not depicted in these details, must be placed at the bottom of each zone. They can be placed in the room utilized by the tank serving the zone below. The hot water tank can operate on a circulating system or feed the floor below without circulation.

In system sizing data elsewhere in this chapter we cover demand, peak demand, and related requirements. For storage tanks the usable volume is the difference in capacity between high and low water levels. This capacity, called net usable capacity, should equal ½-hour

(hr) flow at peak demand. If it is lost in ½ hr, it must be replaced in ½ hr; otherwise during 1 hr of peak demand the tank may be dangerously low or empty. Thus, pump capacity should be equal to replacing what has been used in the same length of time.

Finally, for each of the details submitted, the electric controls that are required to operate the pumps are relatively simple. Each tank regardless of pumps or systems, is the source of water supply. In each case a combination of low water start and high water stop float control devices is all that is required. These level sensing devices start and stop the pumps. On each tank one further device could be added that is similar to the high-level alarm device used to preclude oil spills in oil bulk terminals. This device is set to operate above the normal high-level operating control. If the normal high-level control fails for any reason, the pump would continue to operate and the tank would overflow. The high-level alarm device could, before the tank actually overflows, sound an alarm and perhaps cut the power to the pump.

The pipe to the tanks is sized in accordance with sizing charts and tables which we will present later on in this chapter. The overflow and drain pipe sizing depicted in Table 4.2 is taken from the BOCA plumbing code.

Tankless Zoning

Figure 4.7 presents a different type of zoning. Because of the limited amount of material in this detail, we are also able to depict the hot water piping which we alluded to but omitted, for the sake of clarity, in Figs. 4.2 through 4.6.

For simplicity's sake our detail divides the building into two 12-story sections. It further presumes each section has a floor-to-floor height of 10 ft, making an overall sectional height of 120 ft. If the pressure must be a minimum of 25 psi at the bottom of a 12-story section, the pressure the pump must deliver at the top is the 25 psi plus the static head of 120 ft. This static head is the

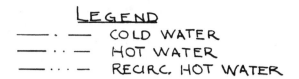

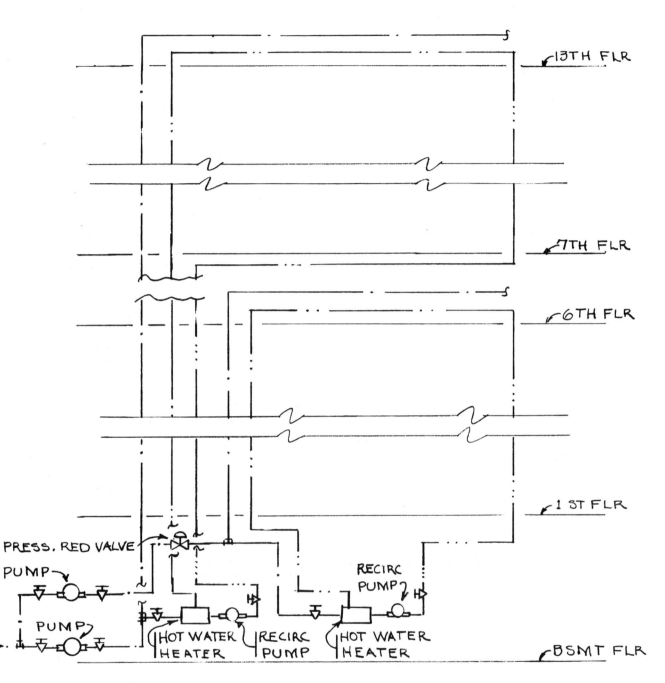

HIGH RISE PUMPED SUPPLY SYSTEM

— NO SCALE —

FIGURE 4-7

equivalent of some 52 psi [120 ft × 0.433 pounds per foot (lb/ft)]. Thus the pressure head of the pump must equal 77 psi (25 + 52). Referring to our previous discussion on zoning, you may note that we are almost at our 80-psi limit. In actuality, we would also have some pipe friction loss which could easily add another 2 or 3 psi to our load so, as a practical result, our pumping head is more or less 80 psi.

Our detail shows two pumps, one operating and one standby. These should be connected via an alternator arrangement so that each time pumping is required the other pump runs, creating equal use of both pumps.

As we have mentioned in previous discussions of well supply, one of the purposes of a storage tank is to preclude on-off operation from constantly occurring. This is a problem in this detail. Thus its application is more suited to reasonably constant demand. Since the multistory building is most likely an office or apartment building, this constant demand ideal is simply not realistic. Thus the pumps will require further engineering analysis and perhaps some form of frequency controller or pressure bypass to resolve low flow pumping requirements.

There is yet another point to note on this detail. The pumps must have the capability of delivering water to the top of the top section. Thus the discharge head of the pumps as shown in the detail is not 80 psi but rather 80 psi plus the load of another 120 ft of static head and accompanying line loss. We already calculated that a head of 120 ft plus some line loss equals 55 psi in round numbers. Therefore that part of the detail supplying the upper section requires pipe and fittings able to withstand a pressure of 135 psi (80 + 55). And the discharge head of both pumps must be rated and set at 135 psi.

This pressure is much too great for the lower zone. As can be seen on our detail, we have easily resolved this problem by adding a pressure-reducing valve to the line feeding the lower zone. This valve can readily be set to reduce the pressure to the required 80 psi.

The water heaters in each case are also arranged so that they have 135 psi and 80 psi properly imposed on the system requiring it. In periods of low flow or no flow the water would tend to cool off over the length of the piping. Consequently some recirculation is required. In each case pump head is very minor because the only load the pumps must overcome is line friction loss. As an energy conservation measure, the pump should be cycled to perhaps 10 minutes (min) of operation per hour. The hot water system, even without a pump, would recirculate, albeit slowly, similarly to the old-fashioned gravity domestic hot water heating system.

System Demand

Paramount to the design of any water supply system using any source or combination of sources is the quantity of water that will be necessary to satisfy the demand of the system. As we have noted in both this chapter and the preceding one, system demand has two values. The first is peak demand. That is the most water that will be used at any particular time. And equally important to well and storage tank supply systems is how much water will be used in a given 24-hr period. The utility source may, in general, be presumed to be constantly available. The well, as we have previously pointed out, has its limits.

Sources of Information

There are a number of sources of information on the calculation of system demand. Some of the better known ones are the American Society of Heating, Refrigeration and Air Conditioning Engineers (ASHRAE) *Fundamentals Handbook;* the BOCA *Basic Plumbing Code;*

Table 4.3 Supply Fixture Unit Values for Various Plumbing Fixtures

Fixture or group	Type of supply control	Supply fixture unit values		
		Hot	Cold	Total
Bathroom group	Flush tank	3.0	4.5	6.0
Bathroom group	Flush valve	3.0	6.0	8.0
Bathtub	Faucet	1.5	1.5	2.0
Bidet	Faucet	1.5	1.5	2.0
Combination fixture	Faucet	2.0	2.0	3.0
Kitchen sink	Faucet	1.5	1.5	2.0
Laundry tray	Faucet	2.0	2.0	3.0
Lavatory	Faucet	1.5	1.5	2.0
Pedestal urinal	Flush valve		10.0	10.0
Restaurant sink	Faucet	3.0	3.0	4.0
Service sink	Faucet	1.5	1.5	2.0
Shower head	Mixing valve	3.0	3.0	4.0
Stall or wall urinal	Flush tank		3.0	3.0
Stall or wall urinal	Flush valve		5.0	5.0
Water closet	Flush tank		5.0	5.0
Water closet	Flush valve		10.0	10.0

Table 4.4 Fixture Unit Values

Name of fixture	Number on system	Supply fixture unit value per fixture (Fig. 4.7)			Total supply fixture units		
		Hot	Cold	Total	Hot	Cold	Total
Bathroom group	24	3.0	6.0	8	72	144	192
Kitchen sink	24	1.5	1.5	2	36	36	48
Total					108	180	240
Demand (gpm)					48	62	72

and the books *Piping Handbook* and *Standard Plumbing Engineering Design* published by McGraw-Hill. All of these texts, and others as well, use the National Bureau of Standards publication *Water-Distribution Systems for Buildings* by Dr. Roy B. Hunter, published in November 1941, as their primary source of data. The basic fixture unit method, the piping flow charts, and the curves relating demand to capacity have been reproduced relatively unchanged in all the above texts on the subject.

Fixture Valves. In designing a system one first has to select some common set of values. Table 4.3 was our selection of a logical source of fixture unit values. This table is a copy of that which appears in the 1984 BOCA *Basic Plumbing Code*. It also closely resembles the table in the ASHRAE *Fundamentals Handbook*. The data in both of these publications closely match the original 1941 National Bureau of Standards report. The BOCA table has added individual hot and cold water demand for each fixture type. Note that the sum of the hot and cold values is larger than the total. This is because the BOCA code is depicting separate probable peak values with the presumption that both do not occur simultaneously. The ASHRAE handbook addresses this hot and cold value by stating that the peak value for either hot or cold use may be calculated as 75 percent of the total.

A fixture unit by definition is a unit of value equal to a flow of 7.5 gpm. When you take a shower it is possible, but unlikely, that you can use, as per Table 4.3, four fixture units or 30 gpm of water (4×7.5). Theoretically, assuming a typical residential well, an 80-gal storage tank, and a 10-gpm well pump, the system could run out of water in a little more than 3 min. We also call your attention to another piece of data. A bathroom group with a flush tank water closet has a total value of six fixture units. Taken individually the bathtub has a value of two, the lavatory has a value of two, and the water closet has a value of five. This totals nine units. Obviously the table reflects the simple logic that it is unlikely that all three fixtures will be used simultaneously.

Sample Calculations. To use a simple example, assume you have a 24-unit apartment house with 24 bathroom groups and 24 kitchen sinks. A simple table of values could be constructed as in Table 4.4, using Fig. 4.8 and Tables 4.3 and 4.5.

Demand Curves. The value in gallons per minute is the numbers taken from Fig. 4.8. These two curves are also items of data taken from the same three sources: BOCA, ASHRAE, and the National Bureau of Standards. The curves originally appeared in the Bureau of Standards report and are probability curves, that is, curves relating fixture units to probable rather than peak demand. Using these curves and going back to the example of taking a shower, the well isn't going to run dry because the shower isn't going to use 30 gpm but rather something on the order of 3 to 5 gpm.

Daily Consumption. For those who have, as part of their site utilities, a source of water from the local utility company there is not a great deal of concern over the total quantity of water used in a given 24-hr period. But, as was pointed out in our previous discussion on water supply systems fed by a well, the daily use as well as the peak hourly use are both items of serious concern.

Table 4.5 is taken from the 1981 BOCA *Basic Plumbing Code*. Values as given in this table have also appeared from time to time in other publications. No one pretends that this sort of data is absolutely accurate. Although through both testing and calculation reasonable demand data can and has been secured for the peak output of a given type of plumbing fixture, it is far more difficult to measure total use in a meaningful fashion. The data are simply being presented as some of the best that can currently be found.

As an example of what we mean, let us take a typical example from this author's experience as a practicing professional consulting engineer. In Table 4.5 there are three different values given for water requirements in schools. These values, as given, vary from 15 to 25 gal per person per day for the school's occupants. These are fairly standard values that, again based on the au-

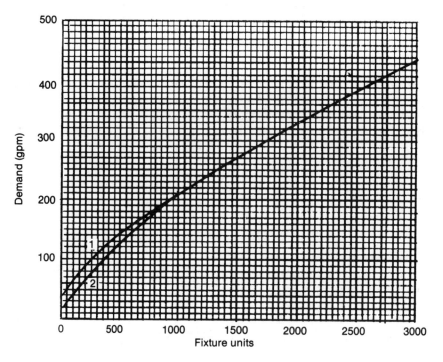

No. 1 for system predominately using flush valves

No. 2 for system predominately using flush valves

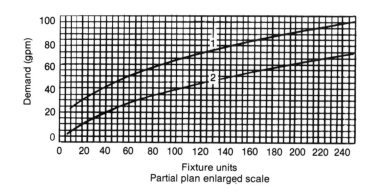

Fixture units
Partial plan enlarged scale

FIGURE 4-8
Fixture units/demand chart.

Table 4.5a   Design Criteria for Daily Water Requirements Based on Building Occupancy

Type of occupancy	Minimum quantity of water per person per day (gal) (or as indicated)	Type of occupancy	Minimum quantity of water per person per day (gal) (or as indicated)
Small dwelling and cottages with seasonal occupancy	50	Day schools, with cafeterias, but no gymnasiums or showers	20
Single family dwellings	75	Day schools with cafeterias, gymnasiums, and showers	25
Multiple family dwellings (apartments)	60	Boarding schools	75–100
Rooming houses	40	Day workers at schools and offices (per shift)	15
Boarding houses	50	Hospitals (per bed)	150–250
Additional kitchen usage for nonresident boarders	10	Institutions other than hospitals (per bed)	75–125
Hotels without private baths	50	Factories (gallons per person per shift)	15–35
Hotels with private baths (2 persons per room)	60	Picnic parks (toilet usage only, gallons per picnicker)	5
Restaurants (toilet and kitchen usage per patron)	7–10	Picnic parks with bathhouses, showers, and flush toilets	10
Restaurants (kitchen usage per meal served)	2½–3	Swimming pools and bathhouses	10
Additional for bars and cocktail lounges	2	Luxury residences and estates	100–150
Tourist camps or trailer parks with central bathhouse	35	Country clubs (per resident member)	100
Tourist camps or mobile home parks with individual bath units	50	Country clubs (per nonresident member)	25
Resort camps (night and day) with limited plumbing	50	Motels (per bed space)	40
Luxury camps	100–150	Motels with bath, toilet, and kitchen range	50
Work or construction camps (semipermanent)	50	Drive-in theaters (per car space)	5
Camp (with complete plumbing)	45 (ind. w.s.)	Movie theaters (per auditorium seat)	5
Camp (with flush toilets, no showers)	25 (ind. w.s.)	Airports (per passenger)	3–5
Day camp (no meals served)	15	Self-service laundries (gallons per wash, i.e., per customer)	50
Day schools, without cafeteria, gymnasiums, or showers	15	Stores (per toilet room)	400
		Service stations (per vehicle serviced)	10

Table 4.5b   Daily Water Requirements for Common Farm Animals

Animal	Minimum daily water requirements (gal)
Horse, mule, or steer	12
Dairy cow (drinking only)	15
Dairy cow (drinking and dairy servicing)	35
Sheep	2
Hog	4
Chickens (100)	4
Turkeys (100)	7

thor's personal experience, have been around as such for at least 30 yr and have been used both as well design criteria and septic tank and field design criteria. Based on actual tests of four schools designed by the author with pupil loading between 500 and 800, no school came close to these values. The junior high with a gym and shower facility ranked the highest at 12 gal per occupant per day.

This personal-experience record of four schools is certainly not statistically representative. For all the author really knows, it may be simply four variations from the norm. The point in all this discussion is to be certain that the user understands that these are representative numbers and not guaranteed values.

Using the data in Table 4.5 is a perfectly straightforward exercise. One simply *and carefully* defines the number of occupants and multiplies this number by the value given. Some values are easier to define than others. For example a 100-bed hospital, according to the table, has an estimated water consumption of 15,000 to 25,000 gal per day (100×150 up to 100×250).

This, however, is not the total final answer on water requirements. To the overall value obtained from the table one must add *separately* the special-use quantities. The common definition of special use is those items not normally part of the expected plumbing system in the building in question. Usually these are specialized machinery, process or equipment demands, or extraneous items such as lawn sprinklers which are normally calculated at 5 gpm per head. Your problem here is not only to find the special-use value but to determine how long per day the item is in use. The values, as determined, must be *added* to the total daily use as calculated from Table 4.5.

This total flow determination is critical in individual projects supplied by a well. If, for example, a judgment is made that a worst-case scenario applies to our above-noted 100-bed hospital, we need 25,000 gal (100×250) of water per day as we pointed out in our previous chapter on wells. Six hours of pumping per day is a reasonable value drawdown for a well, which is equal to 360 min (6×60). Dividing 25,000 gal by 360 min equals 69.44 gpm or 70 gpm rounded. If your well, or wells, cannot safely be pumped at that rate every day of the year, you have a problem no amount of storage will resolve. On the advice of experts in well performance you may be able to pump the required quantity at a lower rate over a longer period of time. Do not attempt to use your own judgment in this situation. Be certain you consult a proper expert in this field and be guided by the advice you receive.

Charts and Tables

There are a number of possible items that affect plumbing water supply design. All of them are available in the texts noted previously. Since some items are used only once in our pipe sizing discussion that follows we felt they could simply be given with a specific source reference. However, in pipe sizing there are two items to which we will refer repeatedly and thus should readily be available. These data are covered in Figs. 4.9 through 4.13 and Tables 4.6 through 4.10. All these data really originated in the more than 40-yr-old report by Roy B. Hunter entitled *Water-Distributing System for Buildings*. Since that time these data in part and in total have appeared in many codes and textbooks. Our source, reprinted and reproduced by permission, comes from the 1984 BOCA *Basic Plumbing Code*. The data are self-explanatory and are briefly described in the paragraphs that follow.

Figure 4.9 and Table 4.6 give flow data on various types of meters. Each was selected for ease of use for solutions to design problems. The fairly common meter, the disk type, is presented graphically in Fig. 4.9 since there can be several choices of meter size especially in the 30-gpm to 150-gpm flow range. For example, at 30 gpm the meter can be ¾ in, 1 in, 1½ in, or 2 in depending on the amount of pressure loss your system can absorb. Since one usually tries to hold pressure losses to 10 psi, the two logical choices are 1 in and 1½ in. Table 4.6 lists other types of meters and specific gallons per minute and pressure loss data as per the American Water Works Association (AWWA) standards C-700-77, C-701-78, and C-702-78. These data are set up in a specific typeset format and the maximum values as stated should not be exceeded. For a very large project requiring flows in excess of the values given, the incoming service should be divided into two or more parallel meters. Further along in this book a detail will be presented showing this sort of installation.

There are a variety of water distribution piping materials permitted by the plumbing code. Figures 4.10, 4.11, 4.12, and 4.13 show friction loss versus flow graphs for types K, L, and M copper, fairly smooth pipe, fairly rough pipe, and rough pipe, in that order. Related to these four figures are Tables 4.7, 4.8, 4.9, and 4.10. These four tables try to make your graph reading easier and less error prone by presenting the data in tabular form. For a given flow in gallons per minute the related pressure loss for a given size has been organized for ease of selection.

While this tabular format may seem easier to use, it does have one drawback. You only are made aware of values that exceed 10 or 15 feet per second (fps) in flow velocity. As you will see in our pipe sizing discussion that follows, the ideal flow velocity rate is between 4 and 8 fps. However, once you have become familiar with both the graphs and the tables, you will develop a fairly accurate sense of what the related velocity is. In doubtful cases you should always verify the velocity situation by double-checking the data on the related graph.

The graph and tabular data for copper pipe are clearly identified as such in Fig. 4.10 and Table 4.7. Figure 4.11 and Table 4.8 depict fairly smooth pipe which is commonly clean screwed or welded steel and brass. Figures 4.12 and 4.13 and Tables 4.9 and 4.10 depict corroded or seriously corroded steel or brass pipe. As we noted previously, not all types of pipe are covered. Plastic piping is a glaring example. However, the difference in values between plastic and the copper piping in Fig. 4.10 and Table 4.7 are not substantial. For our system explanation purposes we really already have more pipe flow than we actually need.

Pipe Sizing Criteria

In the flow of fluids, including water which is our area of concern, there are two controlling criteria which at times are seemingly in conflict with each other. These two items are pressure loss described as the friction loss per square inch per foot or per hundred feet and velocity in terms of feet per second. Using Fig. 4.10 for example, a constant friction loss of 5 psi/100 ft is a reasonable design ideal. At low flows, say 10 gpm, the velocity is 5

fps. If we maintain that constant loss of head due to friction, we find that at 220 gpm of flow the velocity is 10 fps, and at 850 gpm it is 15 fps. This becomes significant when we also are made aware of the fact that the ideal noise-free flow of water should be in the range of 4 to 8 fps. And further for branches of piping serving quick closing valves a velocity of 4 fps is needed to reduce wear and tear on the quick closing device as well as to reduce the effect of back pressure caused by the quick closing, which is commonly called water hammer. Thus, any proper pipe sizing design effort should consider both pressure and velocity.

Difference in elevation between the city or utility supply main frequently is a major factor in piping designs. Again looking at Fig. 4.10 one can follow the 8-fps line and find that we can get 10 gpm through a ¾-in line at a pressure loss of some 15 psi/100 ft. Commonly these pipe sizes are involved in fairly short runs of pipe. Where, for instance, we have 50 psi in the street main, a meter,

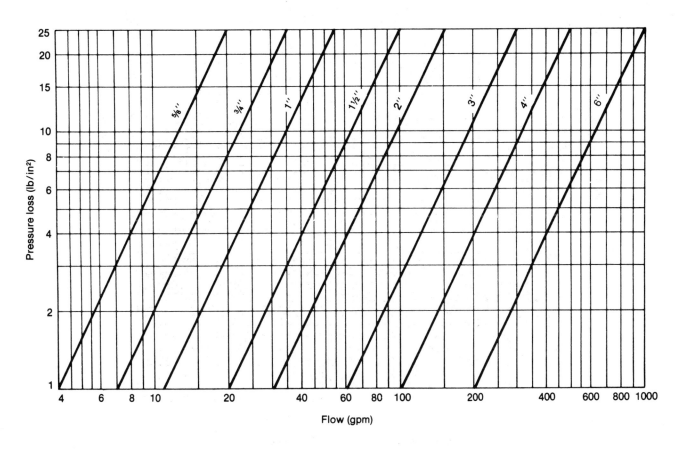

Loss of pressure through disk-type meters (lb/in²)

FIGURE 4-9
Disk-type meter criteria.

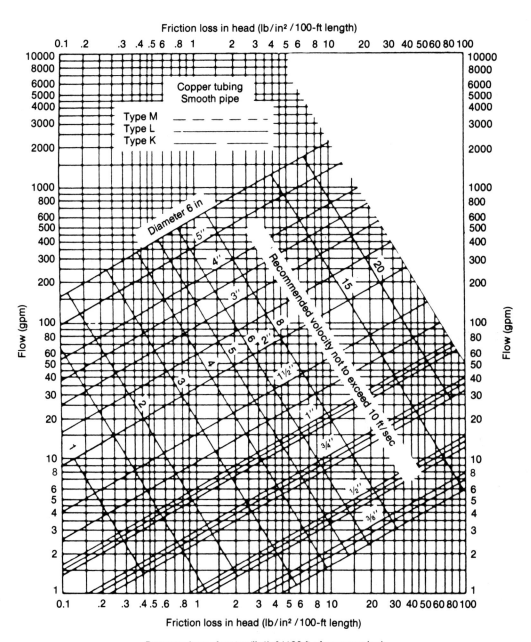

FIGURE 4-10
Copper pipe sizing chart.

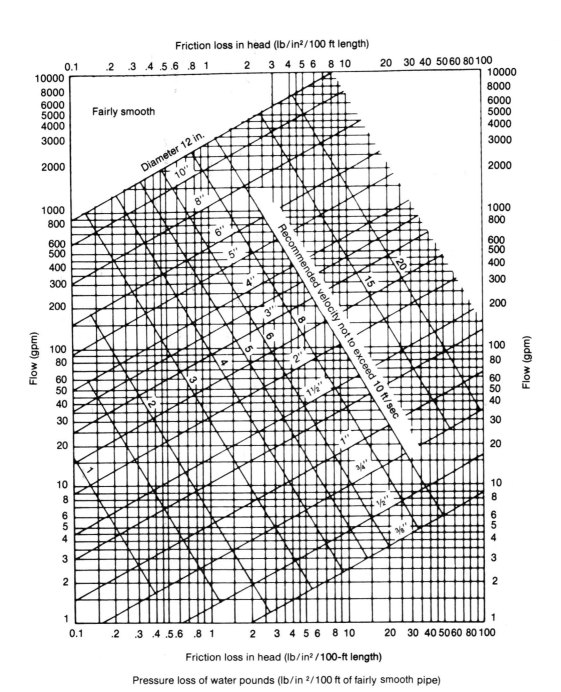

Friction loss in head (lb/in²/100 ft length)

Pressure loss of water pounds (lb/in²/100 ft of fairly smooth pipe)

FIGURE 4-11
Fairly smooth pipe sizing chart.

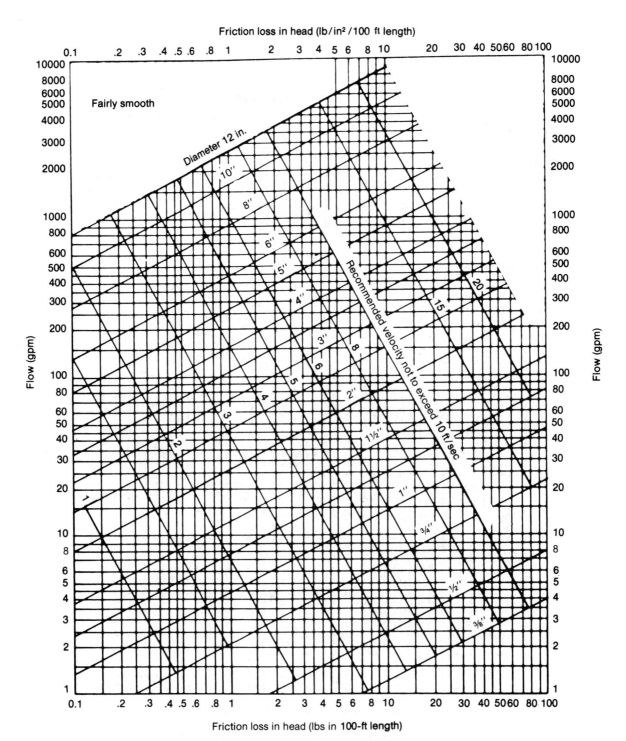

Friction loss in head (lb/in² /100 ft length)

Pressure loss of water (lb/in²/100 ft of fairly rough pipe)

FIGURE 4-12
Fairly rough pipe sizing chart.

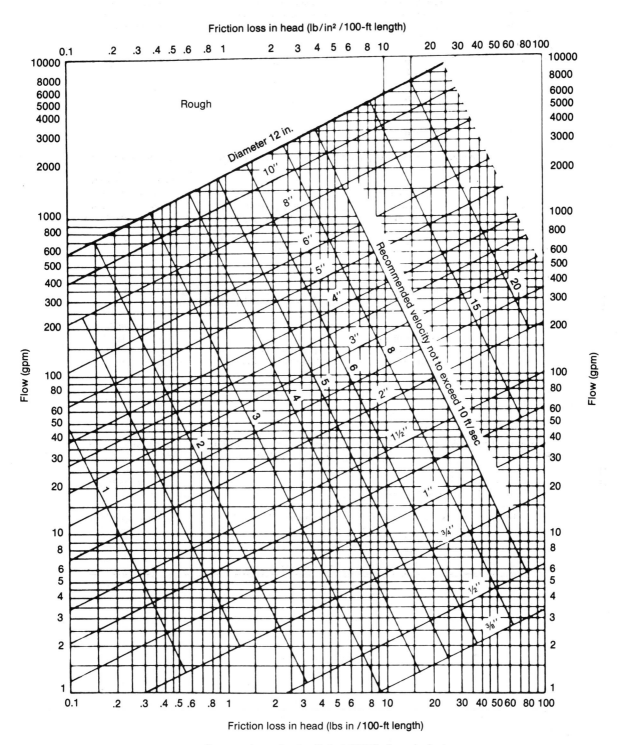

FIGURE 4-13
Rough pipe sizing chart.

Table 4.6 Operating Characteristics for Water Meters

Displacement Type
(AWWA Standard C-700-77, Cold Water Meters, Displacement Type)

Meter size (in)	Safe maximum operating capacity (gpm)	Maximum pressure loss at safe maximum operating capacity (psi)	Recommended maximum rate for continuous operations (gpm)
⅝	20	13	10
⅝ × ¾	20	13	10
¾	30	13	15
1	50	13	25
1½	100	15	50
2	100	15	80
3	300	15	150
4	500	15	250

Compound Type (AWWA Standard C-702-78, Cold Water Meters, Compound Type)

Meter size (in)	Safe maximum operating capacity (gpm)	Maximum rate for continuous duty (gpm)	Maximum allowable loss of head at safe maximum operating capacity (psi)
2	160	80	20
3	320	160	20
4	500	250	20

Turbine Type (AWWA Standard C-701-78, Cold Water Meters, Turbine Type for Customer Service; Class 1, Vertical Shaft and Low-Velocity Horizontal Type)

Meter size (in)	Safe maximum operating capacity (gpm)	Maximum rate for continuous duty (gpm)	Maximum loss of head at safe maximum operating capacity (psi)
1½	100	50	15
2	160	80	15
3	350	175	15
4	600	300	15

Turbine Type (AWWA Standard C-701-78, Cold Water Meters, Turbine Type for Customer Service; Class 11, In-Line (High-Velocity Type)

Meter size (in)	Safe maximum operating capacity (gpm)	Maximum rate for continuous duty (gpm)	Maximum loss of head at safe maximum operating capacity (psi)
2	160	100	7
3	350	240	7
4	630	420	7

Table 4.7   Pressure Loss of Water in Pounds per Square Inch (psi) per 100 ft of Copper Pipe

Gallons per minute	Inches									
	¾[a]	1[a]	1¼	1½	2	2½	3	4	5	6
	0.17									
	0.21									
1	0.27									
	0.56	0.16								
	0.66	0.18								
2	0.84	0.21								
	1.15	0.31								
	1.30	0.37								
3	1.70	0.42	0.13							
	1.80	0.51								
	2.20	0.61								
4	2.80	0.69	0.22	0.10						
	2.70	0.76								
	3.30	0.90								
5	4.20	1.05	0.34	0.15						
	3.80	1.10								
	4.50	1.25								
6	5.70	1.40	0.47	0.21						
	5.00	1.40								
	6.00	1.70								
7	7.50	1.80	0.61	0.27						
	6.10	1.70								
	7.20	2.10								
8	9.60	2.40	0.68	0.33						
	7.60	2.20								
	8.90	2.50								
9	12.00	2.80	0.93	0.42	0.12					
	9.10	2.60								
	11.50	3.00								
10	14.50	3.50	1.20	0.50	0.14					
	18.50	5.20								
	23.00[b]	6.10								
15	28.00[b]	7.10	2.40	1.10	0.27					
	32.00[b]	8.90								
	37.00[b]	9.90								
20	45.00[b]	12.00	3.80	1.70	0.44	0.16				
	46.00[c]	13.00								
	53.00[c]	15.00								
25	67.00[c]	17.00[b]	5.80	2.50	0.68	0.23				
		18.00[b]								
		21.00[b]								
30		24.00[b]	8.00	3.50	0.91	0.32	0.13			
		24.00[b]								
		27.00[b]								
35		32.00[b]	11.00	4.60	1.25	0.42	0.17			
		30.00[b]								
		33.00[c]								
40		38.00[c]	13.00[b]	5.80	1.50	0.52	0.22			
		37.00[c]								
		43.00[c]								
45		48.00[c]	17.00[b]	7.10	1.85	0.66	0.28			
50			19.00[b]	8.70	2.30	0.79	0.33			
60			27.00[c]	12.00	3.10	1.20	0.46	0.12		
70				16.00[b]	4.20	1.40	0.62	0.16		
80				19.00[b]	5.20	1.80	0.79	0.20		
90				24.00[c]	6.20	2.25	0.96	0.24		
100					7.60[b]	2.75	1.20	0.30	0.11	
150					17.00[c]	5.80[b]	2.50	0.62	0.22	
200						9.30[b]	4.00	1.10	0.36	0.15
250						14.00[c]	6.10[b]	1.60	0.52	0.22
300							8.40[c]	2.10	0.72	0.31
350							12.00[c]	2.80	0.98	0.41
400								3.50	1.25	0.52
450								4.30[b]	1.60	0.63
500								5.20[b]	1.80	0.76
600								7.20[c]	2.70[b]	1.15
700									3.40[b]	1.40
800									4.40[c]	1.80
900									5.20[c]	2.20[b]
1,000										2.70[b]

[a]For the ¾-in and 1-in pipe sizes, the three values shown opposite each flow figure are, reading from the top, for types M, L, and K copper tubing respectively.
[b]Velocity at or exceeding 10 fps.
[c]Velocity exceeds 15 fps.

Table 4.8   Pressure Loss of Water in Pounds per Square Inch (psi) per 100 ft of Fairly Smooth Pipe

Gallons per minute	¾ᵃ	1ᵃ	1¼	1½	2	2½	3	4	5	6	8	10	12
							Inches						
1	0.16												
2	0.57	0.17											
3	1.20	0.37	0.10										
4	2.00	0.61	0.17										
5	3.00	0.95	0.25	0.12									
10	11.00	3.50	0.90	0.43	0.13								
15	22.00	7.10	1.80	0.90	0.26	0.11							
20	39.00ᵃ	13.00	3.00	1.50	0.45	0.18							
25	58.00ᵇ	18.00	4.70	2.30	0.68	0.28	0.10						
30		25.00ᵃ	6.60	3.20	0.93	0.40	0.13						
35		35.00ᵃ	8.50	4.30	1.20	0.53	0.18						
40		43.00ᵇ	11.00	5.50	1.60	0.63	0.22						
45			14.00	6.70	2.00	0.80	0.30						
50			17.00ᵃ	8.10	2.40	1.00	0.35	0.10					
60			23.00ᵃ	12.00	3.30	1.30	0.50	0.13					
70			32.00ᵃ	15.00ᵃ	4.40	1.80	0.63	0.17					
80				19.00ᵃ	5.70	2.30	0.83	0.23					
90				24.00ᵃ	7.00	2.90	1.10	0.27					
100				30.00ᵇ	8.50	3.70	1.80	0.35	0.12				
150					17.00ᵃ	7.80ᵃ	2.60	0.70	0.23				
200					30.00ᵇ	30.00ᵃ	4.50ᵃ	1.20	0.40	0.16			
250						18.00ᵇ	6.30	1.80	0.59	0.23			
300							9.00ᵃ	2.40	0.80	0.34			
350							13.00ᵇ	3.30	1.10	0.45	0.12		
400								4.20ᵃ	1.30	0.59	0.15		
450								5.10ᵃ	1.70	0.70	0.19		
500								6.20ᵃ	2.10	0.85	0.23		
600								9.00ᵇ	2.90	1.20	0.32	0.11	
700									3.90ᵃ	1.60	0.42	0.14	
800									4.90ᵃ	2.00	0.56	0.18	
900										2.50ᵃ	0.69	0.23	
1,000										3.00ᵃ	0.81	0.23	0.12
1,500										6.50ᵃ	1.80	0.59	0.24
2,000											3.00ᵃ	0.98	0.40
2,500											4.50ᵇ	1.50ᵃ	0.61
3,000												3.00ᵇ	0.89

ᵃVelocity at or exceeding 10 fps.
ᵇVelocity exceeds 15 fps.

Table 4.9 Pressure Loss of Water in Pounds per Square Inch (psi) per 100 ft of Fairly Rough Pipe

Gallons per minute	Inches												
	¾[a]	1[a]	1¼	1½	2	2½	3	4	5	6	8	10	12
1	0.26												
2	0.91	0.22											
3	2.00	0.47	0.17										
4	3.30	0.82	0.30	0.12									
5	5.20	1.30	0.45	0.18									
10	20.00	4.90	1.70	0.67	0.17								
15	43.00[a]	12.00	3.70	1.40	0.36	0.12							
20	80.00[b]	18.00	6.20	2.50	0.62	0.20							
25		29.00[a]	9.90	3.90	0.97	0.31	0.13						
30		42.00[a]	14.00	5.60	1.30	0.45	0.18						
35		55.00[a]	18.00	7.30	1.80	0.60	0.25						
40		70.00[b]	24.00[a]	9.30	2.30	0.75	0.32						
45			30.00[a]	12.00	3.00	0.96	0.42						
50			37.00[a]	15.00	3.70	1.20	0.51	0.12					
60			52.00[b]	21.00[a]	5.20	1.70	0.70	0.17					
70				28.00[a]	7.00	2.20	0.92	0.22					
80				37.00[a]	9.00	2.90	1.30	0.29	0.10				
90				45.00[b]	122.00	3.70	1.50	0.36	0.12				
100					14.00[a]	4.60	1.80	0.44	0.16				
150					30.00[b]	10.00	4.20	1.00	0.34	0.13			
200						17.00[a]	7.00[a]	1.70	0.59	0.23			
250						26.00[b]	11.00[a]	2.60	0.90	0.35			
300							15.00[a]	3.60	1.30	0.50	0.12		
350							21.00[b]	4.90	1.70	0.69	0.17		
400								6.10[a]	2.20	0.88	0.22		
450								7.60[a]	2.70	1.10	0.27		
500								9.40[a]	3.30	1.30	0.33	0.11	
600								13.00[b]	4.90[a]	1.80	0.46	0.15	
700									6.20[a]	2.50	0.61	0.20	
800									8.10[a]	3.30	0.80	0.26	0.11
900									11.00[a]	4.10[a]	1.00	0.33	0.13
1,000									13.00[a]	5.00[a]	1.25	0.40	0.17
1,500										12.00[b]	2.80	0.90	0.37
2,000											4.70[a]	1.60	0.63
2,500											7.20[b]	2.40[a]	1.00
3,000												3.40[a]	1.30

[a]Velocity at or exceeding 10 fps.
[b]Velocity exceeds 15 fps.

Table 4.10 Pressure Loss of Water in Pounds per Square Inch (psi) per 100 ft of Rough Pipe

Gallons per minute	\(\tfrac{3}{4}\)[a]	1[a]	1¼	1½	2	2½	3	4	5	6	8	10	12
					Inches								
1	0.31												
2	1.20	0.27											
3	2.70	0.62	0.20										
4	4.70	1.20	0.36	0.15									
5	6.00	1.40	0.46	0.18									
10	30.00	7.00	2.30	0.94	0.22								
15	67.00	16.00	6.20	2.10	0.49	0.17							
20		27.00	9.10	3.70	0.89	0.29	0.12						
25		43.00[a]	14.00	5.80	1.30	0.45	0.18						
30		62.00[a]	21.00	8.50	2.00	0.63	0.27						
35		85.00[a]	28.00	12.00	2.70	0.90	0.36						
40			37.00[a]	14.00	3.50	1.20	0.47	0.12					
45			47.00[a]	19.00	4.50	1.45	0.60	0.14					
50			58.00[a]	23.00	5.50	1.80	0.74	0.18					
60			83.00[b]	33.00[a]	7.90	2.60	1.10	0.25					
70				48.00[a]	12.00	3.50	1.40	0.35	0.12				
80				60.00[a]	14.00	4.70	1.85	0.45	0.15				
90				76.00[b]	18.00	5.90	2.30	0.58	0.19				
100					23.00[a]	7.20	3.00	0.71	0.23				
150					50.00[b]	17.00	6.60	1.70	0.53	0.21			
200						29.00[a]	12.00	2.90	0.95	0.37			
250						45.00[b]	18.00[a]	4.50	1.49	0.58	0.13		
300							26.00[a]	6.40	2.20	0.80	0.19		
350							36.00[b]	8.90	2.90	1.20	0.27		
400								12.00[a]	3.80	1.45	0.35	0.12	
450								15.00[a]	4.70	1.80	0.44	0.14	
500								18.00[a]	6.00	2.30	0.55	0.18	
600								25.00[b]	8.30	3.20	0.78	0.26	0.11
700								25.00[b]	8.30	3.20	0.78	0.26	0.11
800									16.00[a]	6.00	1.40	0.47	0.19
900									20.00[b]	7.70[a]	1.80	0.60	0.24
1,000										9.40[a]	2.30	0.75	0.31
1,500										22.00[b]	5.10	1.70	0.70
2,000											9.00[a]	3.00	1.25
2,500											14.00[b]	4.70[a]	2.00
3,000												6.80[a]	2.70

[a]Velocity at or exceeding 10 fps.
[b]Velocity exceeds 15 fps.

and a two-story structure, a loss through the meter of 10 psi on average, a minimum pressure of 15 psi, and the loss caused by static head of 20 ft of elevation (20 × 0.433 or 8.66 psi), the sum of these three items is 33.66 psi, leaving 16.34 psi to overcome friction. If, because of fixture location, the final run of pipe were 100 ft of ¾-in pipe serving 10 gpm with a 15-psi final pressure requirement, one could assume a potential problem existed. Our previously selected ¾-in pipe at 8 fps had a built-in frictional loss of 15 psi/100 ft. The 100-ft run would utilize all but 1.34 psi (16.34 − 15) of the available pressure.

Included in the above discussion of the effect of elevation was the loss through the water meter. Most plumbing design situations involve a choice of at least two different meter sizes, both of which have the capability of serving the load. Unless good design data reasons are given to the utility company, their standard approach is to install a meter one or most likely two sizes smaller than the service pipe main. Commonly the one-size-smaller-than-the-main selection results in a 4-psi meter loss and the two-sizes-smaller selection results in a 10-psi to 12-psi loss that could become a serious problem in poorly designed, marginal installations.

Special loss-producing items such as backflow preventers should be carefully accounted for in the piping design. The backflow preventer can create a very large loss which may well be as large at the loss through the meter. A value of 9 psi of loss through the preventer is not uncommon. Other items such as losses caused by street connections and losses through water heaters should also be noted. Because only the hot water line includes the heater pressure loss, it is not uncommon for the controlling pressure drop due to line, both length and equipment, to be the hot water line and not the cold water line.

Table 4.11 System Pressure Criteria

Item	System losses (psi)	Available source (psi)
Street pressure available		80
Highest pressure required at a fixture	15	
3-in meter loss (3-in meter from Fig. 4.10)	15	
4-in main tap-in loss (4-in tap in 8-in street main, negligible loss, assume 1 psi)	1	
Special system losses (none; special item such as backflow preventer entered here)	0	
Static head loss (0.43 × 80-ft elevation difference = 34.54 or 35 psi rounded)	35	
Totals	65	80
Available pressure		15

Total actual length (measured length of longest run)	180 ft
Total equivalent length (run length plus allowance for fittings and connections)	300 ft
Pipe allowable pressure loss/100 ft (15 psi/300 ft × 100 ft = 5 psi/100 ft)	5 psi

Typical Toilet Room Loads

Fixture	Quantity	Hot	Cold	Total	Quantity	Hot	Cold	Total
Water closet	2	—	20.0	20	3	—	30.0	30
Urinal	2	—	10.0	10				
Lavatory	2	3.5	3.0	4	3	4.5	4.5	6
Service sink	1	1.5	1.5	2	1	1.5	1.5	2
Totals	7	4.5	34.5	36	7	6.0	36.0	38

Restaurant kitchen demand is based on actual demand which is primarily the dishwasher demand for hot water and an estimated demand based on use for cold water. In this instance it is estimated as 20 gpm for hot water and 10 gpm for cold water.

Sizing the Piping System

Table 4.11 and Figure 4.14 are companion pieces. Figure 4.14 is a typical riser diagram for a two-stack plumbing system of an office building, 60 × 200 ft rectangular, containing 12,000 square feet (sq ft) of gross area per floor. Near each end, along the 200-ft dimension, are male and female toilet rooms. On the ground, or first, floor at one end is a small restaurant that serves breakfast and lunch to the building occupants and the neighborhood.

As would be expected, all male toilet rooms as well as all female toilet rooms are identical to each related room on the various floors. Each male toilet room has two flush valves, water closets (WC), wall urinals (UR), and lavatories (LAV) and one service sink (SS). The female toilet rooms have three WCs, three LAVs, and one SS.

Table 4.11 depicts two tabulations: one for the basic system pressure loss and one for the typical toilet room fixture count. The fixture counts from Table 4.3 are 10 WCs, 5 URs, and 1.5 hot, 1.5 cold LAVs for a total of 2, and 1.5 hot and 1.5 cold SSs for a total of 2.

In the type of building depicted in Table 4.11 it is far more economical to have small individual electric water heaters to serve each toilet area. Thus there is no separate hot water heater in the basement with long runs of recirculating hot water lines. The building is a slab-on-grade structure. The incoming service is in an equipment room in a partial basement area. In this equipment room there is a separate hot water heater for the restaurant.

There are two cold water stacks depicted in Fig. 4.14. Since these lines each carry water to be used for both hot and cold water purposes, the line loads are based on the total fixture unit demand. The first step therefore is to depict the fixture counts at each branch. Our detail shows 74 fixture units which is the sum of the 36 male

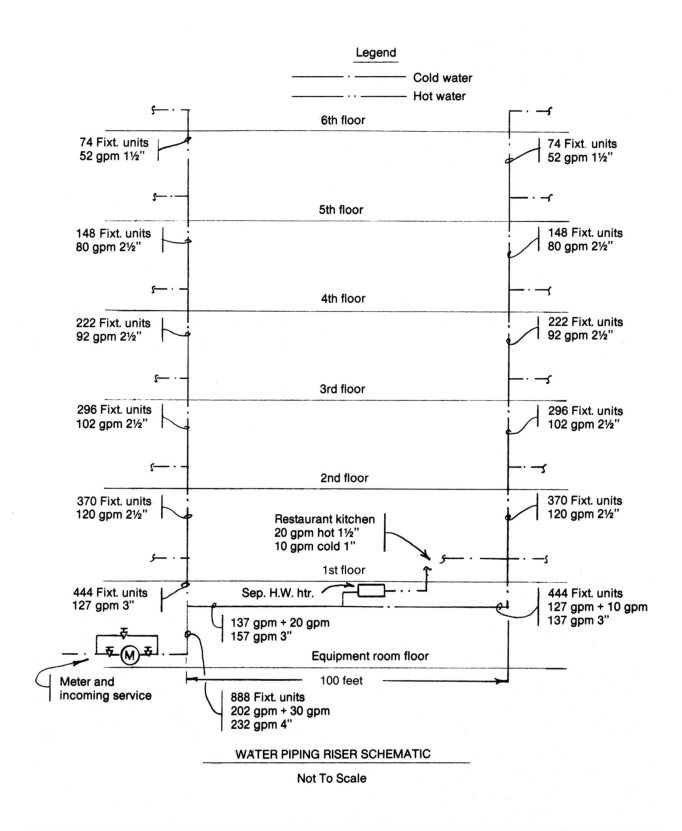

Legend

— · — Cold water
— ·· — Hot water

6th floor

74 Fixt. units
52 gpm 1½"

74 Fixt. units
52 gpm 1½"

5th floor

148 Fixt. units
80 gpm 2½"

148 Fixt. units
80 gpm 2½"

4th floor

222 Fixt. units
92 gpm 2½"

222 Fixt. units
92 gpm 2½"

3rd floor

296 Fixt. units
102 gpm 2½"

296 Fixt. units
102 gpm 2½"

2nd floor

370 Fixt. units
120 gpm 2½"

370 Fixt. units
120 gpm 2½"

Restaurant kitchen
20 gpm hot 1½"
10 gpm cold 1"

1st floor

444 Fixt. units
127 gpm 3"

Sep. H.W. htr.

444 Fixt. units
127 gpm + 10 gpm
137 gpm 3"

137 gpm + 20 gpm
157 gpm 3"

Equipment room floor

100 feet

Meter and
incoming service

888 Fixt. units
202 gpm + 30 gpm
232 gpm 4"

WATER PIPING RISER SCHEMATIC

Not To Scale

FIGURE 4-14

and 38 female total units shown on Fig. 4.14. We then go to Fig. 4.8 and, using the enlarged scale plan and curve 1, get our related gallons per minute, which is 52.

Our pipe size is based on two related criteria, friction loss per 100 ft and velocity in feet per second. Going back to Fig. 4.14, which we will shortly discuss in detail, we find our allowable friction loss is 5 psi/100 ft. In this case, we also have a limiting factor of 8 fps since we have no quick closing devices. Finally our pipe sizing depends on the type of pipe we are using. For this particular detail the pipe is presumed to be type L hard drawn copper tubing with soldered joints. Thus our friction chart is depicted in Fig. 4.10 and Table 4.7. Our task is to size the system based on 5-psi/100 ft loss and to *limit* any size selected to a velocity of 8 fps.

Using the above criteria and Fig. 4.10, we can start at the horizontal base of the chart on the 5-psi line and proceed upward to approximately 65 gpm. At that point we run into our 8-fps limit line. Therefore any value *above* 65 gpm *must* be read by following along the diagonal 8-fps limit line.

Going back to Fig. 4.14 and starting at the top, we begin to accumulate additional fixture units at each floor. For example where the line is below the fifth floor take-off we have 148 fixture units. This is double the single floor load of 74 fixture units. It is *not* double the single floor value of 52 gpm, which would be 104 gpm. Using Fig. 4.8 as before, we find that 148 fixture units equal 80 gpm. The value of 80 gpm exceeds our 65-gpm, 8-fps limit. We must follow the 8-fps limit line until it intersects with the vertical side value of 80 gpm. This gives us a value of barely more than a 2-in line. In our detail we decide to be conservative and call for a 2½-in line. We could just as easily have selected a 2-in line since the practical difference is so small.

Using this method of adding fixture units and selecting the gallons per minute from Fig. 4.8, we finally reach the total value of 444 fixture units for each riser. However, one riser also has the restaurant load imposed on it. The restaurant load is primarily a dishwasher load, and this load is secured from the dishwasher manufacturer if at all possible. The ASHRAE *Systems Handbook* gives hot water values for all sorts of equipment if one has no factual data. We arbitrarily, for our detail, selected hot and cold water gallons-per-minute values. The proper procedure is to find the gallons per minute required for your normal fixture load and add the actual values for restaurants, lawn sprinklers, and the like directly to it (using Fig. 4.8).

For our purposes we have 888 fixture units, which is 202 gpm from Fig. 4.8 plus 30 gpm for the restaurant making a total of 232 gpm at the meter. This value, using Fig. 4.10, Table 4.7, and the 8-fps limit line, requires a 4-in service line.

As we previously noted, we will now cover the pressure loss values in Table 4.11. We have attempted to make everything completely self-explanatory on the table. At the top of the table we have listed the pressure available and the losses in two columns. The first item, street pressure available, must be secured from the utility company. If there is any doubt, a test should be made at the nearest available fire hydrant. Usually the utility company has a fairly small charge for making such a test. The second item is taken from the BOCA *Basic Plumbing Code* for a flush valve water closet, which is commonly one of the highest fixtures in most buildings. Certainly the lavatory is slightly higher but requires a much lower pressure.

All the remaining items each have their own explanation of source and calculation right on the table. While the actual length can easily be verified by looking at Fig. 4.14, the equivalent length, which is noted as 300 ft, is an arbitrary value. In your design the proper way to determine the equivalent length is to count all the likely valves and fittings from the source to the end of the longest run. This, too, is somewhat arbitrary as you have no way of knowing how many fittings the installer will actually use, but your count, if carefully done, can be fairly accurate. This will make for a fairly accurate equivalent length, which is the actual length plus the added length allotted for all pipes, valves, and fittings. A reasonable presumption frequently used by piping designers is to increase the actual length by 50 percent to allow for valves and fittings. We used a figure of 66⅔ percent simply to give us a rounded value of 300 ft for our sample problem.

As all experienced designers know and beginners quickly learn, design calculations are subject to reasoned judgment and experience. If ever there was a place where we would say "don't fool around," it is with the values at the top of Table 4.11. If you do not have enough street pressure to overcome the basic losses shown in a table similar to ours, no matter what games you play you will not have water flowing at the end of your line. Do not assume differently. Put in a booster pump in your supply line. There is *no* alternative.

Pipe Sizing Notes. In the presentation of pipe sizing we have just made in Table 4.11 and Fig. 4.14 we really covered a large amount of data in a limited amount of space. Certain questions that might naturally arise concern use of the chart in Fig. 4.8, connections to individual fixtures, branch sizing, and individual hot water heaters. Using data from the BOCA *Basic Plumbing Code* and a separate detail, we will present information that will assist in answering these questions.

Table 4.12 contains a series of data values for related flow to fixture unit taken from the curves presented in Fig. 4.8. This simplifies your finding gallons-per-minute flows especially for low fixture count values. As you may easily note in looking at the data, the points taken, common to tables of this type, are fairly far apart for

Table 4.12 Demand Data: Maximum Probable Flows for Various Water Supply Fixture Unit Counts

Water supply fixture units	Maximum probable flow (gpm)		Water supply fixture units	Maximum probable flow (gpm)	
	Tank type water closets	Flushometer type water closets		Tank type water closets	Flushometer type water closets
1	1.0		120	25.9	75.7
2	3.0		125	26.5	76.5
3	5.0		130	27.1	77.3
4	6.0		135	27.7	78.1
5	7.0	27.2	140	28.3	78.8
6	8.0	29.1	145	29.0	79.6
7	9.0	30.8	150	29.6	80.3
8	10.0	32.3	160	30.8	81.6
9	11.0	33.7	170	32.0	82.9
10	12.2	35.0	180	33.3	84.2
12	12.4	37.3	190	34.5	85.3
14	12.7	39.3	200	35.7	86.5
16	12.9	41.2	220	38.1	88.6
18	13.2	42.8	240	40.5	90.5
20	13.4	44.3	260	43.0	92.3
22	13.7	45.8	280	45.4	94.0
24	13.9	47.1	300	47.7	95.6
26	14.2	48.3	400	59.6	102.0
28	14.4	49.4	500	71.2	107.0
30	14.7	50.5	600	82.6	113.0
35	15.3	53.0	700	93.7	117.0
40	15.9	55.2	800	105.0	120.0
45	16.6	57.2	900	115.0	123.0
50	17.2	59.1	1,000	126.0	126.0
55	17.8	60.6	1,500	175.0	175.0
60	18.4	62.3	2,000	220.0	220.0
65	19.0	63.8	2,500	259.0	259.0
70	19.7	65.2	3,000	294.0	294.0
75	20.3	66.4	3,500	325.0	325.0
80	20.9	67.7	4,000	352.0	352.0
85	21.5	68.8	4,500	375.0	375.0
90	22.2	69.9	5,000	395.0	395.0
95	22.8	71.0	6,000	425.0	425.0
100	23.4	72.0	7,000	445.0	445.0
105	24.0	73.0	8,000	456.0	456.0
110	24.6	73.9	9,000	461.0	461.0
115	25.3	74.8	10,000	462.0	462.0

Table 4.13 Minimum Sizes of Fixture Supply Pipes

Fixture	Supply pipe (in)	Fixture	Supply pipe (in)
Bathtub	½	Restaurant sink	½
Clothes washer	½	Service sink	½
Combination fixture	½	Shower head	½
Dishwashing machine	⅜	Wall hydrant	½
Drinking fountain	⅜	Wall urinal, flushometer	1
Kitchen sink	½	Wall urinal, tank	⅜
Laundry tray	½	Water closet, flushometer	1
Lavatory	⅜	Water closet, tank	⅜
Pedestal urinal	¾		

values above 300 fixture units. We would suggest that you always use the curves for values larger than 300 fixture units.

Our typical branch line to back-to-back male and female toilet rooms carried a total estimated value of 74 fixture units. In a typical installation the farthest point of the branch feeds four water closets—two in each toilet room. These total some 40 fixture units or 15.9 gpm using the data in Table 4.12, and going back to Fig. 4.10 and our 5-psi/100 ft pressure drop, this requires a 1¼-in type K, L, or M copper pipe since the 1-in size is a borderline case. Here, some local codes and a lot of experienced designers have another rule that requires a 1-in line for one flush valve water closet, a 1¼-in line for two, and a 1½-in line for three. This rule is also based on good quiet operation of branch piping, which mandates a velocity of 4 fps. We are in general agreement with the proponents of all the above noted rules and reasons. At the end of the line you are really designing very short pieces of pipe. And a 32-in length or so of a 1¼-in versus a 1-in line or a 1½-in versus a 1¼-in line is a saving obviously but a very minor one. Certainly it is not worth the possible noise and flow aggravation of the smaller line.

In Table 4.13 we have again borrowed from the BOCA *Basic Plumbing Code* to present a very simple table of pipe sizes to individual fixtures. This table should be on the plumbing detail drawing or in the plumbing specification document. There's really no need to draw riser details covering every line to every fixture. Both the plans and specifications should call for the connections to be made, and no plumbing contractor could seriously debate this work requirement. Furthermore the plumbing contractor knows the plumbing code. Rather than trying to infer one document into another by merely saying "connect as per code," it is far clearer to simply put this table on the plans or in the specifications.

There is another item in pipe sizing that should be especially noted. Carelessly sizing of some parts of the piping can create situations in which the water velocity suddenly increases say from 4 to 10 fps. This can create turbulent flow. Since friction varies directly with the square of velocity, the pressure drop at 10 fps is 6.25 times that at 4 fps. This energy loss creates a scouring action on the pipe, eroding its surface. It also creates noise and, if the pipe system harmonics happen to fall correctly into place, it creates considerable noise in the whole system. Similar abrupt velocity changes through fittings and valves can produce a physical phenomenon known as "cavitation" with very severe noise. Cavitation occurs, for example, at an elbow where the velocity change produces a subatmospheric pressure on the inside of the bend creating air bubbles that form and collapse. This rapid creation and dissipation of air bubbles is the source of the noise.

Water Hammer Arrestor (Shock Absorber). As we have seen in all of the calculations for pipe sizing, the basic design adopted usually resulted in water flow velocity of 6 to 10 fps. To give you a mental picture of the speed of 6 to 10 fps, 60 miles/hour (mph) equals 88 fps. So a speed of 8.8 fps equals 6 mph.

The standard simplified equation that translates velocity into pressure (psi) equals 63.6 × velocity (fps). If you could instantaneously close a valve on a line flowing at 10 fps, you could create 636 psi (10 × 63.6). This pressure would easily blow apart your piping system.

The facts are that the potential of this enormous pressure buildup cannot occur in the practical sense simply because you cannot instantaneously close any valve, pipe is somewhat elastic, and all water contains some air and dissolved gases that act as a pressure cushion.

A standard formula for pressure buildup in a straight run of pipe is

$$P = \frac{0.027LV}{T}$$

where P = compression pressure, psi
L = length of straight pipe, ft
V = flow velocity, fps
T = valve closing time in seconds

In 80 ft of straight pipe, with a velocity of 8 fps and a reasonably fast closing time of 0.2 seconds the result would be as follows:

$$P = \frac{0.027 \times 80 \times 8}{0.2} = 86.4$$

Notice that the above equation has nothing to do with system pressure. The force created depends entirely on how far, how fast, and how quickly the flow stops. The 86.4-psi back pressure creates a force that easily overwhelms the probable 25-psi fixture supply pressure and reverberates throughout the system.

The logical solution is to replace the quick closing valve with a slow closing valve. This is not always practical. Figures 4.15 and 4.16 depict two practical alternatives. Either put an expansion chamber at each fixture connection or put a standard manufacutred water hammer arrestor device on the branch lines feeding each group of fixtures. In the trade these devices are also known as shock absorbers.

Figure 4.15 represents the time-honored old-fashioned way of handling the expansion problem. At each fixture there is a pipe which, at least in the beginning, has a cushion of air which acts as the absorber in the expansion of water caused by sudden changes in pressure. There is nothing wrong with the premise of the detail *except* that over time the air in expansion pipe is absorbed by the water and the cushion no longer exists.

If the system is drained and refilled, the expansion chamber will again function as intended. This required draining is seldom done in actual operating practice.

Figure 4.16 depicts four typical applications of a commercially available water hammer arrestor. In most of your projects you will require more arrestors than the ones depicted. For many commercial applications such as dishwashers and laundry equipment you will need to install an arrestor on each hot and cold supply line just ahead of the final connecting points at the equipment quick closing valves. If a large toilet or shower room has, for example, more than six fixtures with quick closing valves, you may need two arrestors, one at the midpoint of the branch line serving the fixtures and another at the end of the branch main serving the final fixture on the branch line. The catalogs of manufacturers supplying shock absorbers usually supply the data on proper shock absorber selection.

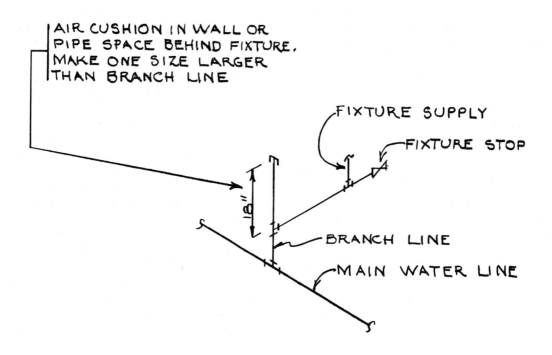

INDIVIDUAL FIXTURE SHOCK ABSORBER

NOT TO SCALE

FIGURE 4-15

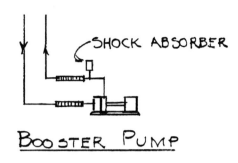

SHOCK ABSORBER

BOOSTER PUMP

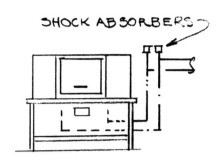

SHOCK ABSORBERS

DISHWASHER

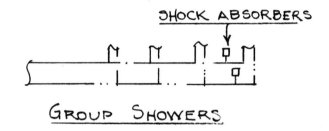

SHOCK ABSORBERS

GROUP SHOWERS

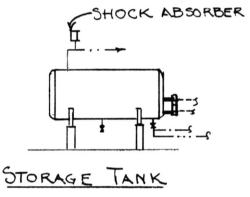

SHOCK ABSORBER

STORAGE TANK

SHOCK ABSORBER

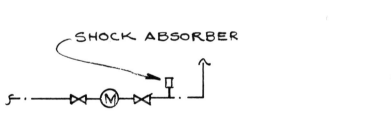

SERVICE

TYPICAL INSTALLATION
OF
WATER HAMMER ARRESTORS

NOT TO SCALE

FIGURE 4-16

FIVE

Water Heating

Until the drastic increase in energy prices of the 1970s the heating and distribution of domestic hot water was not considered very seriously. When serious thought was given to the subject, the emphasis was on space requirements and equipment cost. Everyone "knew" that electrical energy was expensive and electric hot water heaters had a slow rate of recovery. The choice usually involved gas or oil as the fuel. When a boiler was the building heat source, the domestic hot water could be obtained from a coil in the boiler water with or without a storage tank or from a separate heater and storage tank or a tank and heater combination. In forced warm air systems, or when domestic hot water was to be separately generated, a separate oil or gas fired water heater was specified. The electric industry countered with off-peak rates, special rates for all electric systems, and other inducements. Frequently little or no thought was given to the distance between source and use. To resolve any problem that might be caused by line losses one merely specified a mechanical, sometimes gravity, re-circulating system. This casual, almost offhand view of the domestic hot water service does not imply that the systems were not properly designed. In fact many were overdesigned to be certain of ample hot water supplies.

In this chapter we will first cover, in an overview fashion, how to size the system. Then we will present pertinent installation details depicting the proper methods of connecting the system components. As we proceed through the connection details, we will comment on the effect on energy consumption that is inherent in each detail.

The days of cheap energy are definitely a thing of the past. And the days of abundant supplies of water are rapidly running out in many sections of the country. Even in the past when energy was low in cost and water supplies seemingly endless not everyone nor every project treated the supply of hot water casually. Specialized systems for certain institutions, hospitals, schools, colleges, photographic and x-ray labs, and industrial processes have always been concerned with specific control of temperature volume and quality. But even in these cases the emphasis was on control not operating cost.

Consumption

All domestic water heating products use energy to both heat water to the desired temperature and to maintain water in any storage vessel at the required temperature. If the storage vessel and its heater are too small to handle peak demand, there will be times when the system is too small for the load imposed. One might say that is an obvious conclusion. To which might be added that it is also a common problem of overzealous energy saving or just a plain unvarnished design error.

Far more common is the oversized system. Here, in many cases, both the output of the water heater, called the recovery factor, and the size of the storage tank are far in excess of actual need under any circumstance. Systems over 15 years old commonly have this problem. Their designers had some bad experience in the past and were never going to have that occur again.

System Load Values

The first problem is to properly define the domestic hot water supply requirement. Probably the best, although certainly not the only, source of hot water demand data is in the American Society of Heating, Refrigeration and Air Conditioning Engineers' (ASHRAE) *Systems Handbook*. Return for the moment to Chap. 4 and Fig. 4.7 where we presented some data on hot water consumption. Those values came from the 1984 Building Officials & Code Administrators International (BOCA) *Basic Plumbing Code*. This, as we shall see, is best used in instantaneous heater design. For storage tank and heater design, gallon-per-hour (gph) data should be used.

Since we feel that nothing is to be gained by simply repeating all the data contained in the ASHRAE *Systems Handbook* on service water heating, we are going to refer you to that book for detailed information. However, so that this book and especially this chapter provide proper design and detail information, we will present in the data that follow hot water requirements as composite values using the ASHRAE data as a source. The following is our first composite table which, in effect, shows average values for a variety of buildings:

Hot Water Demand (gph) per Fixture at 140°F Final Temperature

Fixture	Quantity
Private lavatory	2
Public lavatory	4
Bathtub	20
Dishwasher	15–225[a]
Shower	30–225[b]
Service sink	20
Circular wash sink	20
Semicircular wash sink	60
Kitchen sink	0.4
Storage factor	0.6–2.0[c]

[a]The value of 15 is residential. A minimum commercial value would be 150. Many restaurant machines have values of 225 to 400. If at all possible, get actual data from the equipment manufacturer.

[b]Residential use is usually 15, commercial use might reasonably be 100, and heavy use for an athletic facility may well be 225.

[c]Most storage factors vary from 0.6 to 1.0. The exceptions are apartment houses at 1.25 and offices at 2.0.

Sizing Water Heating Equipment

To illustrate the problem of sizing equipment let us assume we have a small industrial plant with a fairly dirty manufacturing operation located in a fairly isolated area. The plant contains two sets of toilet rooms, male and female, with a total of 20 lavatories. These are obviously public, not private, lavatories. In addition the plant provides two wash-up rooms, male and female, with a total of four showers, two circular and two semicircular wash-up sinks with associated drying, and locker spaces. Finally, because the plant location is not near any restaurant, the company provides a small cafeteria with a food preparation space containing a dishwasher. The calculation would appear as follows, using previous data:

Item	Quantity	Rate	Total
Lavatory	20	4	80
Shower	4	150	600
Circular sink	2	20	40
Semicircular sink	2	10	20
Dishwasher	1	280[a]	280
Kitchen sink	3	60[b]	180
Total			1,200
Demand factor 0.4 (0.4 × 1,200)			480
Storage factor 1.0 (1.0 × 1,200)			1,200

[a]Data from equipment manufacturer.

[b]Typical double pot sink.

Based on the above answers, the probable maximum demand is 480 gallons (gal) of 140°F water per hour, the maximum hourly demand is 1,200 gph, and the storage tank capacity is 1,200 gph. This means that the hot water heater should have a recovery factor (which is another way of saying output capability) of heating 480 gph, and the properly insulated storage tank should have the capacity to hold 1,200 gal. Since a cubic foot of water contains 7.5 gal, the tank would be 160 cubic feet (cu ft) (1,200/7.5), which is a tank with a 4-ft 6-in diameter and 10 ft 0 in long.

Checking Your Answer. The step taken in our last sentence in the above description should be the first thing you do once the calculation is made. That is, size the storage tank. Several good things will result from this action. First it can and should be done mathematically as was shown. While it is usually easier to get an appropriate size from a tank manufacturer's catalog, the mathematical step will serve as a check against any catalog size selection error, which does happen on occasion. Second, you are instantly aware of the space problem. Maybe you are working on an existing building or there is some other space restriction. There's no rule that the storage has to be provided by one tank. It could be done by two, three, or more tanks if they are properly connected. Third, you may be experienced enough to sense something is wrong. You have done other similar designs that were successful in the past and the tanks in those cases were much larger or perhaps much smaller than the recently calculated tank. Is there a mistake in the calculations? Now is the time to recheck your figures.

As we have noted, the data used were composite data. While they are certainly reasonably accurate, they are not truly definitive for various types of buildings and operational conditions. In the beginning of this discussion we referred to the ASHRAE *Systems Handbook* as a source of complete detailed information on sizing a service water heating system. There is also a wealth of sizing information from various water heating equipment manufacturers.

In the ASHRAE book there is also a presentation of

an alternative method of load and heater calculations. This approach produces load calculations based on occupants rather than installed fixtures. It also, in a series of curves for a variety of installations, addresses the relationship of recovery to storage. In our example we used a fixed factor of 0.4 which resulted in a rating of 480 gph for the heater and 1,200 gal of storage. Clearly a 960-gph heater would not require 1,200 gal of storage. However, the relationship is not a straight line. There is a point at which the ever-larger heater does not eliminate storage unless the heater size is carried to some ridiculous extreme value. That would be beyond anyone's data.

It is this author's opinion based on personal experience that except for simple straightforward office buildings that have no other water-using devices, the controlling items are showers, clothes washers, and dishwashers. Utensil washers could be included with dishwashers. People really control shower use. Preset machinery controls clothes and dishwasher operations regardless of type.

You will also find there is a seeming disparity in the various data from various sources on hot water demand. Part of the variation is caused by the fact that the ASHRAE publication uses 140°F as final delivered hot water temperature and many equipment manufacturers use 120°F as the value. Let us take a simple case in which ASHRAE states the hot water output is 2 to 3 gallons/minute (gpm) and another manufacturer gives the answer as 4 gpm. Using simple algebra in which the hot portion of the water is x and the cold is $(1 - x)$, the following values solve for the percentage of hot water in each case, presuming the cold water is always 40°F and the mixed desired result is 100°F. Using 140°F water, the calculation is

$$140x + 40(1 - x) = 100$$
$$100x - 40 = 100$$
$$x = (100 - 40)/100$$
$$= 0.6$$

Using 120°F water, the calculation is

$$120x + 40(1 - x) = 100$$
$$80x - 40 = 100$$
$$x = (100 - 40)/80$$
$$= 0.75$$

Thus the manufacturer's 4 gpm contains 2.4 gpm (4×0.60) of 140°F water, and at the 120°F value it contains 3 gpm (4×0.75). On the basis of ASHRAE, using 140°F as the standard value, if we assume an average value of 2.5 gpm, the hot water portion is 1.5 gpm (0.6×2.5).

Taking these data a step further, using the ASHRAE data for an apartment, as our example notes, one would

obtain a 20-gph value per shower using 140°F water. Using our calculated 1.5-gpm rate, the use time is 20/1.25 or almost 14 minutes per hour (min/hr). Our manufacturing source notes a use of 18 min/hr, but with water at 120°F. Using our previously calculated values of use, we can multiply $18 \times 0.60/0.75$ and get 14.4 min. In using this system we found considerable correlation between figures from various sources.

Thus to check your answer you should utilize both the occupant-based method and the fixture-based method and compare the two. In addition you should use both fixed demand and storage factors and comparable curve data noted in the ASHRAE text and compare those two. Since data based on occupancy can vary, we prefer and recommend data based on actual fixtures modified only by use when such use can reasonably and logically be defined.

Specialized Loads

The requirements of hot water heating that pose special load calculations are usually for a kitchen. The basic problem is the need for water at two different temperatures. While there can be some variation, the usual temperatures in the kitchen are 140°F for everything but the dishwasher. The dishwasher has a requirement of 180°F for the rinse cycle.

While you can supply water at the higher temperature from a single source, the better way is to supply water at the lower temperature and use a booster in the kitchen application.

Let us assume you need 20 gph at 140°F and 300 gph for the dishwasher at 180°F. The recommendation of the National Sanitation Foundation is that for a heater with sufficient internal storage capacity to meet the *flow demand* in *gallons per minute* the total consumption of the dishwasher can be 70 percent of the demand, which is 210 gph (0.70×300). What you propose to use is a 500-gal storage tank and a gas heater plus a booster for the dishwasher which would be a common set of equipment. Finally we will assume a 2-h kitchen operating time frame.

First the quantity of water in the 500-gal tank is presumed to be up to 140°F before the peak load occurs. Next the standard rule is that only 70 percent of the stored water is usable, which amounts to 350 gal (0.70×500). But this must last 2 hr, so you really have 175 gph (350/2) available. Thus your heater must heat 245 gph ($200 + 210 - 175$) from the standard 40°F to 140°F. The booster must be able to heat 245 gph from 140°F to 180°F. This is equal to slightly more than 4 gpm (245/60). Keep clearly in mind that the booster has no storage capability. It must meet the criteria stated above directly, minute by minute.

Similarly to our composite water heating example, the water heating solution just cited is not the only way to heat water for a kitchen. You could have had a steam or hot water supplied storage tank and heater as a source. But this requires that you have a boiler or some source of steam or hot water year round. While this could be done, it is not energy efficient. Hot water heating, especially kitchen hot water heating, is more economical when it is supplied from a special dedicated source.

There are other specialized uses for hot water. A commercial laundry is a typical example. Here almost 100 percent of the water may be needed at 160 or 180°F. There is no choice except to generate and supply it at the temperature required. What little water is needed at a lower temperature can be supplied using appropriate mixing valves.

However, there are industrial applications in which dual-temperature water supplies are required. In these cases always first investigate the possibility of boosters, as in the previous kitchen example. Secondly investigate the possibility of two separate dedicated systems. It *almost never* pays to heat all water to the higher temperature and to supply a substantial portion via mixing valves at a lower temperature.

Tankless Heater. A tankless heater source of domestic hot water must always have two guiding criteria. Since by definition and by design it has no storage capability, it must have a rated output capability to handle the peak demand load for the duration of the peak load. Equally important, the heater must have a source capable of generating the necessary input to meet the tankless heater's requirements. In the usual installation the heating source, steam or hot water, flows through the shell of the heat exchanger, exchanging its heat with the domestic water in the coil of the heat exchanger.

Generally tankless heaters are used wherever there is a demand for a relatively steady, continuous supply of hot water. These heaters are of the high-demand type, and it is recommended that circulators be installed in both the source supply lines and the domestic supply lines. In residential systems and some commercial applications the tankless heater is installed in a tapping provided in the boiler water area for it. A simultaneous demand for heating and hot water will invariably lower the boiler water temperature for a short period of time.

The heater output to domestic sources must take this fact into consideration and be designed accordingly or else its output will frequently be at unsatisfactorily low temperatures.

The methods used for sizing instantaneous heaters are *not* the same as those used for sizing storage water heating equipment. Thus our composite sizing data previously given in this chapter *do not* apply. As we noted in the beginning of this chapter, we provided in Chap. 4, Fig. 4.7, data based on fixtures. There we depicted typical hot water demand for various types of fixtures. We have referred to the ASHRAE *Systems Handbook* before and in the example that follows we will refer to their approach again. Using the same fixture count

as before but this time using values directly from ASHRAE, the following is the calculation for our same mythical industrial plant building:

Item	Quantity	Value	Fixture units
Lavatory	20	1	20
Shower	4	3	12
Circular sink	2	4	8
Semicircular sink	2	3	6
Total			46
Kitchen sink	3	3	9

We now have 46 fixture units for nonkitchen use as well as 9 fixture units for kitchen use. The dishwasher demand is a separate manufacturer's value of 280 gph. It will actually be in use for only 40 min/hr. As far as we are concerned, the demand of the dishwasher is 7 gpm (280/40).

This tankless heater sizing has been researched by the ASHRAE people and they have come up with a curve similar to the curve in Fig. 4.8, Chap. 4, that shows system demand in gallons per minute versus fixture units. Using the ASHRAE data, the 46 fixture units equate to 24 gpm. The nine kitchen units equate to some 25 gpm. In addition we have the dishwasher demand of 7 gpm. Thus the tankless heater size is 56 gpm (24 + 25 + 7).

The 56 gpm equals 3,360 gph (56 × 60). In our previous calculation, the answer was 480 gph for the heater with a 1,200-gal storage tank. This was because of the application of the 0.40 demand factor. *Do not apply a demand factor to tankless calculations.*

Pipe Sizing. The methods of sizing hot water piping are exactly the same as those we outlined in Chap. 4 for cold water piping. Insofar as distribution methods are concerned there are three distribution methods that apply to multistory buildings and that are relatively similar.

In the high-rise or multistory installation it is common practice to provide a recirculating line. It is also recommended to install a recirculating line in long horizontal one-story and two-story structures when the horizontal distance to the farthest fixture is more than 100 ft. For three multistory high-rise buildings, one method is to use an up-feed system for each riser and at the top of the riser provide a recirculating line. The second method is to take two nearly adjacent risers and make one of the two, in effect, the recirculating return. The third method is to run a feed line to the top of the fixtures and down feed through several branches. These are all schematically depicted in Fig. 5.1. Note that a vent is supplied in the third case. There is always some air in the supply system. If you use the first or second method, your air-relieving problem is solved every time someone opens a fixture on the top floor. The third method tries to get a more even delivery temperature and precludes air release. Thus a vent is required.

System Loading. The domestic hot water load must come from some source. It could be directly from the utility fuel such as an oil, gas, or electric heater dedicated to the domestic hot water supply requirement. In some communities it could be a separate device fed by a site or off-site central steam or hot water plant. Or, in many cases it could be directly or indirectly supplied by the building boiler plant.

As we have been discussing, the domestic hot water load is a very real load that occurs at the rate and value

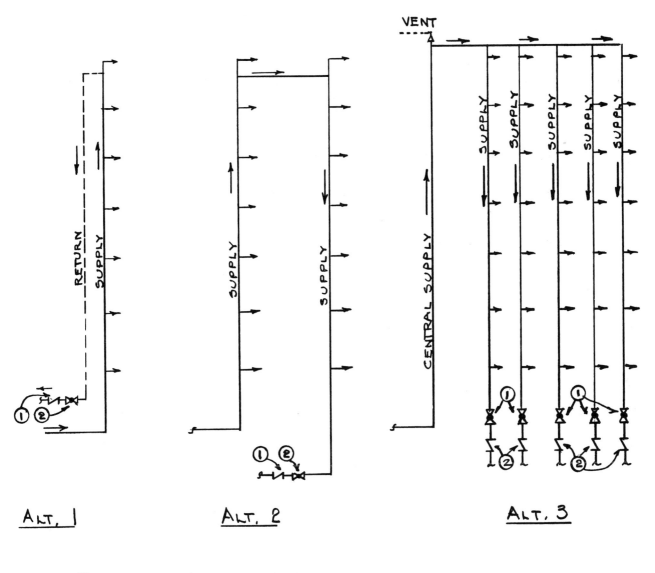

ALT. 1   ALT. 2   ALT. 3

① BALANCING VALVE

② CHECK VALVE

RECIRCULATING PIPING SYSTEMS

NO SCALE

FIGURE 5-1

that our calculations have indicated. In many types of installations in which a central boiler plant serves both as a source of heating and a source of hot water, the peak demand for domestic hot water may well occur at or near the simultaneous peak demand for heat. Certainly we all know that usually the coldest part of the day occurs in the hours before dawn. However, these are not the peak load-time hours on the boiler. For the boiler the 6 to 9 A.M. period represents the changeover from night setback to day operation at 72°F as well as the time the ventilation system usually goes from total recirculation to full or partial fresh air intake. In many cases the domestic water peaks occur around noon. Never assume that peak demands for heat and hot water won't coincide or nearly coincide. The ASHRAE handbook and many manufacturer's catalogs advise that boiler selection should be made on the sum of the two peak values. ASHRAE has some data indicating that when the total domestic load is greater than 50 percent of the heating load, some diversity may be taken into consideration.

Generally the usual project will not have a domestic water heating load greater than 50 percent of the heating load. There is one final point to consider. The water load occurs year round and may well be larger in the summer than in the winter for a variety of reasons. When the domestic load is a small percentage of the heating load, the total system is obviously working in the low end of its range in the summertime. As such, it is not very efficient. In this case and in many others it definitely may prove worthwhile to consider a separate source dedicated solely to the generation of domestic hot water.

Manufacturer's Literature

On a number of occasions in this chapter we have referred to the ASHRAE *Systems Handbook* as well as to literature and engineering data available from equipment manufacturers. We have also pointed out in our discussion that equipment manufacturers commonly use 120°F as delivered water temperature while the ASHRAE text uses 140°F. Because this constant referencing of other texts may create problems for the reader we have secured permission from PVI Industries Incorporated, P.O. Box 7124, Fort Worth, Texas 76111, to reproduce their methods of calculation of hot water requirements for a large variety of buildings; they appear in the Appendix at the end of this chapter and include 10 separate calculations pertaining to hot water supply systems for a large variety of buildings including apartment houses; hotels; motels; food service establishments; barber shops and beauty salons; churches; hospitals; rest homes; nursing homes; orphanages; convents; dormitories; fraternity and sorority houses; schools; commercial, industrial, and coin-operated laundries; office buildings; swimming pools; and baptistries. We have also included basic design and equipment information provided by PVI Industries on design criteria as well as

water-use values for showers, dishwashers, and the like.

This assemblage is, in our view, an excellent compilation of data. The only item deleted from the PVI-supplied data is their statement at the end of each worked-out example that says "Transfer to Selection Guide." Since we are not reproducing their entire equipment catalog, these words would have no usable meaning.

Swimming Pool Heaters. The era of the unheated swimming pool is rapidly becoming a bygone one. Whether the pool is inside or outside, pool water heating is rapidly becoming a standard item in the pool system installation. Outdoor pool usage can be considerably lengthened by heating pool water even though the surrounding air may be cooler than ideal. In some areas this heating of water may nearly double the number of days of pool use.

There are two ways of calculating pool water heating—the area method and the volume method. The most commonly used of these two methods is the area method because it takes into account the climate surrounding the pool. A value based on considerable research states that the heat loss on the surface of the pool equals 12 Btu/hr (BTUH) per degree of temperature difference between water and air. Thus a 50×100 pool has 5,000 square feet (sq ft) of surface. Depth is not a factor since the loss is at the surface. If the pool water is held at 80°F and the surrounding air is 60°F, the water-to-air-temperature difference is 20°F (80 − 60). The above values equate to 1,200,000 BTUH ($5,000 \times 20 \times 12$). Using the standard efficiency of 80 percent for either an oil or gas heater, the input value is 1,500,000 BTUH (1,200,000/0.80).

The volume method is primarily used to heat the water to the desired temperature and does not take into account heat losses at the surface which occur both while the water is being heated and after it has reached the desired temperature. Customary rate-of-rise values are ½ to 1°F/hr for residential pools and per ½ hr for commercial pools. Using the same 50×100 pool and assuming an average depth of 5 ft and 7.5 gal/cu ft, the pool contains 187,500 gal ($50 \times 100 \times 5 \times 7.5$). Using the standard factor of 8.33 and ½°F rise (temperature difference), the load to heat the water is 780,938 BTUH ($187,500 \times 8.33 \times 0.5$). At 80 percent heater efficiency this equals 976,173-BTUH input (780,938/0.80).

The above answer poses a problem. The incoming water temperature to fill the pool is most likely in the range of 40°F to 55°F with an average value of 50°F. If the surrounding air is 80° or higher, the pool will not have appreciable, if any, surface loss. Further since it is being heated at the rate of ½°F/hr, going from 50°F to 80°F takes 60 hours [(80 − 50)/0.5]. Thus it takes 2½ days (60/24) of constant heating to bring the pool up to its 80°F operating value.

While one would not normally try to heat up the pool, indoors or out, when the temperature is 60°F, it is entirely possible to have the pool heating effort occurring when the surrounding air is 70°F at least during a part of the heating cycle. If we assume a value of 70°F during a 24-hr period of the 2½-day heating cycle, we have a surface loss of 750,000 BTUH [(50 × 100 × 10 × 12)/ .80].

If the pool heater were selected to produce 976,173 BTUH, which is the value required to heat the water during our 24-hr period of 70°F, there would be 226,173 BTUH available for water heating [976,173 − 750,000 (surface loss at 10°F differential)]. This value would raise the pool water 0.12°F/hr [(226,173/976,173) × (0.50)], which is slightly less than a 3°F rise in a 24-hr day. At this rate the pool could take 6 to 10 days to come up to temperature. This has happened and pool heater designers have been shocked to discover how long it takes to heat a pool.

If the pool is indoors, the surrounding atmosphere can be maintained at 80°F until the pool is heated. Or a lower temperature can be used and a very long pool water heating time can be accepted. For the larger indoor pool, use the volume method. Once the water is heated, the air-to-water-temperature difference will always be 5°F or less, and the surface loss of 375,000 BTUH at 5°F is less than the 976,173-BTUH heater capacity. In the outdoor pool the best solution would be to heat the water during warm weather and size the heater based on the lowest outdoor air temperature acceptable for pool use utilizing the area or surface calculation method.

Individual Electric Heaters. The selection of the type of water heating depends to a large degree on the demand for domestic hot water. In Chap. 4 we depicted a multistory water pipe sizing problem and its resolution. Our example was predicated on the use of individualized electric water heating. While it may appear that we are selling the concept of electric water heating, this definitely is not the case. As can be seen in most of the preceding discussion in this chapter, the basic thrust of our examples is on central sources using steam or gas as the primary fuel.

In Figs. 5.2 and 5.3 we are presenting two details on the piping of electric heaters to which we alluded in our

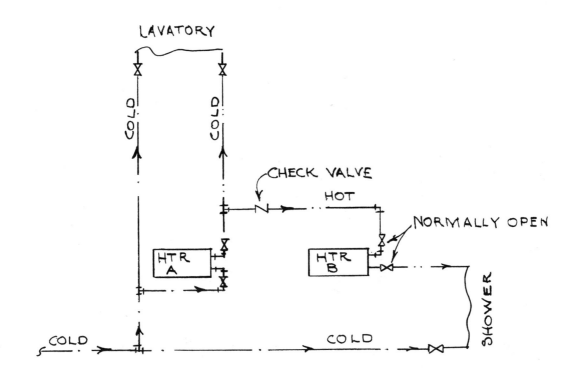

SERIES INSTALLATION — ELECTRIC HEATERS

NOT TO SCALE

FIGURE 5-2

Chap. 4 discussion. Figure 5.2 shows the heaters serving a typical lavatory and shower connected in a series arrangement, and Fig. 5.3 depicts the connection in a parallel arrangement. The basic concept of the series arrangement is that the second heater is effectively a standby heater.

When taking a shower, not everyone uses a large quantity of hot water. It is probably safe to say that the majority do but that a large minority do not. Thus if the two heaters in Fig. 5.2 have properly arranged controls, the second heater, shown as heater B in the series detail, will only be activated when heater A cannot keep up with the temperature demands at the rate of use. Even when called into action, the output of heater B is frequently not used to capacity.

For this energy-saving method of operation there is sometimes a limitation that precludes the operation. The two heaters in series offer a greater resistance to flow. Your system may not have the capability of handling this added resistance. Further there is not one set of controls operating two heaters, which would defeat the economics of the system. Rather there are two sets of controls and some form of step controller is required to activate the staging of heater response. This somewhat more costly installation can be paid off only in energy savings over time.

An alternative arrangement is depicted in Fig. 5.3. This parallel arrangement allows both heaters to feed the shower, and the check valve prevents both heaters from feeding the lavatory. Here the heaters can be preset to a given temperature which is the same for both cases. This arrangement allows a different, more economical in cost, type of heater that has no thermostat at all. In this type of heater the turning on of the electric element is flow actuated. That is, whenever a hot water faucet in either the lavatory or shower opens, flow is created and the moving water through the heater actuates an electrical switch. The heater is instantly on to its full capacity. There are two things to note in using a flow-actuated heater. One is its fixed limit capacity and two

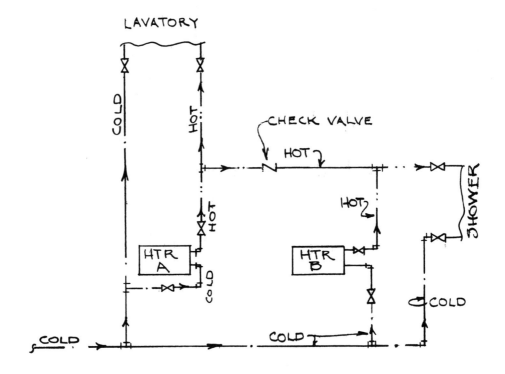

PARALLELING ELECTRIC HEATERS

NOT TO SCALE

FIGURE 5-3

is its full on position. This type of heater installed in series in a low flow operation could dangerously overheat the water since flow of any quantity puts it full on. In a parallel situation with proper flow control on the lavatory faucet and shower heat the flow-actuated heater is a very economical installation. The individual flow-actuated heaters are usually rated at 0.5 gpm.

Heavy use of showers would require further analysis of fuel costs. Situations in which the electric hot water heating has its best and most common application are in office buildings and certain types of light industrial applications that have a water use that is less than 2 gal per person per day and peak use that is probably 1 gph. This type of situation also occurs in many medical office buildings.

As we depicted in our example of a building in Chap. 4, there may well be a restaurant, health club, or other major hot water user, on the premises. Much depends on the location of this facility. Our example presumed the common location on the ground floor. In that case, as we depicted, the restaurant had a separate source of hot water with or without a recirculating line, depending on the use requirements. The volume of water used almost always requires oil or gas, directly or indirectly, as the source of supply.

Circulation Systems

There is the general rule of domestic hot water heating piping which requires that any domestic hot water line over 100 ft in length be provided with a recirculating line to ensure that hot water is available at the fixture. Contrary arguments to this rule usually note that in many cases running the hot water supply 20 seconds (sec) or less will provide the same result without a recirculating line and, in the same vein, considerable electrical energy can be saved by reducing or eliminating the circulator.

Both the general rule and the contrary arguments noted above can be supported by proper calculations. As in many situations, engineering decisions can be realistically adjusted to satisfy most of the situations most of the time. If there is infrequent use of a fixture either because of its location or because of the fact that the building is unoccupied for long periods of time, the hot water supply to a fixture will cool down when the fixture location is at a considerable distance from the supply source. This is, under the best of circumstances, a very annoying situation, and water is wasted draining the hot water line of its cool stagnant supply.

Most of the arguments, except for the one based on a first-installed cost saving, can be overcome by the installation of a timing device to allow the circulating pump to operate, for example, only for 10-min periods in each hour in which the building is occupied and to stop operation during the unoccupied period.

Figure 5.4 illustrates the pertinent connections of two common types of connections using a circulating hot water line. The upper part of this detail depicts the common arrangement of a three-way valve to a usually separate kitchen hot water heater. In this situation water is delivered at a temperature of 140°F to the dishwasher which has its own booster to raise the temperature to 180°F for the dishwasher rinse cycle. As is well-known, this temperature of 140°F is much too high for use in the other kitchen sinks. A mixing valve, usually in the kitchen area, is provided as shown so that the water may be reduced to 120°F or less for kitchen sink usage. The circulator is not depicted.

In the lower half of Fig. 5.4 is the single mixed water temperature system. The essentials of the system are the storage tank, which may be a gas, oil, or electric heated tank or part of a steam or hot water storage tank, and heater system. The tank delivers hot water, usually 10 to 20°F hotter than the system requires, to a three-way mixing valve. This three-way valve mixes hot water and cold water to create a safe, satisfactory delivered water temperature to the system.

The system has a recirculating line which commonly is ½ in or ¾ in and a circulator controlled by an aquastat. When the system usage is small, the water tends to cool down and the aquastat calls the circulator into operation. Heavy usage tends to keep the water in circulation, and at a preset temperature the aquastat will stop the circulator. Some plumbing designers prefer to add a second, reverse acting aquastat which has the ability to stop the circulation if the recirculating water gets too cold. This situation could arise when the source of hot water fails completely and no hot water is available but the system is still trying to function. In that case the three-way valve, under control of a system supply aquastat, would be full open on the hot water inlet part and closed on the cold water part, trying to deliver nonexistent hot water.

The circulated water is returned to the tank as is usual and common in all circulated lines. What is different in this detail is a bypass connection to the service part of the three-way valve. This appears to make the three-way valve into a four-way valve. This is a bypass part on the three-way valve that allows circulated return water to not only go to the tank but to the valve where it is blended with the hot and cold supplies.

The function of the ½-in bypass is to allow previously circulated water to pass through the mixing valve during periods of no draw without entering the hot water tank to be reheated. This secondary source to the valve helps reduce the buildup of undesirable hot water in the system, and it must be installed.

In selecting thermostatic mixing valves there is one note of caution. Designers and engineers have a tendency to always be on the safe side and make equipment selections slightly or considerably oversized depending on the particular specifying individual. Note that all thermostatic water mixing valves have limitations. They simply are not accurate outside of their flow range ca-

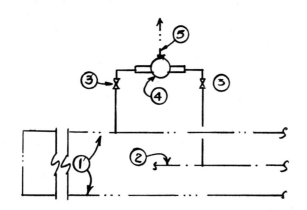

TEMPERED WATER TO SUPPLY RISERS
USING 140° RECIRCULATING SUPPLY SOURCE

NOT TO SCALE

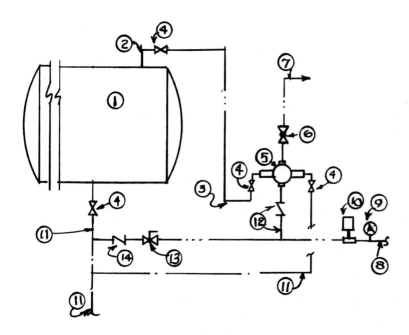

SINGLE SOURCE TEMPERED WATER SUPPLY

NOT TO SCALE

CIRCULATED HOT WATER PIPING DETAILS

FIGURE 5-4

pacity. Do not oversize these valves as most valves have a stated requirement that the flow range must be no less than 15 percent of rated capacity. This is a serious situation in which the oversized valve simply does not function as intended at minimum loading situations. It also relates, rather humorously, to the knowledgeable circulating pump manufacturer who invariably tells the designer "if you must make a mistake in pump selection, make it too small not too large."

Mixing Valves. In Fig. 5.4 the mixing valves depicted frequently represent the end of the designer's thought process. This should not occur. It is of equal and possibly greater importance to understand that delivering a properly selected mixed water temperature is not the end of the process. In ever-increasing numbers the individual hot and cold water faucets on fixtures are being replaced with single lever, automatically controlled devices, thermostatically or pressure actuated. While similar, these two devices employ very different operating mechanisms.

The thermostatic valve uses a bimetallic coil or a liquid- or gas-filled bellows which senses temperature and adjusts installed parts to control delivered temperature. The pressure-actuated valve uses a pressure-sensitive piston or diaphragm which equalizes hot and cold pressures and indirectly controls temperature.

While both types have an antiscald feature, reaction times can vary in the thermostatic units and normally the bimetallic coil reacts faster than the bellows. And, the pressure-acutated valve reacts even faster than the bimetallic element.

The real problem is that the hot water delivered from the central source can and does vary, sometimes substantially, during periods of high and low demand and especially after long idle, nonuse periods. In these varying periods the thermostatic valve can compensate for the temperature change. The pressure-actuated valve cannot, and its outlet temperature will vary with the supply temperature. Where supply temperatures are always varying and the user requires comfort and safety, thermostatic end-use valves should be specified. Where the supply temperature is relatively stable, pressure-actuated valves may be specified. The thermostatic valves are more expensive and are commonly used on shower and bathing units in prisons, mental institutions, nursing homes, group shower control, whirlpool, and handicapped bathing, as well as other specialized applications. The common low-cost solution would be the mixing valve installation as shown in the lower half of Fig. 5.4 plus pressure-actuated valves at the point of use.

Heat Exchangers. Heat exchangers customarily are shell and tube configurations that are used primarily for heating a liquid with steam or water of a higher temperature. The customary piping arrangement introduces the heating medium, steam or water, into the shell of the exchanger which surrounds the tube bundle. The liquid, in our case water, to be heated is circulated in the tubes. Heat passes through the tube walls to heat the water.

In outward appearance the water-to-water or steam-to-water exchangers appear almost to be twins. The primary difference is that the inlet and outlet of the water-to-water heat exchangers are of the same pipe diameter while in the steam-to-water exchanger the inlet connection is much larger (to handle steam vapor) than the outlet, which allows condensed steam (condensate) to flow. The condensate outlet line is always arranged so that it is on the bottom of the installed exchanger. In water-to-water exchangers, especially in installations in which the connecting lines are into a steam boiler and below the water line, it is mandatory that the flow move from the bottom of the boiler into the exchanger, using a circulator, and out the top and back into the steam boiler below the boiler water line. In the case of hot water boilers the direction of flow depends on the prevention of air bending, and the recommended method is also the same as that previously described for steam. Normally the plumbing work does not include the heating medium portion of the piping nor the exchanger in new construction of either hot water or steam source systems. In retrofit work the steam piping may become part of the plumbing work and, by customary trade practice, the steam installation is sublet to a steam filter.

Steam Fed Heat Exchangers. Figure 5.5 depicts the typical piping of a domestic hot water supply using a heat exchanger and a steam source. The basic control, as shown in the detail, is a fixed sensor in the domestic hot water line leaving the exchanger. The detail does not describe this source since there are a number of adjustable sensors, aquastats, and other controllers that may be employed. In all cases they perform the same basic function, which is to control the rate of steam delivery through the pressure-temperature regulation. Controlling the rate of steam fed to the heat exchanger controls the temperature of the water supplied to the system.

The detail also depicts a hot water recirculating pump. The standard plumbing specification states that the pump should be controlled by two aquastats on the pump discharge. One aquastat should act normally—that is, it should start the pump at probably 100°F and stop it at 120°F. The reverse second aquastat shall be reverse acting—that is, to start the pump at 100°F and stop it at 80°F. There are times when the steam supply could fail or be manually shut off. The job of the reverse-acting aquastat is to stop the recirculating pump when the water is not being heated.

In today's energy-conscious environment constantly circulated water when use is small or nonexistent is being recognized as a source of electrical and heat energy waste. The common recommendation is to eliminate the recirculating pump wherever possible and, if that is not possible, to reduce its operation to a few minutes per hour by the addition of an operating single timing device.

Water-to-Water Heat Exchangers. Figure 5.6 presents the use of the water-to-water heat exchanger both as an instantaneous heater and as a storage tank and heater application. In general most designers and engineers prefer the storage tank system. Our detail pulls the heat exchanger and storage tank apart so that this slightly more involved application could be depicted. Obviously the heat exchanger could have been installed in the tank. *In that case the water or steam used as a heating medium would be piped into the tubes of the heat exchanger.*

In the application of the tankless heater, which is one of the most popular and commonly used of all indirect water heating systems, the exchanger is depicted and *presumed* to be installed below the water line. Again, larger buildings will normally require a recirculating line. The general rule is that a recirculating line is needed when the fixture served is more than 100 ft from the source of hot water. One controlling aquastat is depicted. Two can readily be installed. And, if needed, an energy saving timer can be added, as previously described. When the demand is fairly constant this is a very workable application.

In the lower part of Fig. 5.6 a detail is depicted for use in situations in which there are heavy periods of demand which are in excess of the boiler capacity. On a demand for hot water the makeup cold water does not enter directly into the storage tank but rather goes first to the heat exchanger. In the design of the system the tank capacity should be equal to the difference between the peak demand and the average maximum demand produced by the heat exchanger.

The recirculating pump can be used to recirculate both the building return and the water in the tank, and, as noted, its pump controlling aquastat is in the tank. The heat exchanger has a pump to circulate the boiler water heating medium and its control is on the domestic hot water discharge line of the heat exchanger.

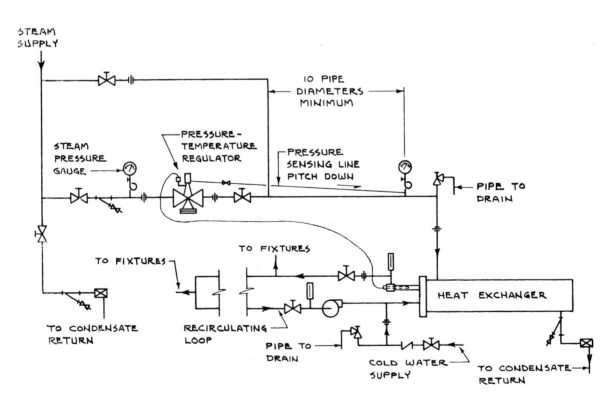

DOMESTIC HOT WATER FROM STEAM

FIGURE 5-5

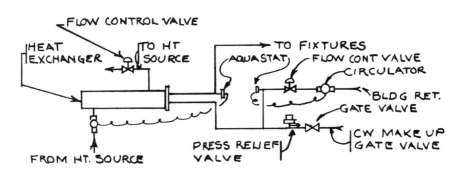

INSTANTANEOUS HEATER PIPING

NO SCALE

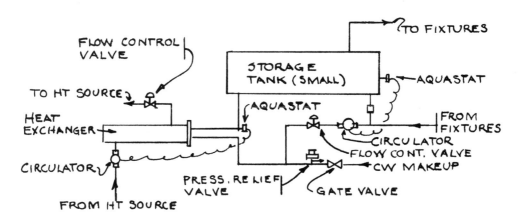

HEAT EXCHANGER WITH STORAGE TANK

NO SCALE

DOMESTIC HOT WATER FROM HOT WATER SOURCE

FIGURE 5-6

Heat Exchanger Dual Temperature Supply. In many buildings two different temperatures of domestic hot water are a common requirement. Usually there is a very large volume required in the 120 to 140°F temperature range and a smaller quantity at 180 to 200°F for dishwashers, sterilizers, and other special applications.

Figure 5.7 is a typical installation of two tankless heaters to satisfy this requirement. Out detail is typical in that the actual location of both tankless heaters is not noted. In practice the lower temperature heat exchanger would commonly be located in the building equipment or boiler room.

Note that the piping is arranged so that the second heat exchanger takes water already heated to the lower operational temperature and further boosts it to a higher one. This is the standard energy efficient design. This second booster should, whenever possible, be located at the point of high-temperature water use to avoid adding recirculating lines for this system.

Gas Fired, Single Temperature, Heater. The unusual impulsive reaction to the mention of a gas fired domestic water heater is the visualization of a heater commonly found in residential basements or heated rooms. This

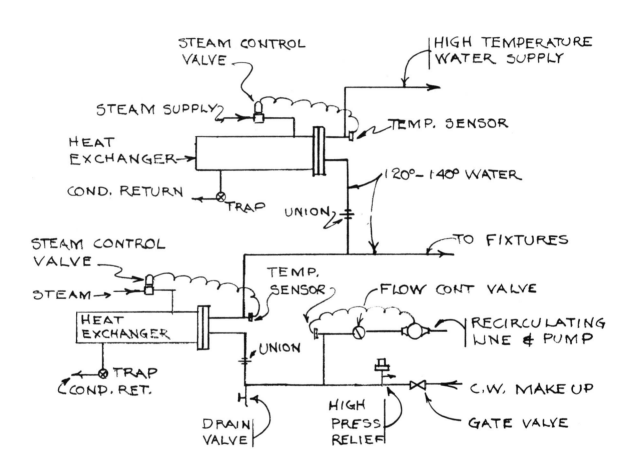

DUAL TEMPERATURE WATER FROM HEAT EXCHANGER

—NOT TO SCALE—

FIGURE 5-7

heater is a combination heater and storage tank rated at 30 to 60 gal. Commercially sized and applied gas water heaters have a much greater gallon per hour output and invariably have a separate storage tank.

Figure 5.8 is a typical piping arrangement with a separate tank where the required water temperature is at one temperature value. The arrangements are similar to those of the heat exchanger connections that have been previously noted. One item in the piping arrangement may seem different or improper to the first time viewer. That item is the flow of water through the gas heater in a top-to-bottom arrangement. The detail as shown is correct. Gas heater manufacturers commonly provide both up-flow and down-flow coils and some

companies, in their product installation details, prefer the down-flow arrangement depicted in Fig. 5.8.

In every case of a separate heating source and a storage tank there is a circulator, as depicted, in the line between the tank and the heater. This is necessary since the proper controlled circulation cannot be effected by relying simply on the natural movement created by the higher temperature of the water in the heater versus that in the tank. The circulator is usually controlled by the aquastat tank temperature control device.

Most of the devices shown on the piping diagrams aid in the overall system operation by providing visual data, and operating conditions call for connections to clean and isolate system components. Relief valves are pro-

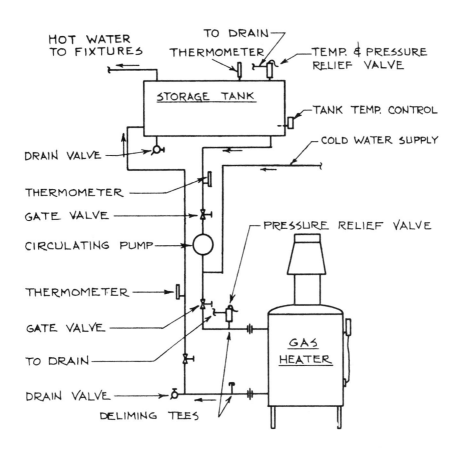

SINGLE TEMPERATURE HOT WATER SYSTEM

GAS FIRED HEATER

NO SCALE

FIGURE 5-8

vided both on the tank and on the circulating line at the heater.

The basic system supplies a single fixed temperature source of domestic hot water. As such the gas fired heater delivers water at, for example, 130°F. Using the circulating pump, water is close to 130°F in the line between the tank and the pump, and the reasonable probability is that the hot water varies slightly in temperature with a probable average of 125°F leaving the tank.

Cold water makeup is connected on the discharge side of the tank recirculating line. This arrangement precludes sudden variations of discharge water temperature to the fixtures by requiring the incoming cold water to be heated before passing into the overall system. This incoming cold water line is the basis of the system's recovery factor. The amount of hot water used by the fixtures must be replaced by cold makeup water. A system with a 200-gph 50 to 140°F recovery factor is a system that uses 200 gph of hot water and has a heater that can raise the temperature of 40°F incoming cold water to 140°F at the rate of 200 gph.

Based on the above data, a storage tank would not be needed. As we know, there are hours when the demand is much larger than an economically sized heater can handle. The tank supplies a standby source of already heated water. With a 500-gal tank whose effective use is say 70 percent we have 200 gph into the heater plus 350 gal (0.70 × 500) in the tank. Thus for the unusual hour we have 550 gal (200 + 350) of 140°F water available.

There is no mixing valve at the tank since the water leaving temperature is preset to meet the design temperature requirement.

Booster Recovery System. When the basic heating system is low temperature (180°F average) forced-circulation hot water or a direct oil, gas, or electrical-supplied forced warm air system, the dishwasher and related kitchen sinks can present a problem. As we have noted in previous discussions in this chapter, some form of boosting of the central source of water temperature is required to meet dishwashing requirements. Using the detail depicted in Fig. 5.8, an electric or gas booster

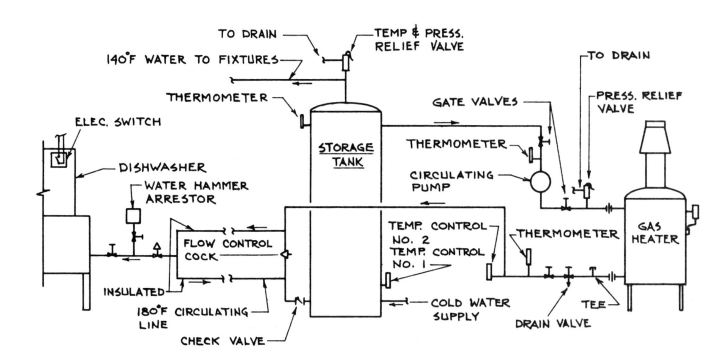

HOT WATER BOOSTER RECOVERY SYSTEM

GAS FIRED HEATER

NO SCALE

FIGURE 5-9

could be added to our standard 125°F delivered water system to resolve the problem.

But, as we also noted, the kitchen is a special case requiring large volumes of high temperature hot water usually for one to three widely separated time periods. The best solution is always, whenever possible, to provide the kitchen and its dishwashing facility with a separate system that is ideally located at or near the source of use. It was also noted that the probable best solution is to heat the kitchen water to 140°F and have a separate electric or steam booster located on or with the dishwasher to raise the rinse water temperature to 180° as required.

Figure 5.9 depicts another way to solve the problem which, in terms of overall cost, may prove economical and certainly should be considered as an alternative solution to the design problem. This detail presumes the heater is gas and there is no booster. To solve the problem of 140°F water and reduce the need for the gas water heater to constantly have to heat water from 40 to 180°F, the storage tank not only is the supply cushion on the system but is also the mixing valve for 140°F water.

There are two temperature controllers. Temperature controller number 1 is an aquastat in the tank that controls the recirculation of water from the tank to the heater and maintains the temperature of 140°F in the tank. It can do this because any use of water at 140°F by the line out of the tank to the fixtures results in a mixture of cold 40°F water and hot 180°F water. Once the water propelled by the circulation reaches the heater, it is raised to 180°F since temperature control number 2, which controls the heater, is set to maintain 180°F.

There is a flow control cock that is manually set so that some 180°F water goes directly to the tank but not all. Part of the water, whether or not the fixtures are used, is recirculated since all the water must be in the tank of the dishwasher. This assures that 180°F water is always available for use. There is no recirculating pump since all lines between the heater, the tank, and the dishwasher are presumed to be 50 ft or less. If they are longer, a recirculating 180°F line pump would be required.

The dishwasher has a quick closing valve, and the water hammer arrestor (shock absorber) is mandatory. It is possible, after long periods of no use of the kitchen equipment, for gravity circulation to heat the tank to more than 140°F although it is not very likely. The circulating pump tank control would be satisfied and the pump would not be running. The idle pump would resist gravity circulation. In addition the heater control number 2 would be satisfied and the heater would be off. Under these conditions a temperature buildup in the tank, which is never a perfectly insulated vessel, is, as previously noted, very unlikely.

As in Fig. 5.8, Fig. 5.9 has two relief valves—one in the tank and one in the tank recirculating line entering the heater. The heater is again a down-flow coil type.

While is would be normal to insulate all hot water lines, neither Fig. 5.8 nor 5.9 calls for this except for the 180°F line. While an obvious requirement, it was felt that 180°F water, commonly exposed in and around the dishwasher, was a real hazard and the note to insulate implies that even if nothing else is insulated, this line must be.

Recirculating Lines. There are many small industrial, institutional, and commercial situations in which dual temperature water is a requirement. Frequently these situations can be solved with a standard storage tank hot water heater, and a lot of the piping can be eliminated. Because these are small, relatively simple arrangements, they frequently end up not getting connected properly. If anything, more mistakes are carelessly made both in design and in installation in the small projects than in the large projects.

Figure 5.10 is a rather elaborate presentation of the so-called simple small job. We could not show, within the confines of the page size, two recirculating pumps with the second being on the 160°F line. Therefore, for the purpose of clarification we're going to presume that the 160°F line is short and doesn't need a pump or, if it is long, that the installation is similar to that shown for the tempered water line.

Again, as in Figure 5.9 the source is at a higher temperature. For other uses (not necessarily a kitchen) we want a controlled source at say 110 to 120°F. For this purpose we have rather clearly depicted a 1-in mixing valve. The tank is an in-the-bottom and out-the-top standard arrangement. Note that we give pipe length so that if you are trying to squeeze the tank in some tight location, you will know the clearance requirement. And the relief valve is piped to near the floor so no one gets hurt when it is used.

The recirculating pump is very clearly tied into the cold water makeup line on the discharge side of the pump. As we've noted before, this is where the connection must be made. The cold water does not go into the suction side of the pump.

The 160°F recirculating line, gravity or pumped, goes into the bottom of the tank, and since this is the lowest point on the system, this is where the tank drain valve is located. The tank has an operating control at the bottom or coldest part of the tank and it has a high limit control near the top or hottest part of the tank. If this tank is gas fired, it will have a flue and draft divertor on top of the tank. If electrically supplied, it will appear as depicted.

The circulator has at least one operating aquastat, as depicted. This keeps the circulating water moving until the temperature reaches usually a preset 100°F. Some designers prefer to add a second reverse-acting aquastat that stops the pump if the water drops to say 80°F since that temperature would presume no hot water because

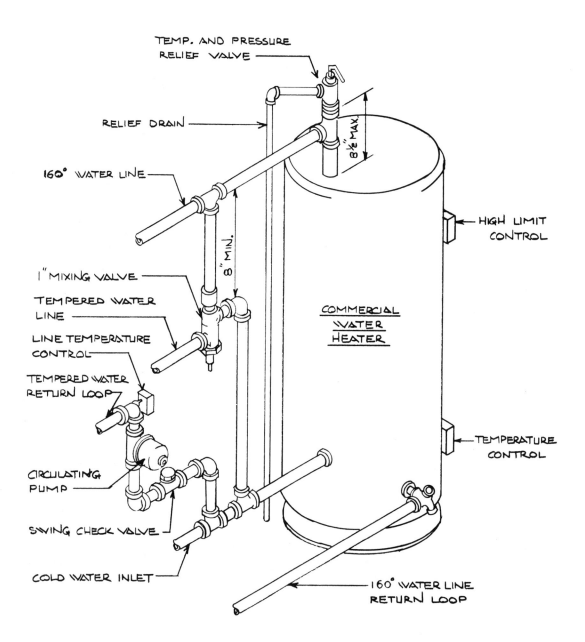

TEMP. AND PRESSURE RELIEF VALVE

RELIEF DRAIN

160° WATER LINE

1" MIXING VALVE

TEMPERED WATER LINE

LINE TEMPERATURE CONTROL

TEMPERED WATER RETURN LOOP

CIRCULATING PUMP

SWING CHECK VALVE

COLD WATER INLET

8½" MAX.

8" MIN.

HIGH LIMIT CONTROL

COMMERCIAL WATER HEATER

TEMPERATURE CONTROL

160° WATER LINE RETURN LOOP

INSTALLATION OF TEMPERED WATER CIRCULATION LINE
— NOT TO SCALE —

FIGURE 5-10

of heater failure, and consequently there would be no need to recirculate cold water.

The check valve is arranged so that incoming cold water cannot flow, contrary to intentions, backward through the pump. All of these seemingly minor details are commonly the ones which, if improperly installed, create the job problems.

Energy Conservation

Except in the area of hot water heating, relatively small amounts of energy are consumed in plumbing systems. There are various pumping operations that, on occasion, with good design and equipment selection can have reduced motor sizes and operational frequency, but this is not a major source of energy saving. In hot water heating there are, however, sources and systems with which energy can be saved.

Three details, Figs. 5.11 to 5.13, present systems that not only resolve problems but also save energy. These details also are good illustrations of plumbing piping in hot water systems.

Figure 5.11 is, in essence, an expansion of Fig. 5.8 in that it depicts diagrammatically and schematically the fixtures served by a single temperature hot water system as described in Fig. 5.8. The expansion of the system illustrated in Fig. 5.11 shows the various branch lines of the distribution system and the extension of the main hot water supply line in a pumped recirculating loop line back to the water tank.

In all of the discussion of previous details it was consistently noted that the recirculation pump was controlled by one operating and sometimes one reverse-acting aquastat. It is well-known that the circulating pump is a fractional horsepower motor probably, in most

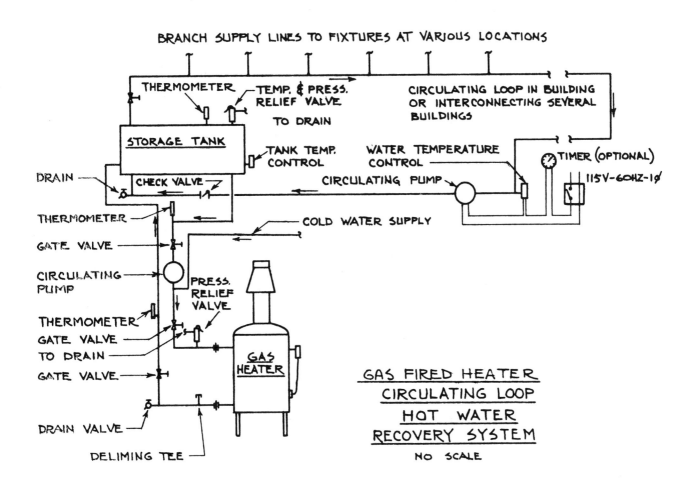

FIGURE 5-11

instances, averaging ¼ horsepower (hp). However, a single control generally results in the pump running constantly. In a majority of the recirculating line installations it really is not necessary for the pump to be running for 8 to 10 hr at night when no one is using the domestic hot water supply. If it were turned off from 8 or 10 P.M. to 6 A.M. some 8 to 10 hr of electrical pump energy could be saved. Granted we are describing a ¼ hp motor. But the shutoff time could be 3,600 hr per year (yr) on that basis. This would be some 900 kilowatthour (kwh) (¼ × 3,600) and at $0.10/kwh would equal $90. A small sum but still a real number.

Figure 5.11 depicts how a simple, inexpensive electrical timer could be placed in the control circuit and set to start and stop the pump. Actually the savings could be much larger. During the occupied cycle the use of the fixtures, as we have noted in other previous discussion, really isn't necessary if the fixtures have fairly frequent use. The pump's only purpose is to preclude cooling down of water in lines for which there is little or no use. A system created by the timer which allows pump operation for only 10 min in each hour has proved to be not perfectly ideal, but it is a satisfactory solution. Now the pump runs only 4 hr per day (10/60 × 24). This doubles the $90 saving noted previously since the idle hours are 7,300 hr/yr (20 × 365) and 7,300 × 0.25 = 1,825 kwh, which at $0.10/kwh is $182.50.

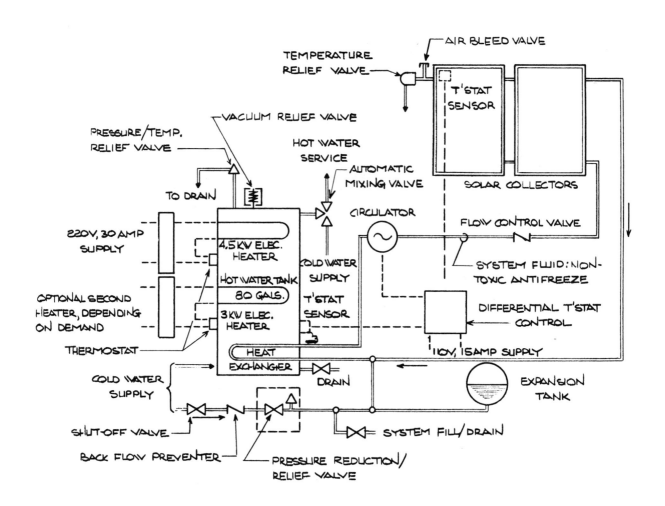

SOLAR HOT WATER HEATER
NOT TO SCALE

FIGURE 5-12

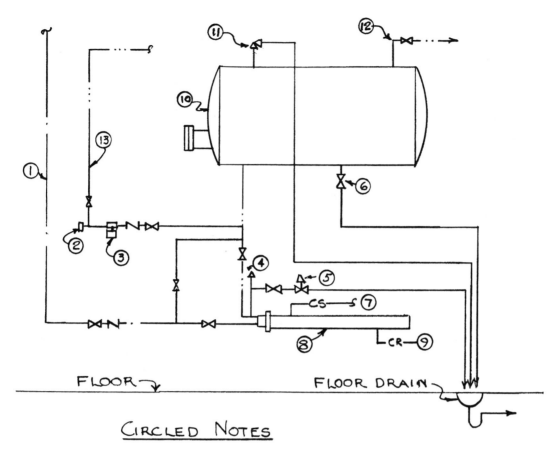

CIRCLED NOTES

1. COLD WATER SUPPLY SYSTEM
2. AQUASTAT
3. HOT WATER RECIRCULATING PUMP
4. RELIEF VALVE
5. AUTOMATIC VALVE
6. TANK DRAIN VALVE
7. HOT CONDENSATE FROM STEAM SYSTEM
8. HEAT EXCHANGER
9. CONDENSATE RETURN
10. HOT WATER STORAGE TANK AND HEATER
11. RELIEF VALVE
12. HOT WATER TO FIXTURES
    RECIRCULATING WATER LINE

HOT WATER HEATER PIPING

WITH CONDENSATE COOLER

NOT TO SCALE

FIGURE 5-13

Figure 5.11 is a schematic arrangement of a solar water heater. It is a well-known energy saving device. The solar water heater, however, is not the quick payback, money saving device that it may seem to be. It obviously can be used anywhere, within reason, that the sun shines. And equally obvious the more hours of sun, unobscured by clouds, the more heat can be generated. This varies, by area, throughout the United States. In the Northeast a rough value is 350,000 Btu/yr/sq ft of solar collector. To heat 1 gal of water from 40 to 120°F, a common range, takes 666 Btu [(120 − 40) × 8.33]. Thus the collector can heat 525 gal/yr (350,000/666). On a daily basis this equals 1.4 gal per day (525/365). Allowing 10 gal per person per day as a hot water requirement, it takes 7 sq ft (10/1.4) of collector (rounded) per person for a daily supply. A standard 4 × 8 collector has 32 sq ft, which is large enough for a family of four.

The detail in Fig. 5.11 depicts a tank of no specified size because there is a problem in this overall situation. In the summer the solar generates more heat than can be used, and in the winter, if you are not careful about your system operating and panel shut-off controls, you will discover that you are cooling, not heating, water. Since in most situations there cannot reasonably be a tank installation big enough to hold a carryover supply from summer to winter, there will be a need to have a supplemental source of hot water heating. The detail depicts electric elements in a tank that logically could be 80 to 100 gal. The control system, not depicted in complete detail, must effectively "lose" excess energy developed in the heat of summer and provide standby heat energy in the winter.

To put some dollar values on the overall situation, the typical family of four might require 40 gal of hot water a day or 14,600 gal/yr, and 100 cu ft of gas in an 80 percent efficiency gas heater could reasonably cost $0.85 per 100,000 Btu and could heat 120 gal of water [(100,000 × 0.8)/666 Btu/gal]. These are all figures subject to calculation and local conditions. However, continuing our example 14,600/120 equals 122 ccf, and 1 ccf equals 100 cu ft of gas or 100,000 Btu. Thus the gas heating cost is $104 (122 × $0.85/ccf). The probable total cost of the residential installation is at least $3,500. Thus the simple payback, if you discarded the gas heater and could totally use solar without any electrical or gas assistance, would take 35 yr ($4,500/$104).

With tax breaks available for energy saving devices and judicious installations that carefully integrate all systems, paybacks of 10 yr and sometimes less have been achieved. Where tax breaks are not available, solar can reasonably compete only in high energy cost situations.

One of the common misunderstandings of solar heating is cleared up in Fig. 5.12. The solar collector does not directly heat the hot water. If it did, the collector plate would freeze and burst in the winter. The collector heats a glycol-water mixture that is recirculated between the collector and the heat exchanger in the storage tank. The mixture is usually in the range of 30 to 50 percent glycol by volume. The system is effectively a hot water heating system with an expansion tank, a recirculating pump, and operating and safety controls. The tank commonly has one or more electric heating elements with thermostats that are only activated when the collector cannot supply all the heat necessary to handle the hot water demand. There can be situations in which extremely hot water is produced by the collector, with temperatures in the summer approaching the boiling point of water. The designer must carefully review this situation. The temperature relief valve in the collector will solve the problem if all else fails but in practice should seldom, if ever, be called into operation.

Figure 5.13 is an ideal way to not only save energy but to also resolve the problem of very hot condensate that is usually found in process systems in industrial plants when higher-pressure steam is a process requirement. One solution is to allow hot condensate to flash into low-pressure steam in the deaerator. But that is a heating solution and may not be readily and economically applied. A plumbing solution is depicted in Fig. 5.13.

Common to many energy conservation solutions, the detail shown has one basic requirement. That requirement is to be able to utilize the available energy. Not every control or piping specialty is depicted, but all of the basic elements of the scheme are presented. Depending on the temperature of the condensate, a reasonable presumption is that the desire of the heating system designer is to cool the condensate which is presumed to be above the boiling temperature (212°F or 100°C) of water to a temperature below 212°F to preclude condensate flashing into steam when the condensate is exposed to atmospheric pressure. This was implied but not depicted in the detail. The steam source was high-pressure steam and the condensate is high-pressure condensate.

No controls are shown since the quantity of condensate is assumed to vary. Since the cold water passes directly through the condensate heat exchanger, it is possible that under low condensate flow and high domestic water use the condensate may be cooled to a temperature too low for heating system purposes. Conversely under low water flow and high condensate flow the domestic water could get overheated. The hot condensate could readily be in the area of 300°F. An automatic temperature-controlled bypass around the heat exchanger of the condensate flow may be required. Commonly the condensate flow is in danger of being overcooled rather than that of the domestic water being overheated.

For many 125-psig steam plants the arrangement shown in Fig. 5.13 is not only an energy saver but also a flash condensate-to-steam problem solver.

Appendix

DRAW CHARACTERISTICS

The percent of an even-temperature water that can be delivered from any given vessel is dependent upon certain factors:

1. **The position of the tank.** Does it lie on the horizontal plane or on the vertical plane? If the tank is placed in the vertical position, a greater number of usable gallons of hot water can be withdrawn. Either forced circulation or an elevated tank with gravity circulation should be used.
2. **The position of the cold water inlet in relation to thermostat and the hot water outlet.** To deliver the maximum number of usable gallons of hot water, the cold water inlet must be exactly opposite the hot water outlet with the sensing element of the thermostat within the line of turbulence created by the flow of the incoming cold water. The cold water inlet must be close to the bottom head, but not so close as to cause thermal shock. Proper location will cause rapid response to the thermostat and excellent draw characteristics.
3. **Flow rate.** Flow rates are determined on a gallons-per-minute basis, calculated from the square root of the tank capacity.

In general, the rules of draw characteristics are:

A. A horizontal tank with a side arm heater with either gravity or forced circulation will deliver 50-60% of the tank capacity at an even delivered temperature.
B. A vertical tank with a side arm heater and forced circulation will deliver 60-70% of the tank capacity at an even delivered temperature.
C. A PVI vertical free standing storage water heater with a STRATA-BAFFLE system is designed to deliver as much as 85% of its capacity, and a PVI horizontal free standing storage water heater with a STRATA-FLO system is designed to deliver as much as 80% of its capacity, both at an even temperature of hot water.

INSULATED LINES

All the sizing in PVI's sizing guides is based on insulated lines with an estimated heat loss of approximately 10%. We find in many installations heat losses in excess of 60% with uninsulated lines. This is dependent upon the amount of air movement over the uninsulated lines; the ambient temperature; and, of course, the gpm flow rate of the circulating system.

We realistically estimate that for installations with uninsulated lines you should expect a 50% heat loss. Heat losses for insulated systems range from 5% to 30% depending on the type and thickness of the insulation. For installations with uninsulated lines, you should multiply the W_T by 1.5. If the lines are to be insulated, depending upon the type of insulation, any multiplier between 1.0 and 1.3 should be used.

HOT WATER

1. **120°F (A range from 105-120°F).** These temperatures are used for personal hygiene, manual dishwashing and other general purpose applications.
2. **160°F (A range from 140-160°F).** These temperatures are used primarily for clothes washing and automatic dishwashing operations. They are too hot for body contact and too cold for sanitizing purposes.
3. **180° plus.** This temperature has become widely accepted in the food service industry as a sanitizing temperature. A dish immersed in 180°F water and allowed to air dry will be relatively free from harmful bacteria.

COLD WATER CORRECTION FACTORS

Cold water correction factors for all selection guides are shown below. PVI does not recommend the use of these correction factors since experience has proven that severe weather in most all sections of the country will at one time or another drop the cold water temperature to approximately 40°F.

Multiply established hot water load (W_T) by:

.88 when incoming cold temperature is 50°F
.75 when incoming cold temperature is 60°F
.63 when incoming cold temperature is 70°F

FORMULA FOR DETERMINING USABLE GALLONS DELIVERED

$$\frac{\text{Stored water temperature} - \text{incoming cold water temperature}}{\text{Desired delivered temperature} - \text{incoming cold water temperature}} \times \text{Percent of vessel usable} \times \text{Actual storage capacity} = \text{Usable gallons delivered}$$

EXAMPLE: How many usable gallons of 120°F water can be delivered from a vertical 500 gallon tank? Stored water temperature 150°F, incoming cold water temperature 40°F.

$$\frac{150 - 40}{120 - 40} = \frac{110}{80} = 1.375 \times .85 \times 500 = 584 \text{ gallons}$$

TABLE 1
WATER CONSUMPTION

ITEM	Gallons Per Hour 120°F	Gallons Per Minute
Vegetable Sink	45	
Single Pot Sink	30	
Double Pot Sink	60	
Triple Pot Sink	90	
Pre-Rinse for Dishes (shower head type, hand operated)	45	
Pre-Srapper for Dishes (salvajor type)	180	
Pre-Scrapper for Dishes (conveyor type)	250	
Bar Sink	30	
Lavatory	5	
Mop Sink	20	
54-inch Bradley Washfountain (full size)	80	8.0
54-inch Bradley Washfountain (half size)	50	5.0
36-inch Bradley Washfountain (full size)	50	5.0
36-inch Bradley Washfountain (half size)	30	3.0
Duo Washfountain	20	2.0
9 and 12 pound clothes washer	40	
16 pound clothes washer	50	

*NOTE: Bradley Washfountain consumption is based on 100% operation for ten minutes. When sizing for other applications use the gpm rating.

TABLE 2
STANDARD & PULSATING SHOWER HEAD FLOW RATES

MANUFACTURER	Model No. & Name	Normal Flow Rate GPM 40 psi	Flow Rate of Recommended Flow Control Valve
American Standard	1414 Stereo	8.5	2.5
	1413 Astro-Jet	10.0	2.5
	1411 Colony	10.0	2.5
Chicago Faucet Co.	620B		2.0
Crane Company	8-1560 Criterion	8.7	3.5
	8-1561 Criterion	8.7	3.5
	8-1563 Criterion	8.7	3.5
	8-2551 Caprice	8.2	2.7
	8-2552 Caprice	3.5	3.5
	8-2556 Rainbeau	6.6	2.6
	8-2557 Rainbeau	6.6	2.6
	8-2558 Rainbeau	3.5	3.5
	R-8595 Riviera	9.5	2.6
	R-8596 Riviera	9.5	2.6
	R-8598 Freshet	8.2	2.7
Dearborn Brass Co.	R-775-Jaclo	4.0	2.5
	R-775-Jaclo	4.0	2.5
	R-777-Jaclo	5.0	2.5
	885 Olderman	3.0	—
	886 Midcor	3.5	—
	888 All American	4.0	—
Dole Valve Co.	*Mark III	3.0	—
	*GS Flow control valve	—	2.0
	*GO Flow control valve	—	2.5
	*GM Flow control valve	—	3.0
	*GL Flow control valve	—	4.0
Milwaukee Faucet Co.	R-3698-1	3.5	2.0
	W-3700-1	3.5	2.0
Sloan Valve Co.	Act-O-Matic	5.5	4.0
Speakman Co.	SS 2250 Model 1	13.0	3.0
	SS 2250 Model 2	12.0	2.5
	SS 2250 Model 3	10.0	2.0
	*SS 2240 Model 1	3.0	—
	*SS 2240 Model 2	2.5	—
	*SS 2240 Model 3	2.0	—
Wrightway Engineering Co.	Bubble Stream	4.0	2.0

*Built-in flow control valves

MANUFACTURER	Model No. and Name	Flow Rate in GPM Pulsating Position	Flow Rate in GPM Spray Position
Chicago Specialty Co.	The Body Shaper™	4.6	7.7
Alsons Corporation	Massage Action™	2.9	3.0
Bowles Fluidics—Aqua Viva Div.	Aqua-Massage®	5.4	6.2
Stanadyne	Touch Control® Model 20915	4.2	6.0
Conair Corporation	Waterfingers™	5.5	5.1
Teledyne Water Pik	The Shower Massage	9.1	7.9

PVI engineering staff have tested the pulsating type shower heads shown above. Flow rate based on 40 psi. Flow control valves are not recommended.

TABLE 3
REQUIREMENTS FOR DISHMACHINES
Conveyor, Door and Hood Types

MANUFACTURER	MODEL NO.	TYPE	GPH @ 190°F & 100% EFFICIENCY	GPH @ 190°F & 70% EFFICIENCY
Alvey-Ferguson	VA, SA-3, KS-88, KW-88	Utensil Washer	85	60
Blakeslee	B7, BC7, D7, DC7	Door	117	82
	1E, 1L, 1M, 1ER	Conveyor	420	294
	FA1E, FA1L, FA1M, FA1ER	Conveyor	420	294
	1LR, 1MR, FA1LR, FA1MR	Conveyor	420	294
	2E, 2L, 2M, 2ER, 2LR, 2MR	Flight	288	202
	3E, 3L, 3M, FA2E, FA2L, FA2M	Flight	288	202
	FA2ER, FA2LR, FA2MR, FA3E, FA3L, FA3M	Flight	288	202
	F1E, F1L, F1M, F1ER, F1LR, F1MR	Flight	420	294
	F2E, F2L, F2M, F2ER, F2LR	Flight	288	202
	F2MR, F3-E, F3L, F3M	Flight	288	202
	XF1E, XF1L, FX1M	X Flight	720	504
	XF1ER, XF1LR, XF1MR	X Flight	720	504
	XF2E, XF2L, XF2M, XF2ER	X Flight	420	294
	XF2LR, XF2MR, XF3E, XF3L, XF3M	X Flight	420	294
General Electric	ST30	Roll Top	51	36
	SK10	Door	75	53
	SK20, SK30	Door	81	57
	*ST50	Roll Top	76	53
	SK40, SK50, SK60, SK501, SK601	Door	100	70
	*SRW, SR14, SS40, SS48, SS70	Single Tank	450	315
	*SS76, SS84, SS62	Single Tank	450	315
	*SS64, SS86, SS100, SS82, SS102, SS116	Multiple Tank	300	210
	*SW80, SW81, SW82, SW83, SW84, SW85, SW 88	Belt Conveyor	420 420	294 294
	*SW80, SW81, SW82, SW83, SW84, SW85, SW 88	Rackless Rackless	540 540	378 378
Hobart	UM4, UM4D, UMC4, UMC4D, UMP4, UMP4D, UMPC4, UMPC4D	Door Door	35 35	25 25
	WM1, WM1D, WMP1, WMP1D, WMC1, WMPC1	Door	35	25
	AM10, AM11, AM11C	Hood	91	64
	SM6T2	Hood	52	36
	LM3T3	Hood	126	88
	AM8T2, AM9T2	Door	91	64
	AM8T3, AM9T3	Door	123	86
	AM9CT3	Door	120	84
	C44, C54, CWS66, CPW80, CRS66	Rack-Conveyor	450	315
	CRS76, CPW90, C64, CRS86	Rack-Conveyor	390	273
	CPW100, C81, CRS103, CPW117	Rack-Conveyor	390	273
	CS100, CS117	Rack-Conveyor	402	281
	FT200 (3RS-8-5, 6, 8, 11)	Flight	468	328
	FT200 (5, 6, 7-8-5, 6, 8, 11)	Flight	468	328
	FT300 (3RS-8-5, 6, 8, 11)	Flight	468	328
	FT300 (5, 6, 7-8-5, 6, 8, 11)	Flight	468	328
	FT400 (88-44-60-72-84)	Flight	426	298
	FT500 (88-44-44-60-72-84)	Flight	420	294
Jackson	JV24A, JV24AF, JC24B, JV24BF	Counter	52	36
	10A, 10AB, 10APR-B	Roll Top	58	41
	JL100, JL100B, JL100PR	Door	100	70
	44CE, 44CERPW	Conveyor	414	290

TABLE 3
REQUIREMENTS FOR DISHMACHINES
Conveyor, Door and Hood Types

MANUFACTURER	MODEL NO.	TYPE	GPH @ 190°F & 100% EFFICIENCY	GPH @ 190°F & 70% EFFICIENCY
Stero	SF1RA	Hood	67	47
	SF2RA	Hood	103	72
	SF2DRA	Door	103	72
	SD2RA, SDRA, SDRA-PACK	Door	92	64
	SCT44, SCT44-10, SCT54, SCT76S, SCT76SC	Conveyor	417	292
	SCT64, SCT76, SCT80	Conveyor	277	194
	SCT94, SCT108, SCT120	Conveyor	277	194
	U31A	Door	92	64
	U31A2	Door	184	129
	SCT94S, SCT108S, SCT120S	Conveyor	277	194
	SCT94SC, SCT108SC, SCT120SC	Conveyor	277	194
	SCT94SM, SCT120SM, SCT150SM	Conveyor	277	194
	STPCW12PS, STPCW15PS	Conveyor	576	403
	STPCW15, STPCW19, STPCW19PS	Conveyor	390	273
	STPCW20, STPCW22, STPCW24	Conveyor	390	273
	STPC12PS, STPC15PS	Conveyor	465	326
	STPC15, STPC19, STPC19PS	Conveyor	330	231
	STPC20, STPC22, STPC24	Conveyor	330	231
	STBUW14	Conveyor	447	313
	SC2-4, SC6-4, SC1-2-4	Conveyor	432	302
	SC1-6-4, SC5-6-4, SC5-2-4	Conveyor	432	302
	SC2-3-4, SC6-3-4, SC2-7-4, SC1-27-4	Conveyor	326	228
	SC1-6-3-4, SC5-2-3-4, SC1-6-7-4,	Conveyor	326	228
	SC5-6-3-4, SC5-2-7-4	Conveyor	326	228
Toledo	TKM-20	Counter	64	45
	TKM-27	Door	98	69
	TKM-44, TKM-62, TKM-62R, TKM-66,	Conveyor	414	290
	TKM-66R, TKM-102R	Conveyor	414	290
	TKM-64, TKM-80, TKM-86R	Conveyor	360	252
	TKM-96, TKM-96R	Conveyor	110	77
	TKM-113F, TKM-113R, TKM-114F	Conveyor	624	437
	TKM-114R, TKM-115F, TKM-115R	Conveyor	624	437
	TKM-215, TKM-216, TKM-217, TKM-317R	Conveyor	624	437
	TKM-318R, TKM-319R, TKM-320R,	Conveyor	414	290
	TKM-321R, TKM-322R, TKM-323R,	Conveyor	414	290
	TKM-324R	Conveyor	414	290
Vulcan	CU-16BTA, R-16BTA	Door	77	54
	3D20T, CD20T	Door	120	84
	A44, A54	Conveyor	480	336
	A64, A81, AO98	Conveyor	510	357
	CP-3, HP-3, CP-2	Conveyor	420	294

Consumptions based on 20 psi flow pressure and 70% mechanical operation.

Rackless types generally are sized on 100% continuous operation.

APPLICATION GUIDE I

APARTMENT HOUSES — HOTELS — MOTELS

In determining proper water heating equipment for any type commercial living establishment, large or small, a diversity factor arises as the number of units increases. Therefore, in determining proper water heating equipment for each of the three types of units, a different diversity factor is used as the number of units is increased.

The major hot water load in an apartment house, motel or hotel will be shower head consumption; consequently, if shower demands can be met, incidentals such as lavatories, kitchen sink, etc. can also be met.

In apartment houses there are two additional factors which must be considered that are not considered in hotels and motels. One is automatic dishwashers; the other, clothes washing facilities. Both must be included over and above the basic shower head sizing.

Many hotels and motels today have central food service and/or commercial laundry facilities. If either or both are to be included, hot water consumption must be determined and added to regular requirements.

IMPORTANT: To conserve energy, insulated hot water lines are mandatory. Many installations, especially apartment houses, do not insulate the hot water lines. The result is greater first costs for the hot water systems because the size of the equipment must be increased. This guide is based on insulated lines. For installations with uninsulated lines, multiple $W_T \times 1.5$.

Many motel and hotel installations are being designed primarily for convention business or other regimented groups of people. These groups of people generally are on a timed schedule and demands are great within a short period of time. DO NOT use the diversity factor (C) in sizing hotels and motels that will have this type clientele.

Also, many hotels and motels are changing from a tourist to a commercial type establishment. When this happens, the demand is increased to a point where it is advisable to eliminate the diversity factor in sizing. The diversity factor is used primarily for apartment houses.

FORMULA I   Apartment houses, hotels and motels
$$W_T = U_T \times FR \times W \times C \times CF$$

FORMULA II   Hotels and motels with central laundry facilities
$$W_T = [(U_T \times FR \times W \times C) + (M_T \times G \times 1.5)] \, CF$$

NOTE: For food service, use Application Guide II and combine with total water demand.

For coin-operated clothes washers, use Application Guide VIII and combine with total water demand.

CLASSIFICATION	FACTOR WATER W	NUMBER OF UNITS U_T	DIVERSITY FACTOR c
Apartment Houses		1-20	.96
1—Bath 4.5		21-40	.92
1—Bath and Dishwasher 4.6		41-60	.88
1—Bath and Clothes Washer 4.8		61-80	.86
1—Bath, Dish & Clothes Washers 4.9		81-100	.82
2—Baths 5.0		101-120	.80
2—Baths & Dishwasher 5.1		121-140	.77
2—Baths & Clothes Washer 5.3		141-160	.75
2—Baths, Dish & Clothes Washers 5.4		161-180	.72
Hotels and Motels		181-200	.70
1—Bath 4.5		201-260	.65
2—Baths 5.0		261-310	.61
W_T = Total Gallons of 120°F Water Required		311-370	.58
U_T = Total Number of Living Units		371-440	.55
FR = Flow Rate in gpm of Shower Heads (See Table 2)		441-520	.53
W = Water Factor		521-600	.50
C = Diversity Factor		601-680	.47
M_T = Total Number Laundry Machines		681-760	.45
G = Hot Water Consumption per Machine		761-840	.43
1.5 = To Convert 160°F Water to 120°F		841-920	.41
*CF = Cold Water Correction Factor		921-1000	.39

EXAMPLE: A 200 unit apartment project with 1 bath and dishwasher in each unit. Shower head flow rate — 4 gpm; 15 coin operated washing machines with a 20-minute cycle; incoming cold water temperature — 40° at the lowest.

$W_T = [(U_T \times FR \times W \times C) + (M_T \times G \times 1.5)] \, CF$ $W_T = 2576 + 698$

$W_T = [200 \times 4 \times 4.6 \times .70) + (15 \times 31 \times 1.5)]$ $W_T = 3274$

*Most areas in the United States should use a 40°F incoming cold to correctly size water heating installation.

APPLICATION GUIDE II

FOOD SERVICE ESTABLISHMENTS

Food service establishments, large or small, have demands for two temperatures of hot water (190°F and 120°F). This demand will vary, not by the number of people served, but by the type of automatic dishwashing equipment installed and general usage fixtures. Maximum demand will peak during a one-hour period.

Field experience has proven that 190°F water must be delivered from the water heating equipment to allow for a line loss between it and the dishmachine where 180°F water is mandatory.

There are two basic types of water heating systems used in food service establishments:

1. **Single System.** Stores 190° water for sanitizing purposes and utilizes a mixing valve for general purpose (120°) water. It is not recommended as a long life system.

2. **Dual System.** Stores 130°-140°F water for general purpose applications and requires a separate booster for 190°F sanitizing purposes. PVI recommends this type of installation for better temperature control and longer life of the central water heating system.

To determine the correct amount of hot water required, use Water Consumption Table 1 for general usage requirements of 120°F and the Automatic Dishmachine Consumption Table 3 for 190°F requirements.

PROCEDURE:

1. List all fixtures and equipment that will require hot water.
2. Set out two columns, one for 120°F water, the other for 190°F water.
3. Turn to Tables 1 and 3 for consumption requirements.
4. Enter in the appropriate column the gallons required (from step No. 3).
5. *Multiply each total by appropriate cold water correction factor.
6. 1.88 = Conversion factor 190°F to 120°F.
7. W_T = Total water.

EXAMPLE:

	Gallons 120°F	Gallons 190°F
2 — Vegetable sinks	90	
2 — Double pot sinks	120	
1 — Triple pot sink	90	
1 — Pre-rinse (hand type)	45	
1 — Mop sink	20	
1 — Bar sink	30	
4 — Lavatories	20	
1 — Dishmachine @ 70%		315
1 — Silver washer @ 70%		45
	W_T @ 120°F = 415	W_T @ 190°F = 360

Single System

W_T @ 120°F = 677   (W_T @ 190°F converted to 120°F; 360 × 1.88 = 677)
W_T @ 120°F = 415
Total W_T @ 120°F = 1092

Dual System

W_T @ 120°F = 415
W_T @ 190°F = 360
†Total W_T @ 120°F = 775
Total W_T @ 190°F = 360

†Adding the requirements of 120°F and 190°F water will size the primary source adequately for the booster.

*Most areas in the United States should use a 40°F incoming cold to correctly size water heating installation.

APPLICATION GUIDE III

BARBER SHOPS AND BEAUTY SALONS

To determine the correct size of water heating equipment for a barber shop or a beauty salon, large or small, there is a diversity factor which applies as the number of chairs increases. For example, it is much more possible for a beauty salon with five chairs to have five women having their hair washed during the same one-hour period than it is for a beauty salon with 15 chairs to have all 15 in use during the same one-hour period.

Barber shops generally use considerably less water to wash a man's hair than a beauty salon uses to wash a woman's hair. The figures shown below in the formula (15 gph for a barber shop and 30 gph for a beauty salon) remain constant. By using the diversity factor, as the number of chairs increases, the gallons of water per chair will decrease.

FORMULA I   For Barber Shops

$$W_T = 15 \times S_T \times C \times CF$$

FORMULA II   For Beauty Salons

$$W_T = 30 \times S_T \times C \times CF$$

W_T	= Total Water Required
C	= Diversity Factor
S_T	= Number of Stations
*CF	= Cold Water Correction Factor

Number of Chairs or Stations S_T	Diversity Factor C
1-5	1.00
6-10	.97
11-15	.94
16-20	.91
21-25	.88
26-30	.85
31-40	.79
41-50	.73
51-60	.67

EXAMPLE: The water heating equipment is to be selected for a modern 14-station beauty salon. Incoming cold water 50°F.

$$W_T = 30 \times S_T \times C \times CF$$

$$W_T = 30 \times 14 \times .94 \times .88$$

$$W_T = 347$$

*Most areas in the United States should use a 40°F incoming cold to correctly size water heating installation.

APPLICATION GUIDE IV

CHURCHES

The greatest demand for hot water in a church will be created in one of two locations — either kitchen facilities or shower room.

IMPORTANT: Both loads must be calculated separately to determine which is greater. The greater load determines the water heating equipment since both loads will seldom occur within the same one-hour period.

Shower Rooms

FORMULA   For church shower loads
$$W_T = S_T \times T \times FR \times CF$$

Shower room facilities should be calculated on a 15-minute basis.

Kitchen Facilities

There is no exact method for determining the hot water load for kitchens without automatic dishwashing equipment; however, an accepted fromula for dishwashing and rinsing purposes is one-half gallon of 120°F water per maximum number of meals to be served. Other equipment such as pot and pan, vegetable and mop sinks should be calculated as shown on Water Consumption Table 1.

W_T = Total Gallons 120°F Water Required
S_T = Total Number Showers
T   = Time in Minutes
FR = Flow Rate in gpm of Shower Head (See Table 2)
*CF = Cold Water Correction Factor

EXAMPLE: A church has a shower room with six showers rated at 2.5 gpm and a kitchen with a dishmachine requiring 64 gph at 190°F at 70% efficiency, one double pot sink, one vegetable sink, two mop sinks and four lavatories requiring 120°F water; 40°F incoming cold. A single system will be used.

Shower Room

$$W_T = S_T \times T \times FR \times CF$$
$$W_T = 6 \times 15 \times 2.5$$
$$W_T = 225$$

Kitchen

Refer to Application Guide II and size for food service requirements (single system).

	Gallons 120°F	Gallons 190°F
1 — Dishmachine at 70%		64
1 — Double Pot Sink	60	
1 — Vegetable Sink	45	
2 — Mop Sinks	40	
4 — Lavatories	20	
	W_T @ 120°F = $\overline{165}$	W_T @ 190°F = $\overline{64}$

Single System

$$W_T @ 120°F = 120 \quad (W_T @ 190°F \text{ converted to } 120°F = 64 \times 1.88 = 120$$
$$W_T @ 120°F = \frac{165}{285}$$

A comparison must now be made. It is noted that more energy and storage will be required for the shower requirements than for kitchen requirements. Therefore the larger unit will be used to size the installation.

*Most areas in the United States should use a 40°F incoming cold to correctly size water heating installation.

APPLICATION GUIDE V
HOSPITALS, REST HOMES, NURSING HOMES, ORPHANAGES AND CONVENTS

The institutions shown above have many common hot water demands. All have general usage requirements of 120°F for labatories, tubs, showers, mop sinks and general clean up. When the total demand is calculated, there will be sufficient hot water available for these requirements.

Most of these institutions will also require 160°F water for central laundry facilities and 190°F sanitizing water for food service.

Because of the very young, the very old and the sickly, a water temperature regulator is recommended for hospitals, rest homes, nursing homes and orphanages. This temperature regulator should be set no higher than 110°F.

FORMULA I   Institutions with no central laundry or food service

$$W_T = FR \times W \times P_T \times C \times CF$$

FORMULA II   Institutions with central laundry facilities—no food service

$$W_T = [(FR \times W \times P_T \times C) + (M_T \times G \times 1.5)] \, CF$$

NOTE: For food service, use Application Guide II and combine the total water demand. Transfer to Selection Guide.

TYPE OF INSTITUTION	WATER FACTOR W	NUMBER OF PERSONS OR BEDS P_T	DIVERSITY FACTOR C
Hospitals	3.7	1-20	.96
Rest Homes		21-40	.92
Nursing Homes	3.1	41-60	.88
Convents		61-80	.86
Orphanages	2.8	81-100	.82
		101-120	.80
		121-140	.77
		141-160	.75
		161-180	.72
		181-200	.70
		201-260	.65
		261-310	.61
		311-370	.58
		371-440	.55
		441-520	.53
		521-600	.50
		601-680	.47
		681-760	.45
		761-840	.43
		841-920	.41
		921-1000	.39

W_T = Total Gallons 120°F Water Required

FR = Flow Rate in gpm of Shower Head (See Table 2)

W = Water Factor

P_T = Total Persons or Beds

C = Diversity Factor

M_T = Total Number Laundry Machines

G = Hot Water Consumption per Machine

1.5 = To Convert 160°F Water to 120°F

*CF = Cold Water Correction Factor

EXAMPLE: A nursing home with 78 beds is to be sized for a water heating system. Shower head flow rate—4 gpm; 2-16# commercial clothes washers; 1 dishmachine that requires 82 gph at 190°F at 70% efficiency; 1 vegetable sink and 1 two-compartment pot and pan sink; 40°F incoming cold water.

$$W_T = [(FR \times W \times P_T \times C) + (M_T \times G \times 1.5)] \, CF$$
$$W_T = [(4 \times 3.1 \times 78 \times .86) + (2 \times 60 \times 1.5)]$$
$$W_T = 1012$$

Refer to Application Guide II and size for food service requirements (dual system).

	Gallons 120°F	Gallons 190°F
1 — Dishmachine at 70%		82
1 — Vegetable Sink	45	
1 — Double Pot and Pan Sink	60	
	W_T @ 120°F = 105	W_T @ 190°F = 82

W_T @ 120°F = 105
W_T @ 190°F = 82
W_T = 187   Combine with W_T above.
W_T Application Guide V = 1012
W_T Application Guide II = 187
Total W_T = 1199

*Most areas in the United States should use a 40°F incoming cold to correctly size water heating installation.

APPLICATION GUIDE VI

DORMITORIES, FRATERNITY AND SORORITY HOUSES

Hot water demand for dormitory type living is based on total number of students and a diversity factor for this number which is directly related to the flow rate of the shower heads installed. When the total demand is calculated, there will be sufficient hot water available for all general purpose applications such as lavatories, showers, tubs, general clean up and small hand laundering.

The exceptions are kitchen facilities, coin-operated laundries and/or large commercial clothes washers. These must be calculated separately and combined with the primary load.

FORMULA I   Institutions with no central laundry or food service

$$W_T = FR \times W \times P_T \times C \times CF$$

FORMULA II   Institutions with central laundry—no food service

$$W_T = [(FR \times W \times P_T \times C) + (M_T \times G \times 1.5)] CF$$

NOTE: For food service, use Application Guide II and combine with total water demand.

For coin operated clothes washers, use Application Guide VIII and combine with total water demand.

TYPE OF INSTITUTION	WATER FACTOR W	NUMBER OF STUDENTS P_T	DIVERSITY FACTOR C
Dormitories, Fraternity and Sorority Houses	4	1-9	1.00
		10-19	.97
W_T = Total Gallons 120°F Water Required		20-29	.94
FR = Flow Rate in gpm of Shower Head (See Table 2)		30-39	.91
W = Water Factor		40-49	.88
P_T = Total Number of Students		50-74	.80
C = Diversity Factor		75-99	.72
M_T = Total Number Laundry Machines		100-199	.65
G = Hot Water Consumption per Machine		200-299	.59
1.5 = To Convert 160°F Water to 120°F		300-399	.54
*CF = Cold Water Correction Factor		400-500	.50

EXAMPLE: A dormitory is to house and feed 125 students. Shower head flow rate—3.5 gpm; 2 - 16# commercial washers; 10 - coin-operated clothes washers, 30-minute cycle; 1 - dishmachine that requires 69 gph at 190°F at 70% efficiency; 1 - vegetable sink and 1 - two-compartment pot and pan sink; 40°F incoming cold water.

$$W_T = [(FR \times W \times P_T \times C) + (M_T \times G \times 1.5)] CF$$
$$W_T = [(3.5 \times 4 \times 125 \times .65) + (2 \times 60 \times 1.5)]$$
$$W_T = 1318$$

Refer to Application Guide II and size for food service requirements (dual system).

	Gallons 120°F	Gallons 190°F
1 — Dishmachine at 70%		69
1 — Vegetable Sink	45	
1 — Double Pot Sink	60	
	W_T @ 120°F = 105	W_T @ 190°F = 69

W_T @ 120°F = 105
W_T @ 190°F = 69   Transfer to Booster Selection Guide.
   W_T = 174   Combine with W_T above.

Refer to Application Guide VIII and size for 10 coin operated clothes washers.

$W_T = M_T \times C_3 \times 1.5 \times CF$
$W_T = 10 \times 28 \times 1.5$
$W_T = 420$ Combine with W_T above.
W_T Application Guide VI   = 1318
W_T Application Guide II   = 174
W_T Application Guide VIII = 420
    Total W_T = 1912

*Most areas in the United States should use a 40°F incoming cold to correctly size water heating installation.

APPLICATION GUIDE VII

SCHOOLS

There are two great demands on a water heating system in a school — showers and food service. Generally, the shower demand will exceed the food service demand in high schools and junior high schools; and the food service demand will exceed the shower demand in grade schools.

Both demands must be calculated separately and added together if one combined system is to be used. However, the trend today is for a separate system for each bank of showers and a separate system for the cafeteria.

There are many other factors (lavatories, mop sinks etc.) that are considered incidental to the overall demand. Therefore, if the shower loads and kitchen loads are calculated, there will be sufficient hot water for the total installation.

To determine the correct water heating equipment for shower facilities, first determine the gpm flow rate of the shower head. Shower heads with small gpm ratings are recommended. (See Table 2 for shower head manufacturers to simplify selection of the correct head size.)

Time is an extremely important factor as all the water required will generally be drawn within a 10-minute period. When shower requirements are met, the other general use requirements such as lavatories and mop sinks will be adequately handled.

When food service facilities are to be added to this water heating system, refer to Application Guide II. The requirements are determined and combined by analyzing the recommendation on Selection Guides.

Stadiums. When water heating equipment is to be determined for stadiums, the time period should be increased to 20 minutes and the W_T transferred to Selection Guide.

Twenty-four hour recovery periods are generally used for stadiums.

FORMULA   For school shower loads
$$W_T = S_T \times FR \times T \times CF$$

> W_T = Total Gallons 120°F Water Required
> S_T = Total Number Shower Heads
> FR = Flow Rate Shower Heads (See Table 2)
> T = Time in Minutes
> *CF = Cold Water Correction Factor

EXAMPLE: A school has 15 showers in the men's locker room and 15 showers in the women's locker room. Shower head—2.5 gpm flow rate; cold water inlet temperature — 40°F.

$$W_T = S_T \times FR \times T \times CF$$
$$W_T = 30 \times 2.5 \times 10$$
$$W_T = 750$$

*Most areas in the United States should use a 40°F incoming cold to correctly size water heating installation.

APPLICATION GUIDE VIII

LAUNDRIES — COMMERCIAL, INDUSTRIAL AND COIN OPERATED

Commercial or Industrial

Most all commercial and industrial clothes washing machines will require between 1 and 6 gallons of 180°F water per pound of capacity per cycle. It is mandatory that this figure be determined from the manufacturer. The figure 3 in the following formula is approximate and may be used as an estimate only.

For a one-hour operation, transfer to Selection Guides. If more than a one-hour operation is required, consult specification sheet. In either case, **the minimum tank capacity shall equal $W_T \times .3$.** For more than a one-hour operation W_T must equal recovery at an 80°F rise.

FORMULA $W_T = M_T \times PC \times 3 \times 1.75 \times CF$

EXAMPLE: 3 — 150 lb. Washing Machines. Cold water 40°F.

$W_T = M_T \times PC \times 3 \times 1.75 \times CF$
$W_T = 3 \times 150 \times 3 \times 1.75$
$W_T = 2363$

Tank capacity minimum $= W_T = 2363 \times .3 = 709$ gallons storage

Coin Operated

The most important single item in determining the correct water heating installation for coin operated laundries is the diversity factor. With one washing machine installed, the diversity factor is 100%; however, this factor drops as the total number of washing machines increases.

The number of machines, total number of cycles per hour and incoming cold water correction factor make up the other requisites for determining the correct water heating load.

FORMULA I   For machines with 20-minute cycle
$\qquad W_T = M_T \times C_2 \times 1.5 \times CF$

FORMULA II   For machines with 30-minute cycle
$\qquad W_T = M_T \times C_3 \times 1.5 \times CF$

EXAMPLE: A coin operated laundry has twenty-six 20-minute cycle washing machines. The cold inlet water temperature drops to 50°F during the winter months.

$W_T = M_T \times C_2 \times 1.5 \times CF$
$W_T = 26 \times 29 \times 1.5 \times .88$
$W_T = 995$

COIN OPERATED LAUNDRY MACHINES				
NUMBER OF WASHING MACHINES M_T	DIVERSITY FACTOR 20-MINUTE CYCLE C_2	DIVERSITY FACTOR 30-MINUTE CYCLE C_3	NO. OF MACHINES	MINIMUM TANK STORAGE
2-4	38	33	2-10	100
5-8	35	30	2-10	100
9-12	33	28	11-20	200
13-16	31	26	11-20	200
17-24	30	25	21-30	300
25-36	29	24	21-30	300
37-52	28	23	31-40	400
53-60	27	22	41-60	500

W_T = Total Gallons 160°F Water Required
M_T = Total Number Washers
C_2 = Diversity Factor 20-Minute Cycle
C_3 = Diversity Factor 30-Minute Cycle
1.5 = To Convert 160°F Water to 120°F Water
1.75 = To Convert 180°F Water to 120°F Water
PC = Pound Capacity, Commercial or Industrial
*CF = Cold Water Correction Factor

*Most areas in the United States should use a 40°F incoming cold to correctly size water heating installation.

APPLICATION GUIDE IX

OFFICE BUILDINGS

Office buildings generally have the greatest hot water demand during a period approximately one half hour before noon and one half hour before quitting time. However, the noon hour and quitting time are not always the same for all businesses; therefore, an office building should be sized on a peak one-hour period.

Usually, lavatories are the key to sizing this type installation. There may also be bar sinks and mop sinks. Apply all fixtures at their full rated capacity (Table 1), then apply diversity factor.

As noted in many other application guides, a diversity factor occurs. As the number of fixtures increases, the chance of all fixtures using their full rated capacity decreases.

If there is food service or laundry facilities, these should be calculated separately and added to the fixture total W_T. (Note Application Guides II and VIII and Table 1.)

FORMULA   Office Building

$$W_T = (L_T \times 5 + B_T \times 30 + M_T \times 20) \; C \times CF$$

	TOTAL NUMBER OF FIXTURES	DIVERSITY FACTOR C
W_T = Total Gallons 120°F Water Required		
L_T = Total Number Lavatories (Table 1)	1-25	1.00
	26-50	.97
B_T = Total Number Bar Sinks (Table 1)	51-100	.92
M_T = Total Number Mop Sinks (Table 1)	101-150	.87
	151-200	.82
C = Diversity Factor	201-300	.75
	301-400	.70
*CF = Cold Water Correction Factor (Page 2)	401-500	.63

EXAMPLE: A 25-floor office building (1 floor for equipment room, 3 floors for parking garage, 1 floor for cafeteria and 20 floors for offices — each with mop sink, ladies and men's rooms with 3 lavatories in each)

$$W_T = (L_T \times 5 + B_T \times 30 + M_T \times 20) \; C \times CF$$

$W_T = (120 \times 5 + 20 \times 20) \; .87$ Total of 140 fixtures

$W_T = 870$

Food service in cafeteria, using example in Application Guide II, dual system:

NOTE: When a dual system is used, a booster is necessary for 190°F sanitizing water.

$W_T = 775$

$W_T = 1645$ Total.

*Most areas in the United States should use a 40°F incoming cold to correctly size water heating installation.

APPLICATION GUIDE X

SWIMMING POOLS AND BAPTISTRIES

Swimming Pools

To heat a swimming pool, the following factors must be determined: Either the total capacity in U.S. gallons or the length, width and average depth.

You must then determine the cold water temperature at the time pool heat up is desired (40°F will help to eliminate a lot of potential unknowns and will offer a fair safety factor. Then determine the water temperature desired in the pool. Most people prefer temperatures between 75°F and 80°F. The difference between the cold water temperature and the pool temperature will determine desired rise. (DR).

The next factor to determine is the number of hours heat up time. Generally 24 to 36 hours should be selected unless the pool is covered. If the pool is covered, you may wish to extend the heat up time to 48 hours. Next, determine the "E" factor (the type of energy to be used—gas, oil or electric).

Heat loss is vitally important in sizing a swimming pool. With the chart below you can determine the heat loss based on ambient conditions. Wind velocity of 0 mph should only be used when the pool is inside or covered.

FORMULA
$$\frac{L \times W \times AD \times E \times DR}{RH} + \frac{L \times W \times HL}{2} = \begin{array}{l} \text{KWH (Electric)} \\ \text{or} \\ \text{Btuh (Gas or Oil)} \end{array}$$

$$\text{Heat Loss} = L \times W \times HL = \text{Btuh or KWH}$$

L	= Length in Feet
W	= Width in Feet
AD	= Average Depth in Feet
DR	= Desired Rise
E	= Gas—Atmospheric 89
E	= Oil or Power Gas 78
E	= Electric .018
HL	= Heat Loss
RH	= Required Hours to Heat Pool

HEAT LOSS FACTORS
75°F POOL TEMPERATURE

AIR TEMPERATURE °F	WIND VELOCITY M.P.H.	HL GAS INPUT BTU/HR.	HL OIL INPUT BTU/HR.	HL ELECTRIC INPUT KW/HR.
40	0	150	131	.031
40	5	200	175	.040
40	10	250	219	.051
40	15	300	263	.062
40	20	350	306	.072
50	0	107	94	.022
50	5	143	125	.029
50	10	179	156	.037
50	15	215	188	.044
50	20	250	219	.051
60	0	64	56	.013
60	5	86	75	.018
60	10	107	94	.022
60	15	129	113	.026
60	20	150	131	.031

Above factors are based on 75°F pool temperature. If temperatures desired are greater or less than 75°F, add or subtract 7% per 5°F to energy calculated for 75°F.

EXAMPLE: A swimming pool is 20' x 35' with an average depth of 5'5". Desired pool temperature — 75°F, 40°F cold water; air temperature — 40°F; wind velocity — 20 mph; energy source — oil.

$$\frac{L \times W \times AD \times E \times DR}{RH} + \frac{L \times W \times HL}{2} = \text{Btuh oil}$$

$$\frac{20 \times 35 \times 5.5 \times 78 \times 35}{36} + \frac{20 \times 35 \times 306}{2} = 399,058 \text{ Btuh required}$$

Transfer to specification sheets and select model number that has a minimum of the required energy.

IMPORTANT: Heat loss should never be greater than total energy required to heat the pool.

Baptistries

Generally, a baptistry can be sized using a 4 to 6 hour heat up time. Heat loss using 60°F ambient with 0 wind velocity. Higher water temperature of 90°F to 100°F are also generally required.

SIX

Waste and Vent Piping

The focus of this chapter is on that portion of the plumbing system that is in close proximity to the plumbing fixture or fixtures. It is concerned with the immediate sanitary piping that accepts the effluent of any fixture permitted by the plumbing code to be drained into the sanitary piping system and to the proper venting of these fixtures and their drain lines to ensure the integrity of their trap seals.

Plumbing systems as such are not very old. Pressurized water systems were begun in the 1840s and the venting of fixtures was only understood and its implementation started in the 1860s. While there were some plumbing fixtures as we know them today prior to 1900, the majority of fixtures currently in use trace their origins to the years just before World War I. Today's modern waste and vent piping traces a lot of its origin to the Hoover Commission study of the early 1930s.

In essence, the design of a waste and vent piping system requires that every trap and trapped fixture shall be provided with a vent. The above statement is modified by the Building Owners and Code Administrators (BOCA) *Basic Plumbing Code* with the words, "except as otherwise approved by this code." While mistakes may be made by the plumbing designer in sizing the waste piping for a group of fixtures, the most common mistake is the one depicting improper and/or incorrect vent sizing and arrangement for a group of fixtures in a typical toilet room arrangement.

The most common mistakes made by the plumbing designer are one or both of the following: Some part of the waste collection piping main or branch is too small; the number, size, and arrangement of the vents are too numerous and too large.

Recently there was a study conducted by the Veteran's Administration, which does *not* have code approval, which has tested and investigated the required size of vent piping. This study depicts vent piping much smaller than permitted by the current code. The vent line sizing, which we will discuss at the end of this chapter, was found to be perfectly workable in these smaller sizes. Again we wish to note that this system has no official approval. It does, however, illustrate that there are continuing, ongoing attempts to improve plumbing system design.

Entire books, such as the *National Plumbing Code Illustrated* by Vincent T. Manas, have been written on the explanation of every facet and interpretation of the plumbing code. In this chapter we will cover the basic design concepts which underline the design and detailing examples of waste and vent piping; they will be presented via a series of examples of typical problems frequently and infrequently encountered in plumbing systems.

Thus this chapter will cover sanitary drainage systems in the area of the fixture grouping as well as the sanitary waste and vent riser portion of the system from the bottom to the top. Our definition covers the connection of the branch waste to the horizontal portion of the main sanitary drain line up through and including the vent through the roof of the building housing the plumbing fixtures.

Waste Pipe Design

The basic value system used in designing waste and vent piping is the fixture unit. This is the same fixture unit we previously used in our water piping design approach

Table 6.1 Drainage Fixture Unit Values for Various Plumbing Fixtures

Type of fixture or group of fixtures	Drainage fixture unit value (dfu)	Trap size (in)	Type of fixture or group of fixtures	Drainage fixture unit value (dfu)	Trap size (in)
Automatic clothes washer (2-in standpipe)	3		Lavatory (barber shop, beauty parlor)	2	1½
Bathroom group consisting of a water closet, lavatory, and bathtub or shower compartment:			Laundry tray (1 or 2 compartments)	2	1½
			Shower, domestic	2	2
Public	8		Shower (group) per head[a]	2	—
Nonpublic	6		Sinks:		
Bathtub[a] (with or without overhead shower)	2	1½	Surgeon's	3	1½
Bidet	3	1½	Flushing rim (with valve)	6	3
Combination sink and tray with food waste grinder	2	1½	Service (trap standard)	3	3
			Service (P trap)	2	2
Combination sink and tray with one 1½-in trap	2	1½	Pot, scullery, etc.[b]	4	2
			Commercial with food waste grinder	—	2
Dental unit or cuspidor	1	1¼	Urinal, pedestal, syphon jet blowout	6	—
Dental lavatory	1	1¼	Urinal, wall lip, washdown	4	1½
Drinking fountain	½	1¼	Urinal, stall, washout	4	2
Dishwasher, commercial	2	2	Washington machine (commercial)	3	2
Dishwasher, domestic	2	1½	Wash sink[b] (circular or multiple) each set of faucets	2	1½
Food waste grinder	—	1½			
Floor drains with 2-in waste	2	2	Water closet, nonpublic	4	—
Kitchen sink, domestic, with one 1½-in waste	2	1½	Water closet, public	6	—
			Unlisted fixture drain or trap size (in):		
Kitchen sink, domestic, with food waste grinder	2	1½	1¼ or less	1	
			1½	2	
Lavatory with 1¼-in waste	1	1¼	2	3	
			2½	4	
			3	5	
			4	6	

[a]A shower head over a bathtub does not increase the fixture unit value.
[b]See plumbing code for method of computing fixture unit value of devices with continuous or semicontinuous flows.

Table 6.2 Horizontal Fixture Branches and Stacks[a]

		Maximum number of fixture units		
			Stacks[b]	
Diameter of pipe (in)	Total for a horizontal branch	Total discharge into one branch interval	Total for stack of three branch intervals or less	Total for stack greater than three branch intervals
1½	3	2	4	8
2	6	6	10	24
2½	12	9	20	42
3	20[c]	20[c]	48[c]	72[c]
4	160	90	240	500
5	360	200	540	1,100
6	620	350	960	1,900
8	1,400	600	2,200	3,600
10	2,500	1,000	3,800	5,600
12	3,900	1,500	6,000	8,400
15	7,000	d	d	d

[a]Does not include branches of the building drain.
[b]Stacks shall be sized according to the total accumulated connected load at each story or branch interval and may be reduced in size as this load decreases to a minimum diameter of one half of the largest size required.
[c]Not more than two water closets or bathroom groups within each branch interval nor more than six water closets or bathroom groups on the stack.
[d]Sizing load based on design criteria.

in Chap. 4. Since our concern is with waste piping and our design must conform with the plumbing code (there is no legal alternative), we are depicting in Tables 6.1 and 6.2 (taken by permission from the BOCA *Basic Plumbing Code*) the fixture unit values in drainage units for various plumbing fixtures and the maximum number of fixtures allowed for horizontal fixture branches and stacks.

Those of you who are familiar with the BOCA *Basic Plumbing Code* may immediately perceive that something is missing or that the values seem strange. In the code, in addition to the above two tables there is a third table for building drains and sewers. We have included the latter set of values in Chap. 7 to avoid confusion between waste and vent piping at the fixture group and basic sanitary drainage system piping. Branch waste piping is usually so constrained that it will only accommodate minimum pitch (fall per foot). Thus the code does not allow for the added capacity available, theoretically, in large pitches.

In the details that follow we will discuss and illustrate the proper sizing of the waste piping. In essence you are adding the values of your fixtures that are taken from Table 6.1 and selecting pipe sizes from Table 6.2. There is one important point in the waste pipe sizing which is a code requirement and which we will note here and repeat elsewhere: No fixture branch connection may be made within 10 pipe diameters downstream from the base of a soil or waste stack. For example, if you have a 4-inch (in) stack taking the waste from one or more floors of plumbing fixtures, you cannot connect the branch of, typically, floor drains within 40 in (10 in × 4 in) of the base of the stack.

Indirect and special waste piping has a separate section devoted to the proper installation of it in the BOCA *Basic Plumbing Code*. Since this is a special subject that

Table 6.3 Size and Length of Vent Stacks and Stack Vents

Diameter of soil or waste stack (in)	Total fixture units connected to stack (dfu)	Diameter of vent (in)[a] (maximum developed length of vent, in feet, given below)										
		1¼	1½	2	2½	3	4	5	6	8	10	12
1¼	2	30										
1½	8	50	150									
1½	10	30	100									
2	12	30	75	200								
2	20	26	50	150								
2½	42		30	100	300							
3	10		42	150	360	1040						
3	21		32	110	270	810						
3	53		27	94	230	680						
3	102		25	86	210	620						
4	43			35	85	250	980					
4	140			27	65	200	750					
4	320			23	55	170	640					
4	540			21	50	150	580					
5	190				28	82	320	990				
5	490				21	63	250	760				
5	940				18	53	210	670				
5	1,400				16	49	190	590				
6	500					33	130	400	1000			
6	1,100					26	100	310	780			
6	2,000					22	84	260	660			
6	1,900					20	77	240	600			
8	1,800						31	95	240	940		
8	3,400						24	73	190	720		
8	5,600						20	62	160	610		
8	7,600						18	56	140	560		
10	4,000							31	78	310	960	
10	7,200							24	60	240	740	
10	11,000							20	51	20	630	
10	15,000							18	46	180	570	
12	7,300								31	120	380	940
12	13,000								24	94	300	720
12	20,000								20	79	240	610
12	26,000								18	72	230	500
15	15,000									40	130	310
15	25,000									31	96	240
15	38,000									26	81	200
15	50,000									24	74	180

[a] 1 ft = 304.8 mm.

is covered by a series of rules and regulations (it is not a waste and vent design element), we refer you to the code for these rules. There is one point in the overall subject which is not noted in our details but does require emphasis. An indirect waste is a waste line that not only has a break (air gap) in the line but also must discharge into a fixture that is properly trapped and vented. Commonly this indirect waste acceptance fixture is a standard sink, a similar built-up unit, or a floor drain. That portion of the requirement is usually readily understood. The portion of the code commonly overlooked or misunderstood is the further requirement that the air gap shall be, in length, at least twice the diameter of the drain served. Finally the code has another requirement that is also frequently overlooked. That is that all indirect waste piping exceeding 2 feet (ft) shall be trapped. For example, the condensate drain line from an air conditioning unit to an indirect waste receptor shall be trapped.

Vent Piping Design

The plumbing code spells out the basic fixture vent requirements in a series of short paragraphs that contain clear, unequivocal sentences. Excerpted from the code are the following sentences.

Every trap and trapped fixture shall be provided with an individual vent, except as otherwise approved by this code. The diameter of the individual vent shall be at least one half the diameter of the drain served, except that a vent pipe shall not be less than 1¼ inches in diameter. The diameter of a relief vent shall be at least one half the diameter of the soil or waste branch served. The diameter of circuit or loop vents shall be at least one half the diameter of the horizontal soil or waste branch served. Every building in which plumbing is installed shall have at least one main stack, not less than 3 inches in

Table 6.4 Minimum Diameters and Maximum Length of Individual Branch Circuit, and Loop Vents for Horizontal Soil and Waste Branches

Diameter[a] horizontal branch (in)	Slope or horizontal branch[a] (in/ft)	Diameter of vent (in) (maximum developed length of vent, in feet, given below)[b]									
		1¼	1½	2	2½	3	4	5	6	8	10
1¼	⅛	NL[b]									
	¼	NL									
	½	NL									
1½	⅛	NL	NL								
	¼	NL	NL								
	½	NL	NL								
2	⅛	NL	NL	NL							
	¼	290	NL	NL							
	½	150	380	NL							
2½	⅛	180	450	NL	NL						
	¼	96	240	NL	NL						
	½	49	130	NL	NL						
3	⅛		190	NL	NL						
	¼		97	420	NL	NL					
	½		50	220	NL	NL					
4	⅛			190	NL	NL	NL				
	¼			98	310	NL	NL				
	½			48	160	410	NL				
5	⅛				190	490	NL	NL			
	¼				97	250	NL	NL			
	½				46	130	NL	NL			
6	⅛					190	NL	NL	NL		
	¼					96	440	NL	NL		
	½					44	220	NL	NL		
8	⅛						190	NL	NL	NL	
	¼						91	310	NL	NL	
	½						38	150	410	NL	
10	⅛							190	500	NL	NL
	¼							85	240	NL	NL
	½							32	110	NL	NL
12	⅛								180	NL	NL
	¼								79	420	NL
	½								26	200	NL

[a]1 ft = 304.8 mm; 1 in/ft = 0.0833 mm/m.
[b]NL means no limit. Actual values in excess of 500 ft.

Table 6.5 Maximum Distance of Fixture Trap from Vent[a]

Size of trap (in)	Size of fixture drain (in)	Fall per foot (in)	Distance from trap (ft)
1¼	1¼	¼	3½
1¼	1½	¼	5
1½	1½	¼	5
1½	2	¼	8
2	2	¼	6
3	3	⅛	10
4	4	18	12

[a] 1 in/ft = 0.083 mm/m; 1 ft = 304.8 mm.

diameter, which shall run undiminished in size and as directly as possible from the building drain through to the open air above the roof. All vent and branch vent pipes shall be graded and connected so as to drain back to the soil or waste pipe by gravity.

These short, terse sentences are a very clear, concise description of what vent piping is all about. The sentences were quoted out of context, and yet, as presented above, form a description of venting that is as good as any author could create. Actually all that remains is to size the vent piping of any system. To accomplish this task we repeat from the BOCA *Basic Plumbing Code* Tables 6.3 to 6.5 which cover the size and length of vent stacks and stack vents; the minimum diameters and maximum length of individual branch, circuit, and loop vents for horizontal soil and waste branches; and the maximum distance of fixture trap from vents.

Primarily the designer's problem comes in two areas of the design. These are areas in which the vent may be omitted and/or wet venting employed and in the circuit venting of a battery of fixtures. In the details that follow we will address these situations. For the newer designer the term "wet venting" may seem confusing. A vent is "wet" and a fixture is "wet vented" when the vent involved is also the waste line of an adjacent fixture. This is a very common occurrence in any bathroom situation in which the tub or shower drain has its trap vented through the waste line of the adjacent sink.

Loop Venting.   In the opening remarks in this chapter there was a reference to the two problems in the plumbing detail of waste and vent piping: some parts of the waste piping being too small, and the vents being too numerous and too large. Usually these problems occur when there is a battery of fixtures to be delineated in a detail.

The BOCA *Basic Plumbing Code* states,

A branch soil or waste pipe to which two but not more than eight water closets (except blowout type), pedestal urinals, fixtures having floor outlet trap standards, shower compartments, or floor drains are connected in battery, may be vented by a circuit or loop vent which shall be taken off downstream from the fixture most distant from the soil stack. In addition,

lower floor branches serving more than three water closets shall be provided with a relief vent installed downstream from the fixture nearest the stack. Where lavatories or similar fixtures discharge into such branches, each vertical branch serving such fixtures shall be provided with a continuous vent.

Figure 6.1 is a hypothetical toilet room which illustrates this section of the code. In this figure we have depicted a men's room containing four water closets, two urinals, two lavatories tied to a single stack, and a floor drain. The first and most common mistake is to become somewhat confused by the code and the vent for every fixture. Commonly the confusion here occurs over two statements in the code. One is the general statement that every fixture must be vented and the other is the statement (quoted above) "except blowout type." The blowout type water closet or urinal is a special type of fixture that eliminates its water with considerably more force than other common types. As such the blowout fixtures must be individually vented or a common vent installed in back-to-back installations.

In Fig. 6.1 we have four water closets, which is more than three, and as the code states we need a vent riser downstream of the fourth water closet. We also have two lavatories tied together in a stack and this stack must be vented. Finally we have eight connections to our branch line. To comply with the rules we need a vent downstream of the farthest fixture from the stack. From the detail we have four fixtures units until we reach the first urinal connection, after which we have eight. There our drain line size is 2 in (floor drain connection standard of 2-in sets beginning size). After the first urinal as per Table 6.3, we need a 3-in line. The 3-in drain continues until we pass the second water closet. Here the rule is that more than two water closets require a 4-in line.

From Table 6.3 our first two single vents are 1¼ in. Where they join together we require a 1½-in vent and the load on our relief vent past the first water closet is 18 fixture units which requires a 2-in vent. The stack, by code, has a 3-in diameter.

Wet Venting.   As implied by the title of this description the vent is wet, which means that it is used as a drain for some other fixture. This is a very common occurrence in residential or apartment waste and vent piping.

The standard bathroom is one that contains a bathtub with shower or a stall shower, a lavatory, and a water closet. As can be seen in the detail shown in Fig. 6.2, the shower drain may be vented through the waste line of the lavatory. Commonly the main waste and vent stack is behind the water closet.

The plumbing code allows not more than one fixture unit to be drained into a 1½-in diameter wet vent or not more than four fixture units drained into a 2-in wet vent and prohibits the kitchen sink from being one of the fixtures involved in this arrangement. Note that the bathtub has a 1½-in drain standard while the shower has a 2-in drain standard. Thus the drain line is 1½ in if the line is from a bathtub and 2 in if the line is from a stall shower. What the code further says is that the horizontal drain line can accept a back-to-back toilet arrangement. If that occurs, there would be an arrangement that is also shown in Fig. 6.2.

Relief Vent. Tall stacks in high-use buildings can have both vacuum and pressure problems. Generally the problem is air flowing down the stack, pushed by liquids from the upper floors, meeting liquids at high rates of flow from the drain lines of a lower floor discharge. As illustrated in Fig. 6.2, the plumbing code requires a relief vent for stacks having more than 10 branch intervals. The counting starts at the top and proceeds downward, not from the bottom up. As stated in the code, the size of the relief vent must equal the size of the vent stack to which it connects. And the connection at the soil stack is made with a wye below the horizontal branch. There is also a wye connection required where the relief vent connects into the vent stack.

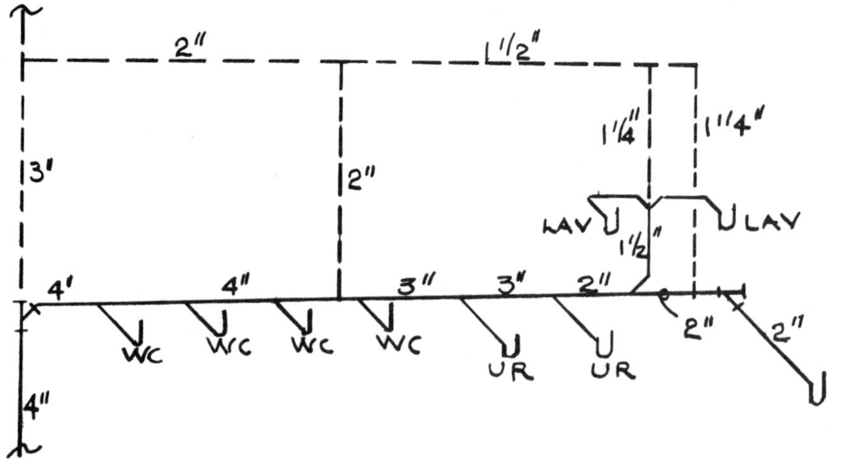

FIXTURE UNIT COUNT

FIXTURE	F.U. VALUE	QUANTITY	TOTAL
FD	2	1	2
UR	4	2	8
LAV	1	2	2
WC	6	4	24
			36

LOOP VENTING

— NO SCALE —

FIGURE 6-1

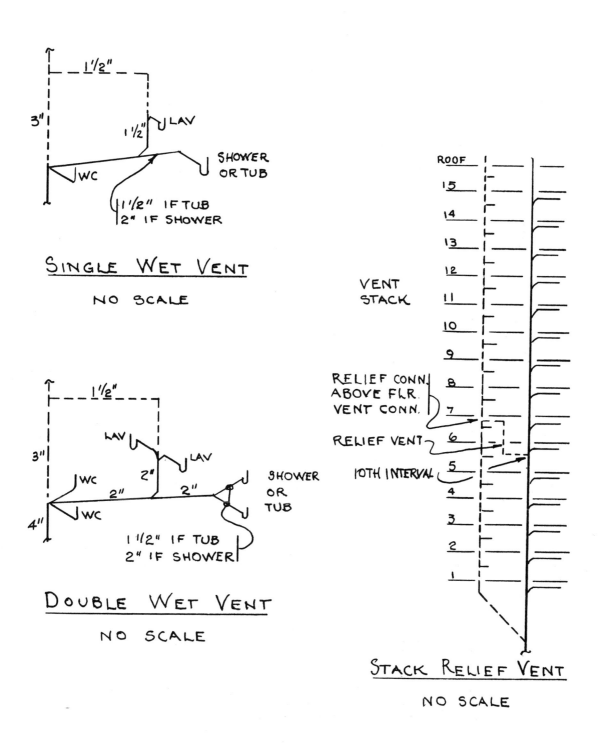

SINGLE WET VENT

NO SCALE

DOUBLE WET VENT

NO SCALE

STACK RELIEF VENT

NO SCALE

WET & RELIEF VENT DETAILS

FIGURE 6-2

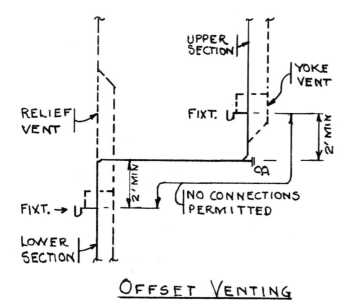

OFFSET VENTING

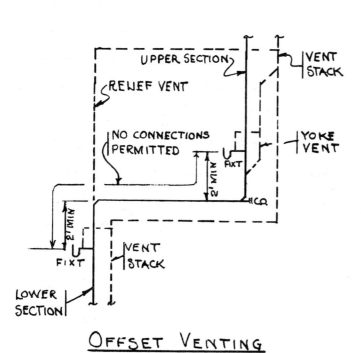

OFFSET VENTING

FRESH AIR INLET

RETURN BEND

GRADE

TRAP CLEANOUTS

FRESH AIR TEE

HOUSE TRAP

NOTE: INSTALL TRAP IN CONCRETE PIT WITH COVER OR BRING CLEANOUTS UP TO FINISHED FLOOR

BUILDING (HOUSE) TRAP

VENT DETAILS

— NOT TO SCALE —

FIGURE 6-3

Offset Connection. There are situations in tall buildings (more than 10 stories in height) in which a horizontal offset in the sanitary stack occurs. If this is your problem, there are two solutions allowed by the plumbing code; they are depicted in Fig. 6.3. One solution would be to have two separate vent stacks with the relief vent becoming a separate stack. The other is to allow the relief vent to cross over and tie into the offset vent stack. Note that no connections are permitted in the horizontal drainage line offset nor are they permitted for a distance 2 ft 0 in above and below the offset. In both cases the vent for the upper fixture ties into a yoke vent, which is effectively a separate vent for the horizontal portion of the drainage stack.

Building (House) Trap. The primary purpose of any trap is to protect health and safety. Originally house traps were required by plumbing officials when sewers to which the building systems were connected were combination storm and sanitary sewers and were unvented. These sewers contained rats and odor-causing decaying matter. In modern times sewers are also plagued by oil and gas lines that leak and seep into sewers. Building or house traps seal off explosive gas mixtures in sewers and act as a shock absorber when an explosion in the sewer does occur. Conversely there are communities that forbid traps *because* the sewer gas is so corrosive that the system needs the ventilation provided by the individual building vent.

Figure 6.3 also depicts a typical house or building trap. The fresh air inlet aids in the integrity of the trap seal and reduces the corrosive atmosphere within the system. House traps may be a municipality code requirement to avoid odor problems in septic tank systems which frequently occur in one-story housing developments. The fresh air inlet also relieves any air pressure rise at the trap caused by the trap's resistance to the flow of air. Finally, it should be noted that the house trap is *not* a code requirement.

Motel Waste and Vent Piping

Figure 6.4 is the first of the figures in this chapter that treat the practical problems which confront all plumbing designers. In this figure and subsequent ones we will present solutions to real, recognizable piping situations beginning with the most simple of these situations.

In many motels one of the niceties that is provided in individual room layouts is the toilet room with a separate countertop lavatory that is exterior to the toilet room. This enables one person to shower or perform other functions while the other washes, shaves, or puts on makeup.

Depicted in the detail is the single-story situation. The primary desire of the building owner is to build the units at the lowest possible first cost. Since plumbing costs are basically controlled by both sizes of materials and numbers of fittings, the placement of the main waste and vent lines is the first consideration. Our main stack is centered on the back-to-back location of the two water closet locations. Using this arrangement, we have reduced the number, size, and length of 3-in piping to an absolute minimum.

Almost invariably the motel provides a combination tub and shower. This allows us to use 1½-in individual tub drains and tie them to a 2-in drain which in turn is wet vented through a 2-in waste from the back-to-back lavatories. As can be seen in the figure, the majority of the vent piping is 1¼ in and 1½ in.

To illustrate the connection to the building sewer we have depicted a short vertical drop from the main stack. Commonly there would be a continuous sanitary main connecting a series of individual sanitary lines to a street sewer or septic tank system, depending on local conditions.

This detail is very nearly a repeat example of our previous sample detail which was presented earlier in this chapter. It is not an accident that this has occurred. In our previous presentation of typical details our intent was to present only those details that commonly were part of real, practical design problems.

In this detail, and all that follow, there are a variety of materials that may be used for the waste and vent piping. We have no particular bias or preference toward any material. Any that the code permits and which will result in a workable, low first-cost solution is acceptable. All of the pipe sizing was obtained from the data contained in Tables 6.1 through 6.5 in this chapter. Regardless of the material used, the size remains the same.

Office Waste and Vent Piping

In the typical office building using a slab-on-grade one-story design there are one or more clusters of back-to-back toilet rooms for male and female occupants. Figure 6.5 is a fairly common version of this typical toilet room solution.

The design approach to this area is the same as for that in Fig. 6.4, which was previously covered. The basic approach is to show on the plan the horizontal portion of the piping, and, in the accompanying riser diagram, to show the pipe sizes. In some offices the horizontally depicted piping of the waste and vent piping is also sized on the plan. While there is no objection to this arrangement of sizing information, considerable care must be used to avoid confusion and sizing errors. In our detail we limited the sizes depicted to those that are of primary concern to the general building contractor. Thus all that our detail depicts are the waste and vent riser and the extension of pipe to the floor drains. Some local codes may require any vent pipe that passes through the roof to have a diameter of at least 4 in. Thus the last few feet of the vent riser must be enlarged from 3 in to 4 in. This was once thought to be a required protection

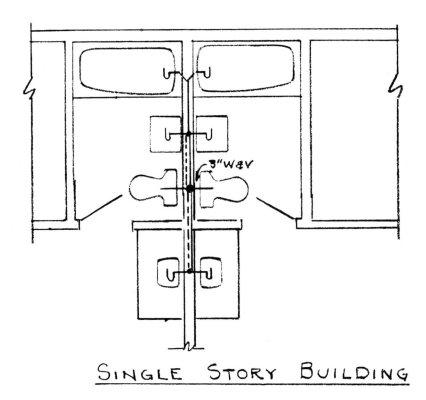

SINGLE STORY BUILDING

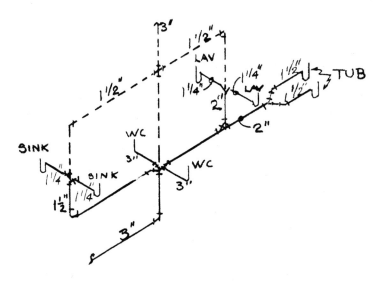

WASTE AND VENT PIPING

TYPICAL MOTEL UNIT

—NOT TO SCALE—

FIGURE 6-4

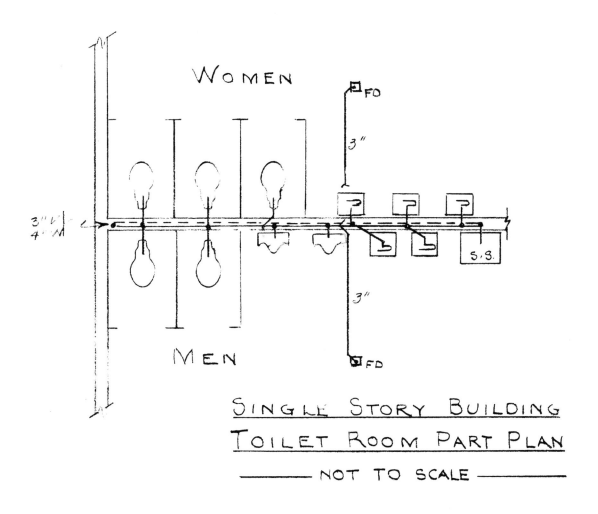

WOMEN

MEN

SINGLE STORY BUILDING
TOILET ROOM PART PLAN
—— NOT TO SCALE ——

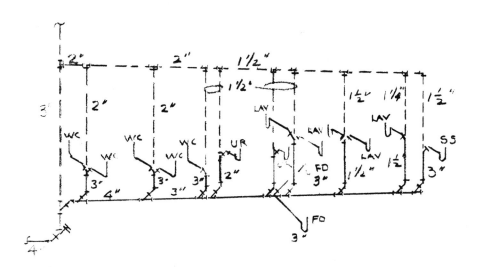

WASTE AND VENT RISER DIAGRAM
—— NOT TO SCALE ——

FIGURE 6-5

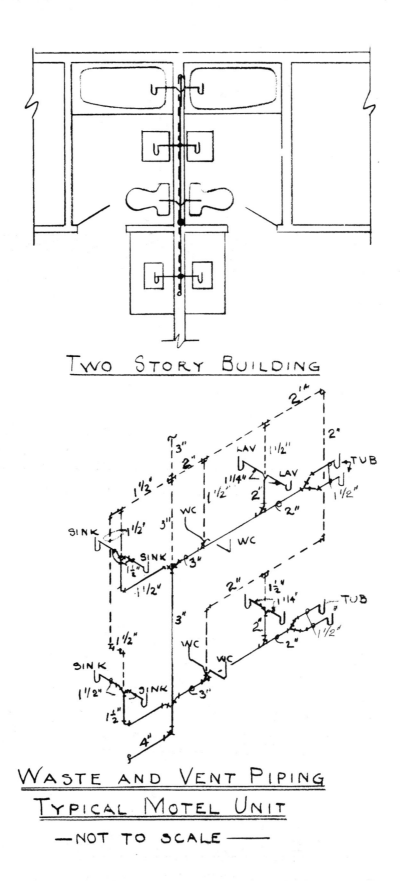

TWO STORY BUILDING

WASTE AND VENT PIPING
TYPICAL MOTEL UNIT

—NOT TO SCALE—

FIGURE 6-6

against vapor freezing and sealing a vent in very cold climates. This requirement has been dropped in many areas of the country.

As we have noted in the previous discussion, the sizing is obtained by using the values in the tables in this chapter and simply adding fixture unit values as you proceed from the start of the line at the service sink until you reach the end at the waste and vent stack. The 3-in floor drains do not, by code, require a vent connection at the floor drain trap. The variety and arrangement of the fixtures create the need for all the vertical vents to be shown as connections to the loop vent. One of the items to note is that the two wall, not stall, urinals are separately vented. The installed physical separation of the two make this the simplest arrangement. In addition the single water closet which is approximately back to back with one of the urinals also requires an individual vent. The back-to-back water closets can be tied to a single vent if they are not of the blowout type. Blowout types must be individually vented.

In Fig. 6.5 and in others that follow the advantage of sizing the piping on the riser diagram should be self-evident. Many building plans are at ⅛ scale. While it is difficult to show even the horizontal waste and vent piping on a ¼-scale plan, it is practically impossible to do so at a ⅛-scale plan. The riser diagram has no scale and can be expanded to any size necessary to depict all sizes clearly.

Two-Story Motel

Having covered the more obvious and simpler forms of waste and vent piping in Figs. 6.4 and 6.5, the purpose of Fig. 6.6 is to expand on one of these situations. Another expansion of systems will be covered in Fig. 6.7. In many areas of the country the two-story motel is a standard of many national hotel chains. Here the hotel's designers concentrate on repetitive sizing and repetitive arrangement. This repetitive arrangement very definitely involves the bathroom and plumbing system arrangement.

On both floors the exterior countertop lavatory is a hotel feature. As can readily be seen in our detail, this creates a separate looped branch in the waste and vent system. Insofar as the detail is concerned, it is simply the placing of the two plumbing situations in Fig. 6.4 on top of each other and finding the most economical way to connect waste and vent piping in accordance with code requirements.

Note that the strategic placing of the fixtures creates the opportunity to install a very cost-effective piping arrangement. Since by code the vent stack must be 3 in, the arrangement permits most of the vent stack to be a waste stack utilized to reasonable capacity. The size of a vast majority of the waste and vent piping is 1½ and 2 in which are less expensive in both material

and labor. The identical arrangement can be used in a motel that has no exterior countertop lavatory. You would simply omit the waste and vent piping to the left of the main stack.

The real solution in these cases is even more dependent on not crossing stacks with vents. It would be possible, as far as the code is concerned, to take the vent from the lower floor countertop lavatory and connect it to the vent from the toilet and bathroom lavatory. However, this would require crossing the 3-in waste stack. Very commonly motel walls are built to minimum requirements, and the units are carefully arranged one over the other. Further, the bathroom walls are no thicker than absolutely necessary. If the party wall on the lower floor were thickened by 2 in to allow for pipe crossing, this would, in the final analysis, add feet to the overall length of the building, increasing foundation, floor, exterior wall, and roof costs. No one would accept that alternative.

Four-Story Toilet Rooms

Commonly the plumbing designer has the problem of designing toilet rooms that are in multistory buildings that are not skyscrapers but are several stories high. Since the toilet rooms are repetitive, the waste and vent detailing is repetitive. In Fig. 6.7 we have selected a four-story building in which the toilet room arrangement creates repetitive piping. Note that Fig. 6.7 shows four stories of the toilet room that was originally depicted as a single-story unit in Fig. 6.5.

One of our objectives, especially for the newer designer, is to use this method to illustrate that multistory piping is not a complex subject. Frequently the first time one sees multistory piping, especially if it is depicted as a series of riser diagrams for a tall building with five or six toilet areas, one feels this is truly complicated. If the toilet rooms remained the same, the piping between the waste and vent risers would remain the same. And if there were 20 floors and 10 toilet areas, as in Fig. 6.7, we'd simply have a big sheet showing the 10 sets of waste and vent piping for 20 floors.

As we previously discussed, we would have, as required by code, a relief vent at each tenth interval beginning from the top floor. Thus the above noted hypothetical case would have a relief line between the waste and vent stack. Figure 6.3 illustrates this situation. But, getting back to Fig. 6.7, the multistory building diagram would look like our detail.

The major and actually only difference between Figs. 6.5 and 6.7 is the separate vent stack. You should carefully note that this is a main stack vent that ties into the main waste system at the base of the main waste riser. Thus the horizontal vent crossovers on the top and bottom floors are larger than those of the floors in between.

Certainly the 3-in vent could have extended straight

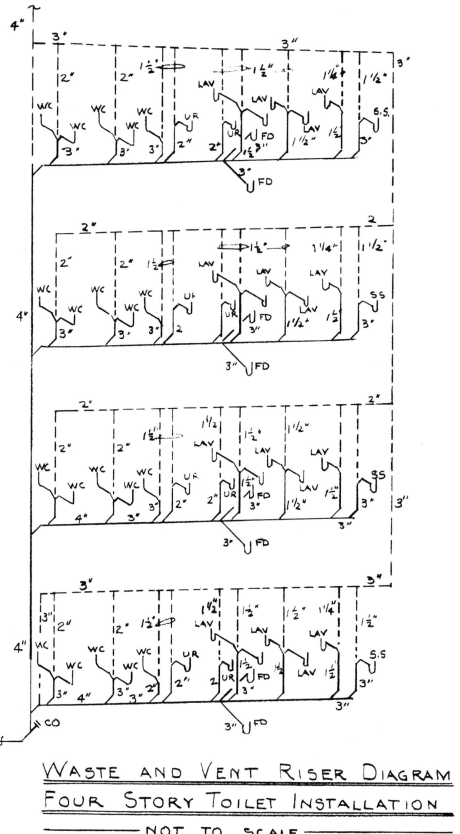

WASTE AND VENT RISER DIAGRAM
FOUR STORY TOILET INSTALLATION
———————— NOT TO SCALE ————————

FIGURE 6-7

through the roof. Thus there would have been two, not one, vents through the roof. This would have reduced the top floor horizontal vent riser to 2 in and, like the two lower floors, the top floor horizontal line would no longer be part of a continuous main vent.

Total installed costs are the determining factor on the subject of one vent versus two. Usually, but not always, the connections as shown in Fig. 6.7 are the least expensive.

Special Waste and Vent Systems

There are two types of waste and vent branch piping that present unusual problems. These are waste and vent piping handling the drainage of acid waste from a laboratory and waste and vent piping handling volatile, flammable oil waste, usually from an automotive service area. The typical piping is depicted in Figs. 6.8 and 6.9. In each detail the neutralizing tank or oil-separating

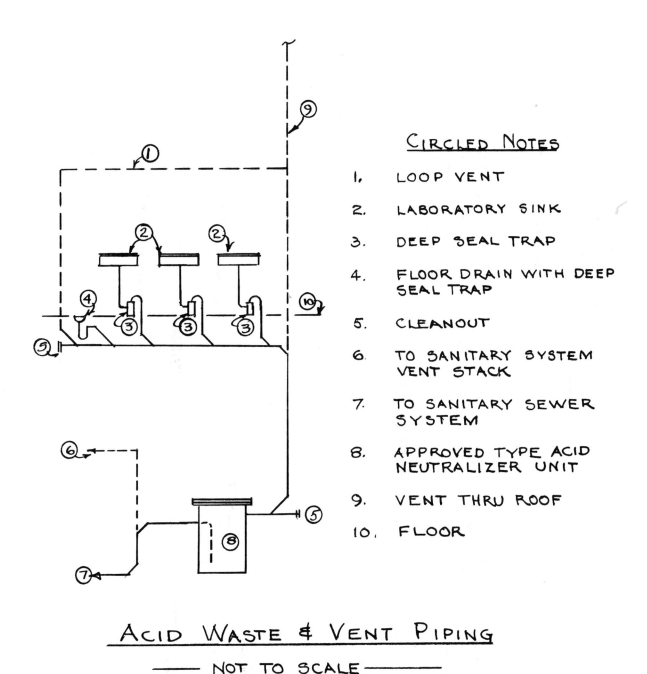

CIRCLED NOTES

1. LOOP VENT

2. LABORATORY SINK

3. DEEP SEAL TRAP

4. FLOOR DRAIN WITH DEEP SEAL TRAP

5. CLEANOUT

6. TO SANITARY SYSTEM VENT STACK

7. TO SANITARY SEWER SYSTEM

8. APPROVED TYPE ACID NEUTRALIZER UNIT

9. VENT THRU ROOF

10. FLOOR

ACID WASTE & VENT PIPING

— NOT TO SCALE —

FIGURE 6-8

tank is shown with typical connection points. These connecting points may vary to some degree depending on the specific, proprietary neutralizer or separator product you select.

These two details cover very special areas of pollution control. In each case pollutants that are harmful to the sanitary system are being removed. This is done by chemical reaction in the case of the acid neutralizing tank and by separation in the oil separator. Both of these devices require maintenance. Over time the chemicals in the neutralizer are used up in the neutralization reaction and must be replaced with fresh chemicals. And, in time, the oil separator becomes filled with oil. Normally the separator is arranged to allow the oil to float to the top and relatively clear water to be drained from the bottom, as we depicted in Fig. 6.9.

Figure 6.8 depicts a typical small laboratory area of three sinks and a floor drain. The sinks do not have the

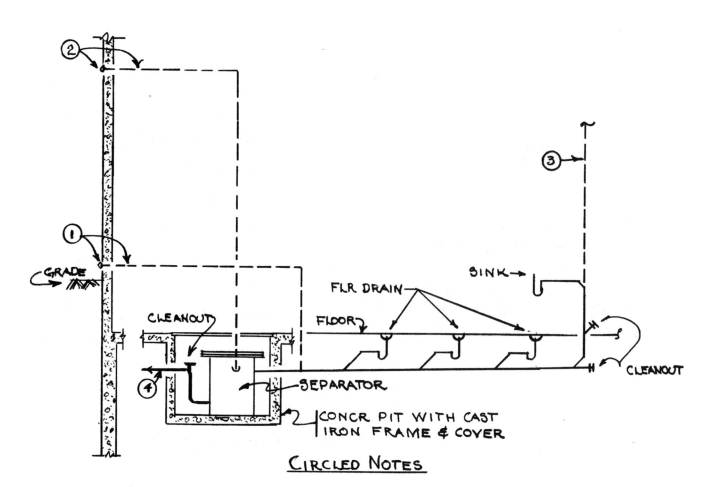

CIRCLED NOTES

1. FRESH AIR INLET THRU WALL AT LEAST 6" ABOVE GRADE.
2. VAPOR RELIEF VENT THRU WALL 12'-0" ABOVE GRADE.
3. VENT THRU ROOF.
4. CONNECT ONLY TO SANITARY SEWER SYSTEM.

FLAMMABLE OIL WASTE & VENT PIPING

——— NOT TO SCALE ———

FIGURE 6-9

usual type of P trap but rather a deep seal trap sometimes called a pot trap. The waste line piping to which these traps and the floor drain are connected should be two sizes larger than the standard sizing tables would dictate because this horizontal portion of the waste line has much more of a problem with stoppage and blockage than an ordinary sink waste line. Commonly local authority requires that its vent stack be separated from other normal vent stacks. In addition the neutralizing tank is also vented on the discharge side of the neutralizer. This vent may be connected into the regular sanitary vent stack system.

Figure 6.9 is quite different from Fig. 6.8 in that it shows a far greater concern with explosive, flammable vapors. The horizontal waste line is again two sizes larger. The line is back vented and also has a fresh air inlet on the line as it enters the separator and a vapor relief vent on the separator itself. The separator is basically a minature oil storage tank and as such requires a relief line similar to that of the very large oil storage tank at any oil terminal area.

The lines in Figs. 6.8 and 6.9 have not been sized since the details should always be reviewed with your local code authority and sized according to local rules prevailing in your area. There may also be variations to these details that are required by your local code authority. In general what has been depicted is reasonably accepted in most locations.

Finally there are industrial applications that may simply empty wastes from a piece of equipment directly into some sort of neutralizer or separator and have a vertical, running trap on the discharge side of the neutralizer or separator. This has not been depicted as it is not a plumbing item and the treatment takes place out of the plumbing system. The vented running trap is simply complying with the code requirement that all fixtures have traps which are properly vented. In this case the trap is the same size as the treatment tank's outlet connection.

Complex Toilet Room Areas

In the various toilet room plans and accompanying waste and vent riser diagrams, the design work undoubtedly did not appear complicated to those with some knowledge of the subject. Even to the relative newcomer the details must have appeared stereotyped and repetitive. From a typical design problem viewpoint this is a reasonable judgment. Many of the situations that will arise are no more complicated or involved than those that have been presented. However, there are complicated situations that do arise and do require thought and ingenuity. Figures 6.10 and 6.11 are companion plan and piping schematics that illustrate this point.

The toilet room plan depicted in Fig. 6.10 is an existing area in a large building complex that is being renovated. Typical of all renovation work the piping design must contend with a number of situations that cannot be altered to permit a simpler solution. The main sanitary 5-in drain line enters the area and veers to the left as shown and is not going to be relocated. That's where it is relative to the two toilet rooms and that is where it will remain. The only three possible vent riser locations are as depicted on the plan. Because of the arrangement of walls and partitions in the spaces above the toilet room, these locations are also fixed. Finally because of building code requirements, handicapped requirements, and client and architectural requirements, the quantity and location of each of the fixtures has been firmly settled.

Following our previous detail procedures, the location of horizontal waste and vent piping is shown on the plan. Only those lines that are out of the congested piping and the vent riser locations are sized on the plan. This is a situation in which the plan is at a large enough scale that, with careful and neat drafting, the horizontal waste and vent lines could have been sized on the plans, and a schedule of vertical vent sizes could have been prepared to complete the sizing information. We believe, however, if the waste and piping detail is put on the same sheet as the plan, the result is far easier to understand and to size and would preclude any misunderstanding.

The waste and vent riser diagrams in Fig. 6.11 are in the same order, top to bottom, as the plan in Fig. 6.10. If even further clarification is desired, a note such as "See elevation 1, 2, 3" could be placed at each of the three piping situations in sequence and the three waste and vent riser diagrams in Fig. 6.11 could be numbered "Elevation 1, 2, 3."

Once the plan is broken up into the three elements as depicted in Fig. 6.11, the design in the top figure is slightly more complex than we have presented so far but, in general, has suddenly become a lot simpler, and the sizing solution is a lot more obvious. In the top diagram in Fig. 6.11 the code requirement of individual venting of the somewhat peculiarly located shower, wall urinals, and water closet becomes clear. In the center detail in Fig. 6.11 the vent requirement of the water closet on the far end of the sanitary drain line also becomes obvious. Finally, the lower portion clarifies that one set of water closets qualifies for loop venting and one set does not. It would be difficult to depict why this is so without the waste and vent riser diagram.

This detail, taken from an actual situation, is the final in our series of waste and vent diagrams.

Roof Penetration. As we noted in the beginning of the previous chapter, there are a variety of plastic, copper, and iron piping materials that are permitted by code for use in waste and piping systems. In addition there are many situations in which the vent riser at the point of roof penetration is smaller than permitted by local code. Usually the local codes require vent risers through the

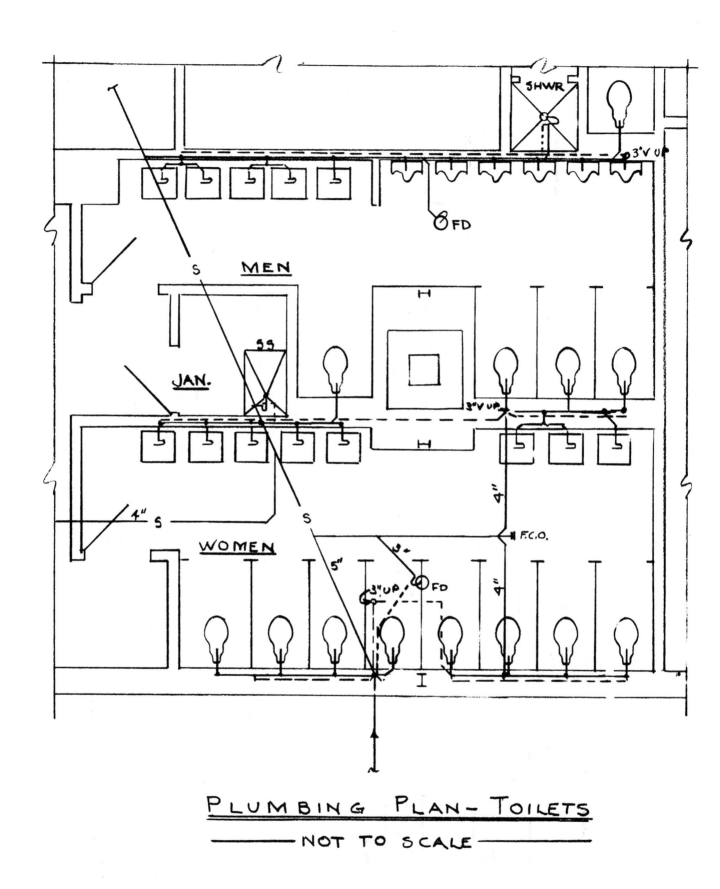

PLUMBING PLAN - TOILETS

NOT TO SCALE

FIGURE 6-10

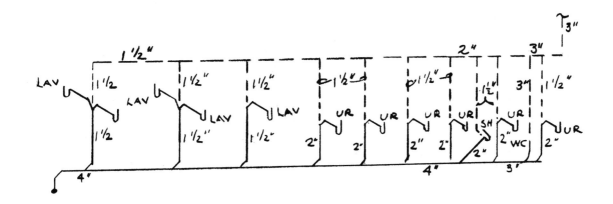

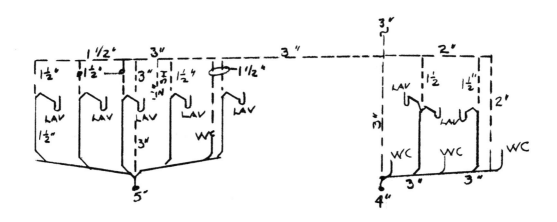

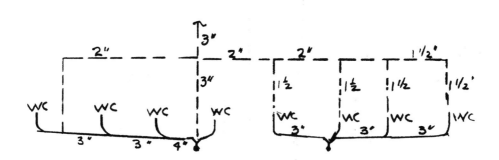

WASTE AND VENT PIPING
TOILET ROOM PLAN
—NOT TO SCALE—

FIGURE 6-11

roof to have a 3-in or 4-in diameter as a minimum value. For the plumbing designer there are two important considerations. The first is the size of the vent pipe as well as its height above the roof and its horizontal distance from walls and windows. The second is to maintain the integrity of the roof.

Penetrating any roof is a serious matter as frequently the owner of the building demands and receives a guarantee, usually for 20 years, against roof failures and leakages from the roofing contractor. As a result of this guarantee, the roofing contractor is very particular that any opening cut through the roof is installed in a fashion that precludes leakage around the opening. Usually certain portions of the flashing of the vent pipe are installed by the roofing contractor to ensure that the result is to the satisfaction of this contractor. Copper and plastic

roof vents, where permitted, have specially designed flashing materials that are soldered or sealed to the vent pipe and then counterflashing is applied over the flashing flat base flange. This is done is a variety of ways by the roofing contractor depending on the type of roof material. In either case the work of the plumbing designer and plumbing contractor is relatively simple. Merely attach the flashing material properly to the vent riser at the roof line.

The iron pipe depicted in Fig. 6.12 is not so simple. In this detail all of the elements of the flashing are shown. The first task of the plumbing designer is to find out which of the elements in the area in which the work is to be installed are to be performed by the plumber and which are to be performed by the roofer. Generally the plumber would provide all the piping, the pipe sleeve,

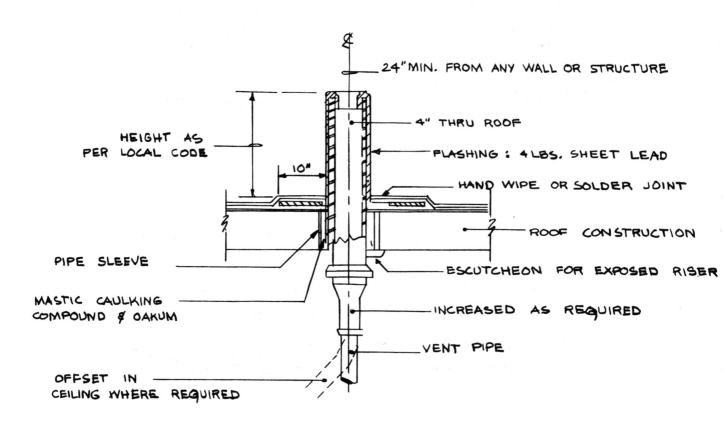

VENT THRU ROOF DETAIL

NO SCALE

FIGURE 6-12

the sealing between the pipe sleeve, and the pipe and the lead (sometimes copper) flashing shown in Fig. 6.12. When iron pipe is involved, there are flashings, usually copper, that may be permitted to rise up a certain distance and be sealed to the iron pipe. No matter how the sealing is done, it is hard to assure that it will remain sealed for the life of the roof-guarantee period. The use of a continuous sheet of lead, as depicted in the detail, that rises up and turns down inside the vent pipe assures that there is no joint to leak. The roofer usually then installs and seals a flashing material as shown which is as weathertight as one can install. This system is more costly but is very successful when properly applied. The actual, final roofing material is applied by being placed on top of the flashing.

Reduced Size Vent Systems. In some respects the sizing of vents for any waste and vent system relates to the discussion of how well you can hear with wax in your ear. Those of you who have gone to a doctor to have your ear passages flushed know there are two reactions. The first is the pleasant feeling of having objects or substances which have caused pain or discomfort removed. The second is the decision of whether or not you can actually hear better. Commonly you are told that so long as the ear passageway is not totally blocked, a totally clear ear passage is no better than a small but clear ear passage.

The long-term steady rise in building costs has spurred investigation in many areas. One of these is vent piping and code-designated vent pipe sizing. The Veterans Administration recently sponsored a National Bureau of Standards research project on vent pipe sizing. The report is too long to discuss in detail here. The basic approach was to trace probable continuous air flow paths through the vent system and to determine overall pressure loss. In many situations vent pipe sizes in the range of ½ to 1 in were entirely satisfactory.

The above brief discussion is not to take sides in the discussion but merely to point out to the engineering-trained reader that while the future of waste and vent pipe sizing may very well be based on more precise mathematical formulas, the present situation is controlled by tables and data in the plumbing code.

Summary

While there are many special situations in waste and vent piping, most can readily be resolved by applying the basic concepts, or versions of them, which were illustrated in this chapter plus a thoughtful and careful reading of the BOCA *Basic Plumbing Code*. For the reader who would like a more complete illustration of what each code-described requirement means, we suggest the purchase of a small book entitled *National Plumbing Code Illustrated* written by Vincent T. Manas and published by his company, Manas Publications, 1868 Shore Drive South, St. Petersburg, Florida 33707.

While mistakes can be made in sizing individual fixture vents, the plumbing designer who carefully reads the code and our discussion in this chapter should not, and usually does not, have a problem with this individual portion of the waste and vent system design and the related detailing. Usually the problem occurs when the individual parts are put together to comprise the total waste and vent system.

To resolve this problem our advice is to take the proper and necessary steps to design the overall system before concerning yourself with the individual fixtures within the system. Using the tables presented in this chapter and in Chap. 7, which covers drainage systems, design the system as follows:

1. Calculate the number drainage fixture units.
2. Size the horizontal branch drains.
3. Size the vent stack.
4. Size the vent header.

If you follow this procedure, you will have properly sized the principal parts of the system. All that remains is the simple task of sizing the individual waste and vent piping. The individual vents as noted in our details and in the code sometimes serve more than one fixture, sometimes they are actually the drain pipe (wet vent) of another fixture, and sometimes, as in batteries of water closets, they may serve a number of fixtures.

SEVEN

Drainage Systems

Drainage systems are of two types—sanitary and storm—or put another way, those that transport rain water and those that transport effluent from plumbing fixtures and other contaminating sources. It is important to clearly understand the two types of systems. Even though the final connection of both systems may be to a street utility combined storm and sanitary street sewer, the plumbing code very clearly states that storm water shall not be drained into sewers intended for sewage only. The plumbing designer cannot create a on-site combined storm and sanitary sewer even if that is what occurs in the street.

While there are two types of drainage, it is important to understand that the plumbing code *does not* cover the fact that there are two types of storm drainage, interior and exterior, which frequently affect plumbing design and plumbing designers. Storm drainage frequently includes building and paved areas that exceed area values that are in the plumbing code tables. For example a one-story factory of 200,000 square feet (sq ft) with 100,000 sq ft of paved area is not uncommon or unusual. Further it is fairly common for storm drainage to be run at ⅛-inch to 1⁄16-inch (in) pitch per foot or 0.50 in/100 ft. The tables in the code stop at a 15-in sewer with 238,000 sq ft of allowable roof area at ⅛-in/ft pitch with a 1-in/hr rainfall.

The above problem is further compounded by site zoning laws which frequently create a requirement that the factory building cover no more than 50 percent of the land on which it is situated. Depending on the site situation and other requirements, the problem may very well involve a 200,000-sq ft factory that is located on a parcel of land totaling 400,000 sq ft of which 200,000 sq ft are for the factory roof, 100,000 sq ft are for the paved parking area, and 100,000 sq ft are for an attractive grassed area that must also be drained.

Correctly addressed, the problem of site drainage is not part of the plumbing system. It is part of the site drainage to be designed by a competent civil-sanitary engineer and/or licensed landscape architect. This is the gray area of plumbing design since the paved area contains the interior building drains as they go to the street. The exterior drainage, depending on local building codes and ordinances, may also be a requested part of the plumbing design.

In this chapter we will present the basic formulas and calculations that are required to properly design the site drainage. This is being done with the basic premise that any extensive site drainage design performed even by an experienced plumbing designer should at least be reviewed and approved by a knowledgeable civil engineer and/or landscape architect. This is especially important when paving is extensive and the pitching of it becomes a factor. This is a much more firm requirement when site grading of the unpaved land is required to effectively drain it. These areas of site drainage expertise are not usually or commonly part of any normal plumbing expertise.

Hydraulic Overview

For the plumbing designer the code has already resolved most of the hydraulic calculations that may be required for most building roofs. The following five tables have been taken from the BOCA *Basic Plumbing Code*. If the plumbing designer knows the number of fixtures to be connected to the sanitary drain and the pitch of the drain line, the proper sanitary drain size can be picked from Table 7.1. If a special gallons-per-minute figure is given for a special fixture, the pipe size can be selected from Table 7.2. Similarly, for a given roof or paved area all you need is to know is *the local maximum hourly rate of rainfall* and the pitch of storm drain line. Note that we have underlined the local maximum rainfall rate. The values in Tables 7.3, 7.4, and 7.5 are based on 1 in/hour (hr).

If the maximum hourly rainfall rate in your area is 3 in/hr, for example, using Table 7.3 a 4-in drain at ⅛-in pitch per foot would handle 2,507 sq ft (7520/3). A 4-

in line, using Table 7.4, would handle 4,613 sq ft (13,840/3), and, using Table 7.5, a 4-in gutter with a ⅛-in/ft pitch would handle 680 sq ft (2,040/3).

Before going into the hydraulic calculations, we would like to make several pertinent points. First, the values for sanitary drainage piping are based on diversity. As you will recall, a plumbing fixture unit equals 7 ½ gallons per minute (gpm). In Table 7.1 a 4-in drainage set at a pitch of ¼ in/ft permits a total of 216 fixture units to be connect to this 4-in line. At 7 ½ gpm per fixture unit this equals 1,620 gpm (7 ½ × 216). In Table 7.2 the same 4-in line at the same pitch has a capacity of 106.8 gpm (2 × 53.4) or which, as noted, is based on the Manning formula about which more will be said later in this discussion. The same value of 216 fixture units has been in plumbing codes for more than 40 years (yr). It is not only a tested value, but it is also a time-tested value. The simple fact is that there is considerable diversity in plumbing sanitary waste discharge. Quite obviously there are at least 1 in 16 fixture units or less (106.8/1,620) which occur as a simultaneous load. The values in Tables 7.1 and 7.2 are both correct.

Second, we would again like to make the point that the plumbing designer is not a civil engineer, and when his or her actual problem is the interior storm drain line picking up interior roof leaders from a 400,000-sq ft factory roof with a 2-in maximum rainfall, the tables in the code do not solve the problem. The answer cannot be an educated guess or extrapolation of the code tables. It has to be a calculation.

Third, diversity in storm drainage can not normally be a safe answer for roofs. While, in theory, you can

Table 7.1 Building Drains and Sewers[a]

Diameter of pipe (in)	Fall per foot (in)[b]			
	1/16	1/8	1/4	1/2
2			21	26
2 ½			24	31
3		36[c]	42[c]	50[c]
4		180	216	250
5		390	480	575
6		700	840	1,000
8	1,400	1,600	1,920	2,300
10	2,500	2,900	3,500	4,200
12	2,900	4,600	5,600	6,700
15	7,000	8,300	10,000	12,000

[a]Maximum number of fixture units may be connected to any portion of the building drain or the building sewer including branches of the building drain.

[b]To convert inches per foot to centimeters per meter multiply by 8.33.

[c]Not over two water closets or two bathroom groups.

Table 7.2 Approximate Discharge Rates and Velocities in Sloping Drains Flowing Half Full[a]

Diameter[c] of pipe (in)	Discharge rate and velocity[b,c]							
	1/16-in/ft slope		1/8-in/ft slope		1/4-in/ft slope		1/2-in/ft slope	
	Discharge (gpm)	Velocity (fps)	Discharge (gpm)	Velocity (fps)	Discharge (gpm)	Velocity (fps)	Discharge (gpm)	Velocity (fps)
1 ¼	1.20	0.63	1.70	0.89	2.41	1.26	3.40	1.78
1 ⅜	1.57	0.67	2.22	0.95	3.13	1.34	4.44	1.90
1 ½	1.96	0.71	2.77	1.00	3.91	1.42	5.53	2.01
1 ⅝	2.40	0.75	3.40	1.06	4.81	1.50	6.80	2.12
2	4.21	0.86	5.96	1.22	8.42	1.72	11.90	2.43
2 ½	7.63	1.00	10.80	1.41	15.30	1.99	21.60	2.82
3	12.40	1.13	17.60	1.59	24.80	2.25	35.10	3.19
4	26.70	1.36	37.80	1.93	53.40	2.73	75.50	3.86
5	48.30	1.58	68.30	2.23	96.60	3.16	137.00	4.47
6	78.50	1.78	111.00	2.52	157.00	3.57	222.00	5.04
8	170.00	2.17	240.00	3.07	340.00	4.34	480.00	6.13
10	308.00	2.52	436.00	3.56	616.00	5.04	872.00	7.12
12	500.00	2.83	707.00	4.01	999.00	5.67	1,413.00	8.02

[a]*For ¼ full* multiply discharge by 0.274 and multiply velocity by 0.701. *For ¾ full* multiply discharge by 1.82 and multiply velocity by 1.13. *For full* multiply discharge by 2.00 and multiply velocity by 1.00. *For smoother pipe* multiply discharge and velocity by 0.015 and divide by n value of smoother pipe.

[b]Computed from the Manning formula for ½ full pipe; $n = 0.015$.

[c]To convert inches to millimeters multiply by 25.4; to convert gallons per minute to liters per minute multiply by 3.785; to convert feet per second to meters per second multiply by 0.3048.

Table 7.3 Size of Horizontal Building Storm Drains and Building Storm Sewers[a]

Diameter of drain (in)	Maximum projected area in square feet and gallons per minute flow for various slopes[b]					
	1/8-in/ft slope		1/4-in/ft slope		1/2-in/ft slope	
	Sq ft[a]	GPM	Sq ft[a]	GPM	Sq ft[a]	GPM
3	3,288	34	4,640	48	6,576	68
4	7,520	78	10,600	110	15,040	156
5	13,360	139	18,880	196	26,720	278
6	21,400	222	30,200	314	42,800	445
8	46,000	478	65,200	677	92,000	956
10	82,800	860	116,800	1,214	165,600	1,721
12	133,200	1384	188,000	1,953	266,400	2,768
15	238,000	2473	336,000	3,491	476,000	4,946

[a]Based on a maximum rate of rainfall of 1 in (25 mm)/hr for a 1-hr duration and a 100-yr return period. The figure for drainage areas shall be adjusted to local conditions by dividing by the local rate in inches per hour.

[b]To convert inches per foot to centimeters per meter multiply by 8.33; to convert square feet to square meters multiply by 0.093; to convert gallons per minute to liters per minute multiply by 3.78.

Table 7.4 Size of Vertical Conductors and Leaders[a]

Size of leader or conductor[b,c] (in)	Maximum projected roof area[c]	
	Sq ft[a]	GPM
2	2,176	23
2 1/2	3,948	41
3	6,440	67
4	13,840	144
5	25,120	261
6	40,800	424
8	88,000	913

[a]Based upon a maximum rate of rainfall of 1 in (25 mm)/hr for a 1-hr duration and a 100-yr return period. The figure for drainage area shall be adjusted to local conditions by dividing by the local rate in inches per hour.

[b]The area of rectangular leaders shall be equivalent to the circular leader or conductor required. The ratio of width to depth of rectangular leaders shall not exceed 3:1.

[c]To convert inches per foot to centimeters per meter multiply by 8.33; to convert square feet to square meters multiply by 0.093; to convert gallons per minute to liters per minute multiply by 3.78.

Table 7.5 Size of Semicircular Roof Gutters[a]

Diameter of gutter[b] (in)	Maximum projected roof area for gutters of various slopes[c]							
	1/16-in/ft slope		1/8-in/ft slope		1/4-in/ft slope		1/2-in/ft slope	
	Sq ft[a]	GPM	Sq ft[a]	GPM	Sq ft[a]	GPM	Sq ft[a]	GPM
3	680	7	960	10	1,360	14	1,920	20
4	1,440	15	2,040	21	2,880	30	4,080	42
5	2,500	26	3,520	37	5,000	52	7,080	74
6	3,840	40	5,440	57	7,680	80	11,080	115
7	5,520	57	7,880	81	11,040	115	15,600	162
8	7,960	83	11,200	116	14,400	165	22,400	233
10	14,400	150	20,400	212	28,800	299	40,000	416

[a]Based upon a maximum rate of rainfall of 1 in (25 mm)/hr for a 1-hr duration and a 100-yr return period. The figure for drainage area shall be adjusted to local conditions by dividing by the local rate in inches per hour.

[b]Gutters other than semicircular may be used provided they have an equivalent cross-sectional area.

[c]To convert inches per foot to centimeters per meter multiply by 8.33; to convert square feet to square meters multiply by 0.093; to convert gallons per minute to liters per minute multiply by 3.78.

presume that the roof can temporarily have a buildup of water that will eventually drain, you run the risk of the weight of the temporary excess water creating a sag or worse in the roof structure. Unfortunately these sags invariable occur away from the usual drain locations, and the next time it rains you will have the serious problem of undrainable ponds of water on the roof. Situations like this invariably create leaks and roof collapses under the worst-case conditions.

Hydraulic Calculations

As noted in the previous discussion, no calculations are required unless the storm drainage situation exceeds the values in Table 7.3. For example, assume a 2.5-in rate of rainfall applies in your local area and you have a 120,000-sq ft roof to drain. Your plan is to have a series of 4-in rain leaders properly spaced that descend in straight lines to a below-floor storm drainage piping system.

Using Table 7.4 you note that a 4-in vertical roof leader can handle 13,840 sq ft of roof based on a 1-in/hr rainfall. Since your locally based rainfall value is 2.5 in/hr, the area your 4-in leader will cover is 4,536 sq ft (1/25 × 13,840). The total number of individual 4-in rain leaders is 26.5 (120,000/4,536). At least 27 drains and leaders are required.

Even before the design is completed, you are requested to provide the size of the storm sewer for the total building. Since Table 7.3 for horizontal drain lines also is based on 1-in/hr rainfall rates and gives maximum areas permitted at this rate, the easiest thing to do is to modify your roof area so that it can directly relate to the area values in the table.

At the stated 2.5-in/hr rainfall rate the 120,000-sq ft roof is the equivalent of a roof area of 30,000 sq ft (2.5 × 120,000) which has a 1-in/hr rainfall rate.

Applying this value to Table 7.3, it may be readily noted that the required storm drain size is 15 in and that the pitch of the horizontal storm drain line is at least ¼ in/ft.

If your building had 200,000 sq ft of roof, the related roof value would be 500,000 sq ft, which is beyond the values offered in Table 7.3. Obviously you could have two separate storm sewers, each taking half of the roof and solve your problem. This may not always be possible. Second, you may be asked to continue your line through a 100,000-sq-ft attached paved parking lot, pick up the lot drains, and come up with the size for a total of 300,000 sq ft of drained area. Obviously a calculation is required.

The Manning formula referred to in Table 7.2 applies to both sanitary and storm drainage systems. In general form the formulas are:

$$V = \frac{1.486 \times R^{2/3} \times S^{1/2}}{N}$$

and

$$Q = A \times V$$

where V = velocity of flow, feet per second (fps)
R = hydraulic radius, ft
S = hydraulic slope of the surface of flow, ft/ft
N = a pipe surface roughness coefficient
A = interior area of pipe, sq ft
Q = quantity rate of flow, cubic feet per second (cfs)

The hydraulic radius is the radius of the cross-sectional area of the flow to the wetted perimeter of the pipe. For pipes flowing half full the cross-sectional area is $3.14\text{-}D^2/8$, the wetted perimeter is $3.14D/2$, and the hydraulic radius equals $D/4$ [$(3.14D^2/8)/(3.14D/2)$]. If the pipe is rated at full flow, the cross-sectional area is $3.14D^2/4$ and the wetted perimeter is $3.14D$. The hydraulic radius is $D/4$ [$(3.14D^2/4)/(3.14D)$]; thus the hydraulic radius is $D/4$ for either half full or full, which are the two common pipe flow valuations.

For all calculations the first step is to change all values to foot equivalents. A 15-in pipe at ¼ in/ft becomes a $\frac{1}{12}$-ft or 1.25-ft pipe. The ¼-in/ft pitch becomes 0.0208 ft/ft (¼ × 12). Flowing full the hydraulic radius is $D/4$ or 1.25/4 which is 0.3125. The roughness coefficient is a standard that varies from 0.013 to 0.015. A conservative value is 0.015. Thus for a 15-in pipe at ¼-in pitch the velocity is

$$V = \frac{1.486 \times (0.3125)^{2/3} \times (0.0208)^{1/2}}{0.015}$$

$$= \frac{1.486 \times 0.461 \times 0.144}{0.015}$$

$$= 6.57 \text{ fps}$$

The area of the pipe is $3.14D^2/4$ or $3.14 \times (1.25)^2/4$ which equals 1.22 sq ft. Therefore since $Q = VA$ the flow quantity becomes 8.00 (rounded) cfs (6.57 × 1.22). Since 1 cubic foot (cu ft) of water equals 7.5 gallons (gal), the 8 cfs becomes 60 gal/sec (8 × 7.5) which equals 3,600 gpm (60 gal/sec × 60). This compares to the 3,491 gpm stated in Table 7.3 for a 15-in pipe at ¼-in pitch. Using this formula and adding up the gallons-per-minute values from Table 7.4, the proper size of horizontal drain line can be calculated for any flow.

Having a method to calculate the size of the horizontal storm drainage line, there remains the requirement of calculating the amount of water that will flow into the system. We are, in this case, limiting our calculation effort to roofs and related parking lots. Commonly parking lots are the paved portion of the overall site and are part of the overall site drainage system which normally is not part of the plumbing system design or plumbing contractor's installation. There are cases, however, in which the total site is limited and zoning has permitted the land to be almost totally covered by building and paved parking spaces. While the parking lot drainage is still a civil engineering design problem, this is a gray area in which the plumbing designer may feel he or she has the necessary expertise, and he or she may be under pressure to include the parking area drainage as part of the plumbing design.

In general the basic premise of the roof drainage system is that the quantity of rain falling in 1 hr is drained off of the roof in that same hour. In addition, the coefficient of runoff, which similarly to the piping coefficient of roughness is related to surface condition, may be taken as unity (10) for modern smooth roofs. The paved parking area is not as smooth as the roof. It generally has a coefficient of 0.85 to 0.90. There is also time to drain called dwell time and many other factors to consider.

On the presumption that being overly conservative about the relatively small parking lot's drainage system will not materially affect the design and will simplify the plumbing designers calculations, we will use a runoff coefficient of 1.0 in the formula for basic runoff calculations, which is

$$Q = C \times i \times A$$

where Q = runoff, cfs
i = intensity of rainfall, in/hr
A = area, acres
C = runoff coefficient

In our previous discussion of hydraulic calculations we remarked that a 200,000-sq-ft roof and a 100,000-sq-ft parking lot in an area in which the rainfall is rated at 2.5-in/hr would be a problem since the values exceed those in the available tables. To resolve the problem we first have to get the total area of 300,000 sq ft into acres. By definition 1 acre is 43,560 sq ft. Thus, 300,000 sq ft equals 6.89 acres (30,000/43,560). The total runoff is

$$Q = 1 \times 2.5 \times 6.89$$
$$= 17.2 \text{ cfs}$$

Our problem is that since we do not know the pipe size nor its pitch, we cannot use our piping formulas. We can assume a pitch of ¼ in but that still leaves us the pipe size which affects both the velocity formula for which the pipe diameter is a required value and the flow formula for which the pipe area is required. Our problem requires a series of trial values which can be handled by a computer or by hand, as we will illustrate.

In looking at Table 7.3 we note that a 15-in pipe can accept almost twice the flow of a 12-in pipe. Our 300,000 sq ft of area with a 2.5-in/hr rain equates to 750,000 sq ft. We also know pipe sizes are in 3-in increments. We just might solve our problem with an 18-in pipe and, if not, surely with a 21-in pipe. Starting with an 18-in (1.5 ft) pipe at ¼-in pitch, the velocity calculation is

$$V = \frac{1.486 \times (.375)^{2/3} \times (0.02087)^{1/2}}{0.015}$$
$$= \frac{1.486 \times 0.52 \times 0.144}{0.15}$$
$$= 7.42 \text{ fps}$$

The pipe area is 1.77 sq ft (3.14 × (1.5)²/4). Therefore flow in the 18-in pipe is 13.13 cfs (1.77 × 7.42), which is not equal to the required 17.2 cfs. Repeating the calculations using a 21-in pipe gives us 8.27 fps, an area of 2.4 sq ft, and a flow of 19.85 cfs. Thus a 21-in line is required.

Having made these two calculations, we are also equipped to quickly answer the frequently asked question, "Someone else is going to design the parking lot and site drainage. How big is your pipe as it leaves the building?" Since we originally knew it was bigger than a 15-in and by simple arithmetic an 18-in pipe handles 229,000 sq ft (13.13/172 × 300,000 sq ft), the answer is a quick and easy 18 in with a flow rate of 11.47 cfs (17.2 cfs × 200,000/300,000).

Hydraulics Summary. The basic Manning formula and the storm drainage formula were used in our calculations since they are the generally accepted standards. While the Manning formula applies to all sanitary and storm drainage, it is very unlikely that a situation would arise in which any calculations are required for sanitary drainage. In the storm drainage calculations the selection of one pipe size larger than the largest value in the table served not only to illustrate the correct procedure but also to show that a shortcut guess is not the way to solve the problem.

To illustrate what is meant by shortcut guessing, the code shows in Table 7.3 that at ¼-in pitch a 12-in pipe carries 1,953 gpm and a 15-in pipe carries 3,491 gpm. Many designers feel that if you know one flow value, you can get the rest by the simple ratio of areas (3.14D)(¾). Since 3.14 and 4 are constants, it is, according to these designers, simply D^2/D^2. If that were true, the 12-in pipe at 1,953 gpm would relate to a 15-in pipe carrying 3,052 gpm [(15²/12)² × 1,953 gpm]. This is very different than the stated value for the 15-in pipe of 3,491 gpm or our calculated value of 3,600 gpm. This shortcut guess is obviously very wrong and so are all the other shortcut methods that you may feel can be employed.

Sizing Sanitary Drainage Piping

When the horizontal portion of the sanitary drainage line is relative short (under 100 ft) or when the difference in elevation between the beginning invert of the sanitary line and each final connection to the street sewer or septic tank is large (4 ft or more), there is usually not a problem in sizing the horizontal sanitary drain line. The basic requirement of the plumbing code is that the horizontal portion of the line must be pitched at a slope that will create a flow of not less than 2 fps when flowing half full.

Tables 7.1 and 7.2 provide two separate pieces of information. Table 7.1 provides for each capacity of the drain line by specifically stating how many fixture units may be connected into a given size at a given slope. Table 7.2 provides the pipe capability in gallons per minute at a given slope. It also provides the answer as to the velocity of flow when flowing half full. It may be noted that while the capacity is doubled at full flow, the velocity remains the same. Further, Table 7.2 is based on a roughness coefficient of 0.015. This happens to be a value for the surface of cast-iron pipe in bad condition. This book does not reproduce a table that is available in the code for many types of pipe in various stages of surface condition. As noted in the footnotes of Table 7.2, both the flow quantity capability and the velocity may be improved by using smoother surface materials such as brass, copper, and plastic.

The starting invert of the horizontal sanitary line is the bottom of the inside of the pipe at the beginning of the first horizontal connection to a fixture or battery of fixtures. Usually the pipe has a 3-in diameter at this point but not always. A single lavatory could be the starting point and the line may well be the 2-in code minimum standard. As can be seen in Table 7.2, a 1 ½-in to 2 ½-in line requires a 1 ½-in/ft pitch to meet the code velocity requirement, while a 3-in or 4-in line requires a ¼-in pitch. The pitch can be reduced to ⅛ in with a 5-in line, and to ¹⁄₁₆-in pitch with an 8-in line. Before you jump to the conclusion that in the case in which your starting point is a single lavatory you can solve your pitch problem by using a 3-in line, note that your lavatory is one fixture unit or 7.5 gpm. A 3-in line handles 24.8 gpm when flowing half full at ¼-in pitch. Your flow is much too small, and your velocity will be below the minimum feet-per-second standard.

The proper design approach requires a selection of the point at which the sanitary line will leave the building to connect to a sanitary sewer or septic tank. This point may be rigidly fixed or you may have some latitude in choosing its location. Having made this leaving-point selection, the sanitary pipe invert at this point must be carefully calculated. This calculation requires knowing what the final invert at the septic tanks or sewer must be and pitching upward, usually at ⅛-in or ¼-in pitch, to the point at which you leave the building. Your line may also terminate in an exterior manhole. Regardless of whether this is a public or private utility sewer, a site manhole, or a septic tank, it is the plumbing designer's responsibility to be certain that the drainage system can flow from its final connection back to its beginning invert in accordance with the plumbing code requirements.

From the leaving point to the starting point the ideal path is a straight line, which is a condition that seldom occurs. The code requires changes in direction to be made by 45° wyes or bends of ¼ to ¹⁄₁₆. Thus the sanitary drainage pipe path commonly must avoid building wall footings and other obstacles and may well have several offsets both to more readily pick up wastes from groups of fixtures and also to avoid obstacles.

Once the path of the sanitary drainage line is set on the plan and the horizontal waste from each fixture group is connected to it, the pipe size is determined by using Table 7.1 with the pitch controlled by using the 2-fps rule and Table 7.2. As the number of fixture units increases, the pipe size increases. In larger systems when the horizontal line becomes 5 and 6 in, the slope is commonly ⅛ in. In a very large system the line may be required to be 8 in, and the pitch may be ¹⁄₁₆ in/ft. It is very unusual to find a situation in which the building horizontal sanitary sewer is required to be larger than 8 in. Cleanouts should be provided at all offsets and at 150-ft intervals in straight piping.

On occasion the required slope in the pipe is such that sewage cannot flow by gravity to the sewer or septic tank. If nothing can be done to raise the pipe at its inception or relocate fixtures to shorten the length of run, there is no alternative to the code requirement that in this case sewage shall discharge to a tightly covered and vented sump, and the liquid shall then be lifted by automatic pumps, ejectors, or any equally efficient method into the exterior sewer system. The code further requires that the sump shall not be larger than the volume of sewage delivered in 12 hr. The details presented and discussed later in this chapter will cover this pumping requirement.

Figure 7.1 shows a typical large industrial building with a series of plumbing branch connections. All data are typical and the fixture units at each connection point have been arbitrarily presumed to exist. Further, the building structural system is presumed to permit the interior horizontal piping to be installed as shown.

Since there is some sanitary branch piping below the floor to each point, our starting invert is presumed to be 1.5 ft (18 in) below the finished floor. Our final point of connection is to a sanitary manhole beyond the parking lot and in the street. Thus by simple inspection the beginning point becomes fairly obvious. As previously discussed, the beginning point is the point farthest away from the street connection.

All of our horizontal sanitary drainage piping is based on Table 7.1 data. Since the distance is very long (almost 900 ft) from beginning to end, we are seriously concerned about not ending up below the street sewer flow line. Thus as soon as the data in the tables permit it, we will switch from a ¼-in slope to a ⅛-in slope in the pipe. Commonly the sewer line in the building is extra heavy cast-iron or a code-approved plastic, although other materials are permitted.

We have shown the accumulating fixture unit count on the plan; we are presuming you will also do this on a separate sheet and save the tabulated values which create the need for ever-larger pipe sizes since fixture counts are not shown on contract drawings. What is shown on contract drawings is what is crucial to the proper installed elevation of the sanitary drainage line. That is, the pipe invert elevations at key points (usually 200 ft or so apart) and the pipe sizes. Regardless of distance, inverts should be shown at all major turning points.

Once the line leaves the building, it is not part of the plumbing contract but rather part of the site utility contract. Regardless of this fact, the line should be shown continuing to the final connection point with its proper invert to ensure that the exterior line will not preclude a satisfactory drain line in the building. In the street the utility data normally given for a manhole are as we have noted. The top of the manhole cover at street level (T.F. or top of frame) and the line invert are the standard items of information. Utility rules vary. One of the common ones is that your line meets the manhole invert at a 45° angle extending back a foot or more. For example, this rule would require, say, the final 18-in length of your line to be approximately 12 in in invert level above the line in each street. In our case we are approximately 24 in above the street invert (86.4 versus 84.5).

The proper design approach to this typical example and to all similar problems is to draw approximately the path of the pipe under consideration and to measure its total length. Then determine if your line at an ⅛-in/ft

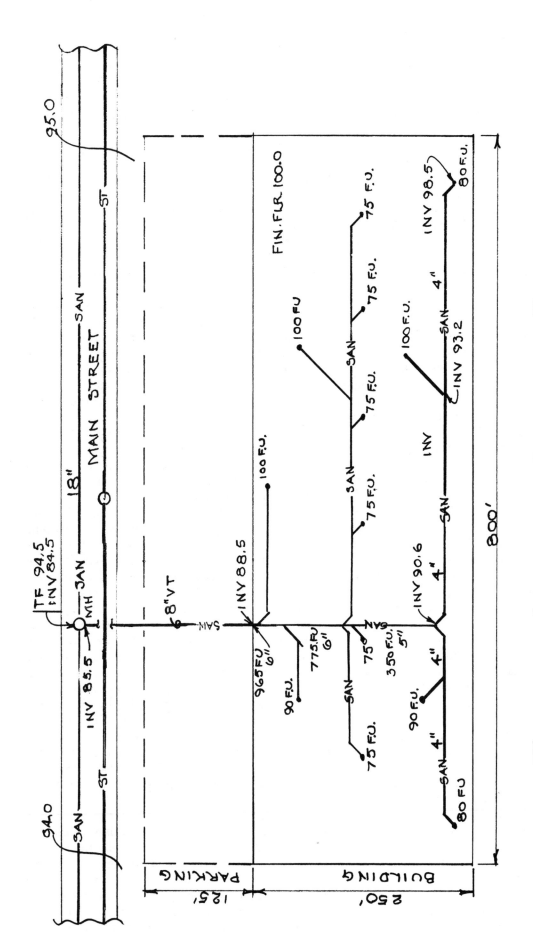

TYPICAL SANITARY DRAINAGE PIPING

— NOT TO SCALE —

FIGURE 7-1

slope or greater can be installed. For example, assume you have 100 ft of exterior run and 380 ft of interior run, which totals 480 ft. At a ⅛-in/ft slope the line drop is 60 in (480/8) or 5 ft. If the building floor covering your pipe has an elevation of 146.0 ft and you begin at a 144.5 invert, your line ends at 139.5 in (144.5 − 5.0). If 139.5 in is below the street sewer invert of 140.6 ft, you may be able to change the routing and "make" this elevation, or you may very well require a sewage lift station. Thus this procedure is always your first step in plumbing design and, if the facts are known early enough in the design process, the architect and structural engineers may be in a position to change the building elevations or fixture locations and resolve your problem for you. But they can't do it if they are unaware of such a condition and, once the design is fixed, changes in the building design are not easy or welcome.

Sizing Storm Drainage Piping

Having demonstrated in our hydraulic calculations how to size the horizontal storm drainage piping using the data in Tables 7.3, 7.4, or 7.5 or by the hydraulic formulas as presented, the task of sizing the storm drainage piping is similar to that outlined for sanitary drainage piping, with one exception. The exception is how many roof drains or parking lot area drains must be provided.

Since there is no doubt that roof drains, interior vertical leaders, and the horizontal storm water piping within the building is part of the plumbing design and plumbing contract, our task begins with the roof drainage. If the roof is pitched and roof drainage is handled by gutters and exterior leaders connected to an exterior storm drainage line, the work involved is not plumbing and the plumbing designer should not get involved in the design. Modern building designs by the architect and the structural engineer usually incorporate some pitch in all roofs including flat roofs. Thus the roof low points created by this roof sloping create the obvious locations for roof drains.

Table 7.4 gives the size of the vertical leader. Again it is important to note that the sizes given are for a 1-in rainfall. In our previous sample hydraulic calculation we used a typical 2.5 in/hr rainfall. Again using that same value, a 4-in leader, shown in Table 7.4 as being capable of handling a 1-in rainfall per hour spread over 13,840 sq ft of room, is reduced to 5,536 sq ft (13,840/2.5).

In the design of the roof drainage system ideally the 4-in leader connects to a 4-in roof drain located in the center of this 5,536-sp ft area which is a square of nearly 75 by 75 ft. From any corner of the square the distance to the drain in the center of the square is approximately 53 ft. This ideal location is seldom achieved. The roof drain locations in smaller roofs are frequently located along the outer edges of the sides of the roof and the first requirement is to be certain that in the water travel to the drain there are no low spots along the way. The

code states any roof of 10,000 sq ft or less must have at least two drains, and an area larger than 10,000 sq ft must have at least four drains. Obviously a roof with an area of 200,000 sq ft needs more than four drains.

The words in the code are somewhat enlightening. The code did not say how you divide up the 10,000 sq ft for the two drains. However, for 10,001 sq ft, to be truly precise, you need four drains. Of course this minuscule difference is really not the point to be made. Simple logic would lead one to conclude that a drain of the proper size should be provided for approximately 5,000 sq ft of roof area. This allows for reasonable assurance that water can flow to the drain.

There is one final point to be made on roof drain location. Owners, architects, and builders are commonly very firm in requesting that drain headers follow along side of vertical columns to their final connection at the horizontal drain line. This is acceptable until the line reaches the roof. Here the line must be horizontally offset to allow the roof drain to not be located next to a column which invariably is a high point, not a low point, on the roof. Secondly, as the line penetrates the final slab-on-grade, the line must be quickly offset to avoid penetrating the column footing and disturbing its structural integrity.

Once all the roof drains and their related vertical leaders are located, the running of the horizontal storm drain line proceeds in the same design concept fashion as was outlined for the sanitary drain line. First, the location leaving the building must be resolved. This location is determined by the street or site storm drainage system requirement. These two factors also determine the invert elevation since one or both are controlling the final connection invert elevation. Once this location and invert elevation have been determined, a line, as straight is possible, is drawn back to the drain leader location at the opposite, farthest end point. From this end point to the point leaving the building becomes the path of the storm drainage line. The areas served in each leader are added and the line is sized based on the data in Table 7.3. Again, in Table 7.3, the line size must be adjusted for actual rainfall. For example, a 12-in line at a ¼-in pitch and a 2.5-in rainfall handle the drainage from a total roof area not exceeding 74,400 sq ft (186,000/2.5), and as we noted in the hydraulic calculations, the table ends at a 15-in pipe, which under the same 2.5-in rainfall conditions and ¼-in pitch equals 134,400 sq ft (336,000/2.5). A roof larger than 134,000 sq ft requires a hydraulic calculation.

The exterior paved parking lot drainage, as we have noted before, is not plumbing work and is not really part of the plumbing design, as we will now illustrate. In our previous hydraulic calculation we discussed a 200,000-sq-ft factory roof with a 100,000-sq-ft parking lot, and we calculated the pipe leaving the building and containing the flow of all the roof drains as an 18-in pipe. And the pipe became 21 in after it picked up the drainage

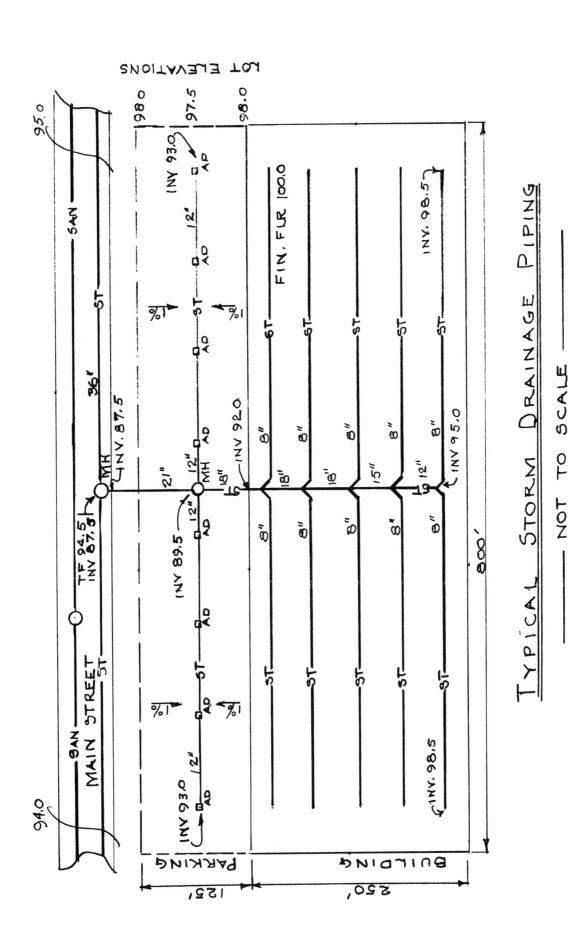

TYPICAL STORM DRAINAGE PIPING

— NOT TO SCALE —

FIGURE 7-2

from the parking lot. This pipe is very large and expensive if it is cast iron. While the pipe in the building will most likely be cast iron, the exterior pipe will commonly be reinforced concrete (RCP). As noted in our tables and in the code, we will have to adjust our calculations since the coefficient of roughness is 0.016 for concrete and we've been using 0.015. This is not a large change, and our pipe capacity is only going to be reduced to 18.6 cfs, which is still larger than the 17.2 cfs shown in our hydraulic calculation.

The sizing calculation noted above is not the primary task. The major problem is locating, sizing, and selecting the paving area drains and their related piping. Figure 7.2 is an illustration of the storm drainage system we have been discussing. The building is 800 by 250 ft, and the parking lot is next to the building and is 800 by 125 ft. Thus the roof area is 200,000 sq ft and the parking lot paved area is 100,000 sq ft. The lot is adjacent to the street and the street storm water line. In older areas of older cities the sewer may be a single combined sanitary and storm sewer.

To simplify the presentation the roof drains and vertical leaders are not shown. At 5,000 sq ft per roof drain there would be 40 drains and vertical leaders. Since there are 10 branch mains, there would be 4 leaders per branch. Again we have only shown main branch pipe sizes selected by using our tables. As may be noted, the 18-in line of our hydraulic calculations begins about midway in the main branch collection line. Commonly the storm drainage system can be installed approximately as shown if your building is similar in size and shape. In the detail we have used ⅛-in pitch throughout. Our line starts at 1.5 ft or 18 in below the floor and ends at almost 10 ft below the floor.

Once we leave the building, the plumbing contract ends and all further work is site utility work that is usually handled by the general contractor or site utility specialists. If you, as a plumbing designer, are going to include the parking lot drainage as part of your design, you must have some expertise in site drainage. Figure 7.2 has deliberately kept the problem simple by presuming there is no other unpaved land involved in the calculations. If added land drainage is part of the design problem, the exterior drainage definitely is not part of plumbing design and will not be covered in this book on plumbing design.

Commonly exterior paving is pitched toward paving drains at a 1 percent slope or 1 in/100 ft. Lesser slopes of 0.5 in/100 ft can be and have been used, but within the context of this book we are not recommending them. Our scheme presumes the drainage must be contained on the lot and that there will be no curbs; it includes a general premise of 12,500 sq ft per drain with the paving at a 1 percent pitch and a 6.25-in drop for the high outside edges to a continuous center low point. This simplifies the paving problems and creates a situation that can be realistically accomplished by the installing

contractor; it also provides a natural bump-free passage for the cars parked in the area. The same 2.5 in of rain that falls on the roof in 1 hr also falls on the parking lot. The area covered by one of the eight area drains is 12,500 sq ft (100,000/8). At a 2.5-in/hr rainfall this is 2,604 cfh of rain which, for removal calculation purposes, is 0.72 cfs (2,604/60 × 60). In overall site drainage calculations that are affected by various slopes and surface coefficient factors, one of the formulas used is the rational formula which is

$$Q = A \times C \times i$$

where Q = runoff, cfs
C = surface coefficient
i = rainfall, in/hr
A = area, acres

Again we will use a surface coefficient of 1.0. Thus $Q = (12,500/43,560) \times 2.5 \times 1.0 = 0.72$ cfs, which corresponds to our simple logic approach.

Now that we know the volume of water to be handled, we have to size the cast-iron inlet to our area drain. A roof drain cannot be located in a parking lot. One of the formulas used to size the area drain grating is based on inlet capacity at a given free area of the grilled openings in the drain. This formula is

$$Q = C A \sqrt{2gh} \times CL$$

where C = orifice coefficient
Q = flow, cfs
A = net free area, sq ft
g = 32.2
h = head (height), ft
CL = clogging factor of the drain

The orifice coefficient may be taken as 0.6 and the clogging coefficient is 0.67. In our case the high point is 6 in, or 0.5 ft. Therefore to find the free opening

$$A = \frac{Q}{X \times 29 h \times CL} = \frac{Q}{C \times \sqrt{2gh} \times CL}$$

$$= \frac{0.72}{0.6 \times (\sqrt{2 \times 0.5 \times 32.2}) \times 0.67}$$

$$= 2.24 \text{ sq ft}$$

Since 2.24 sq ft equals 322 sq in (2.24 × 144), our next step is to consult a manufacturer's table of grille sizes with related free areas. In a typical manufacturer's catalog a 26-in by 38-in grille has a free area of 366 sq in. Thus our louvered grating is 26 by 38 in, covering a concrete area drain box of the same size and at least 3 ft 0 in deep. Our separate detail and related discussion on area drains will cover the drain itself in more detail.

To try to keep the paving area drainage pipe sizing in plumbing terms, it may be noted that all our calculations treated flow as if it were roof drain flow, which is not wrong per se only more conservative because of the 1.0 surface coefficient we used. Table 7.3 gives a capacity of 478 gpm for an 8-in pipe at a ⅛-in pitch. Our calculations call for 0.72 cfs (0.72 × 450) which is 324 gpm. An 8-in line to the drain will be satisfactory. The main to which we are connecting is deep enough to easily allow for this pipe slope.

Again, the line is underground. The pipe is not cast iron. Usually it is some form of concrete or heavy duty plastic. Most buried lines are at least 10 in. One fourth of our system must be 10 in and one fourth, 12 in. To avoid all the changes in size and to anticipate clogging problems, we recommend and show the line to be 12 in over its entire length.

While the calculations and design as discussed and depicted are correct and safe and will work properly, it should be obvious that we have long ago left the subject of plumbing. In all cases we have taken the conservative approach. Our intent has been to provide the plumbing designer with the necessary data to include simple exterior drainage as part of the design and to avoid getting into design areas for which the civil engineer must be employed.

Finally we have simplified the system by piping from area drain to area drain directly, which may not be permitted in your area. Each two area drains may have to be connected separately into a storm manhole and then from the final manhole to the manhole we show on the 21-in main storm drainage line. Thus, Fig. 7.2 could require two additional manholes, one for each branch with 8-in feeders from individual drains to the manholes and a 12-in line between the manholes and the final 21-in storm sewer manhole on the parking lot.

Manholes

Manholes are the means by which exterior sanitary and storm drainage lines may be collected into one line to be delivered to the final connecting point as well as the required means of effecting change in the direction of exterior storm and sanitary sewer lines. They are also used as cleanout points in exterior straight lines over 300 ft in length. In addition they may be, and are, used when an abrupt drop in pipe elevation is required. The details that are being presented in this discussion show typical sizes and shapes. Practically every city and town has local, state, and utility company rules covering manholes that are acceptable to the local authorities and meet local regulations. These local rules regulate manhole size, shape, materials of construction permitted, and other special requirements. These rules should be carefully checked. The details depicted in this book must be modified to satisfy all local requirements. Manholes used for storm drainage systems are primarily used for street collection systems although they may also be used for site drainage systems.

Figures 7.3 through 7.6 illustrate standard manhole details. As noted on Figures 7.3 and 7.4, the manholes are the same for both storm and sanitary sewers. Figures 7.5 and 7.6 are merely additional detailing for these dual-purpose manholes. Both Figs. 7.3 and 7.4 are typical, field constructed brick masonry manholes. The advantage of the field constructed manhole is the ability to construct it to the precise invert elevation required as well as to easily incorporate local regulatory conditions.

The walls of the manholes are 8 in thick when the manhole is not over 12 ft 0 in deep. If it is over 12 ft 0 in deep, the walls must be 12 in thick. There are occasions when very deep manholes are required, sometimes 25 to 30 ft deep. All requirements for these should be carefully checked and, depending on soil conditions, may require the services of a structural engineer to calculate, size, and select both walls and bases.

Both of these details note that the manhole frame and cover data are located elsewhere. In this book the frame and cover data are depicted in Fig. 7.6. Since the manhole has an interior diameter of typically 4 ft 0 in and the frame and cover at the surface are smaller, the upper portion of the manhole is corbelled (sloped) to provide a smooth transition from the larger diameter manhole to the smaller diameter frame and cover.

Figure 7.3 depicts a standard straight-through manhole in which the flow passes in a straight line through the manhole. The open gutter in the bottom is a half-round channel of the same diameter as the entering and leaving piping. If the angle between entering and leaving is not a straight line, the channel must make a smooth curve for the transition from the incoming to the leaving direction. On occasion the manhole may also accept flow from two lines and combine the flow into one leaving line. Within the 4-ft circle it is difficult to effect many smooth transitions to a single leaving point. The entry of one of the incoming lines should be arranged so that the flow is in a straight line, or nearly straight line, and the connection of the second line, which has some 2 ft of interior space in which to change direction, may be accomplished in a smooth curving channel arc. If the leaving line is larger than the entering line, or lines, the channel diameter should be the size of the leaving line throughout.

Manholes must, from time to time, be cleaned and inspected. Steps forming a ladder must be installed as depicted in the detail and on 15-in centers or less when the top-to-bottom depth does not neatly concide with some multiple of 15 in.

The base of the manhole is commonly 9 in or more of reinforced concrete, and the channel is constructed of additional concrete on the base sloped toward the customized and formed channel in this additional concrete. This is a requirement for all manholes.

Splashing in the storm and sanitary sewer creates problems of sewage buildup in manholes and should be avoided, especially in sanitary sewer manholes. When

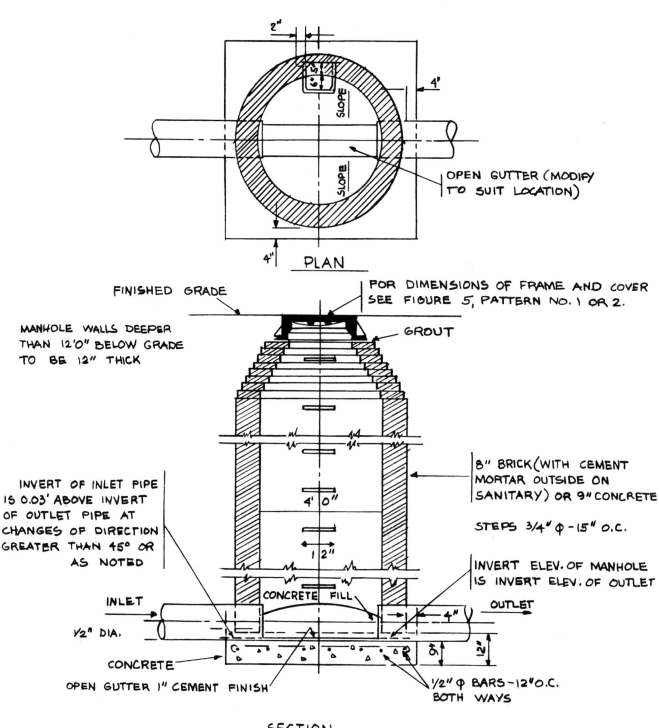

2"

SLOPE

SLOPE

SLOPE

4'

4"

OPEN GUTTER (MODIFY TO SUIT LOCATION)

PLAN

FINISHED GRADE

FOR DIMENSIONS OF FRAME AND COVER SEE FIGURE 5, PATTERN NO. 1 OR 2.

MANHOLE WALLS DEEPER THAN 12'0" BELOW GRADE TO BE 12" THICK

GROUT

8" BRICK (WITH CEMENT MORTAR OUTSIDE ON SANITARY) OR 9" CONCRETE

STEPS 3/4" ϕ - 15" O.C.

4' 0"

INVERT OF INLET PIPE IS 0.03' ABOVE INVERT OF OUTLET PIPE AT CHANGES OF DIRECTION GREATER THAN 45° OR AS NOTED

1' 2"

INVERT ELEV. OF MANHOLE IS INVERT ELEV. OF OUTLET

CONCRETE FILL

INLET

OUTLET

4"

1/2" DIA.

12"

CONCRETE

OPEN GUTTER 1" CEMENT FINISH

1/2" ϕ BARS - 12" O.C. BOTH WAYS

SECTION

STANDARD MANHOLE, SANITARY & DRAINAGE

FIGURE 7-3

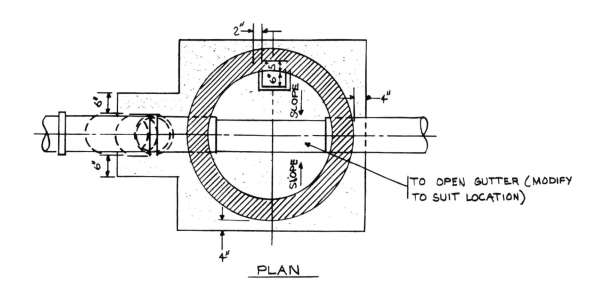

PLAN

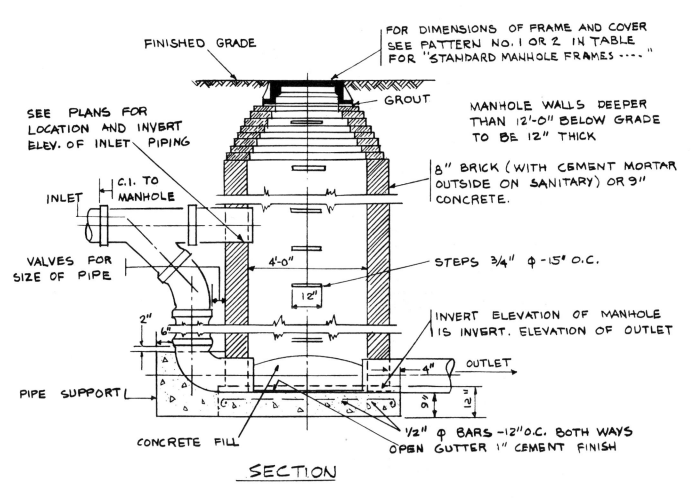

FINISHED GRADE

FOR DIMENSIONS OF FRAME AND COVER
SEE PATTERN NO. 1 OR 2 IN TABLE
FOR "STANDARD MANHOLE FRAMES ----"

GROUT

MANHOLE WALLS DEEPER
THAN 12'-0" BELOW GRADE
TO BE 12" THICK

SEE PLANS FOR
LOCATION AND INVERT
ELEV. OF INLET PIPING

8" BRICK (WITH CEMENT MORTAR
OUTSIDE ON SANITARY) OR 9"
CONCRETE.

C.I. TO
MANHOLE

INLET

VALVES FOR
SIZE OF PIPE

4'-0"

STEPS 3/4" φ -15' O.C.

12"

INVERT ELEVATION OF MANHOLE
IS INVERT. ELEVATION OF OUTLET

2"

6"

4"

OUTLET

PIPE SUPPORT

9"

12"

CONCRETE FILL

1/2" φ BARS -12" O.C. BOTH WAYS
OPEN GUTTER 1" CEMENT FINISH

TO OPEN GUTTER (MODIFY
TO SUIT LOCATION)

SECTION

STANDARD DROP MANHOLE SANITARY AND DRAINAGE

FIGURE 7-4

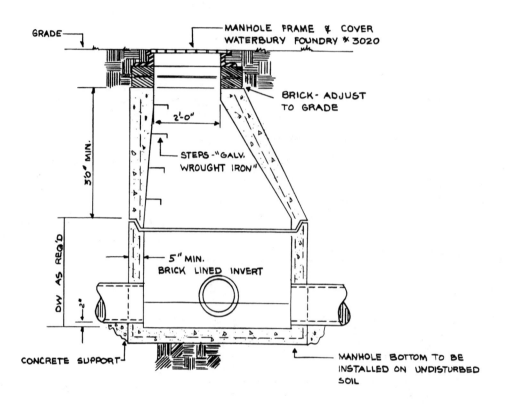

GRADE

MANHOLE FRAME & COVER
WATERBURY FOUNDRY #3020

BRICK- ADJUST
TO GRADE

2'-0"

STEPS -"GALV.
WROUGHT IRON"

30" MIN.

DW AS REQ'D

2"

5" MIN.
BRICK LINED INVERT

CONCRETE SUPPORT

MANHOLE BOTTOM TO BE
INSTALLED ON UNDISTURBED
SOIL

NOTES:

1. MANHOLE CONE & RISER SECTIONS TO CONFORM
    TO A.S.T.M. DESIGN C-478-61T.

2. PROVIDE LIFTING HOLES ON ALL UNITS.

3. ABSORPTION NOT TO EXCEED 8%. AS PER
    A.S.T.M. C-76.

4. FIELD BUILT MANHOLE WILL BE ACCEPTABLE-
    FURNISH SHOP DRAWINGS SHOWING CONST.
    DETAILS.

MANHOLE   DETAIL

FIGURE 7-5

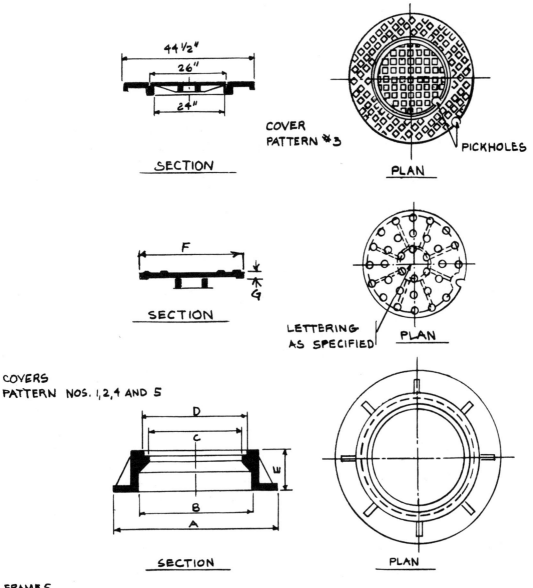

44 1/2"

26"

24"

COVER
PATTERN #3

SECTION

PLAN

PICKHOLES

F

G

SECTION

LETTERING
AS SPECIFIED

PLAN

COVERS
PATTERN NOS. 1, 2, 4 AND 5

D

C

E

B

A

SECTION

PLAN

FRAMES
PATTERN NOS. 1, 2, 3, 4, 5

PATTERN	WT., MIN.	A	B	C	D	E	F	G
1	440 LB	36"	26"	20 1/2"	23 1/4"	9"	1-3/4"	23"
2	330 LB	36"	25 1/2"	20 1/2"	23 1/4"	9"	1-1/2"	23"
3	1075 LB	56"	46"	42"	44 3/4"	10"	1 3/4	44
4	435 LB	38"	27"	23"	25 1/4"	9"	1 1/2"	25"
5	350 LB	35"	25 1/2"	23"	25 1/4"	7 1/2"	1 1/2"	24"

NOTE: COVER FOR STORM MANHOLES TO BE SOLID OR PERFORATED
AS SPECIFIED
PATTERN NOS. 1 AND 4 FOR USE IN PAVED AREAS.
PATTERN NOS. 2 AND 5 FOR USE IN UNPAVED AREAS

STANDARD MANHOLE FRAMES & COVERS, PATTERNS #1, 2, 3, 4 & 5

FIGURE 7-6

the difference in entry and leaving invert elevations is greater than 24 in and cannot be otherwise changed, the drop manhole connection depicted in Fig. 7.4 is a requirement since steep angles also create problems. In essence, the steep angle provides a path of horizontal air and sewage relief through the straight line continuation of the sewer line into the manhole. The ebb and flow of sewage, especially from taller structures, creates rapid changes of flow from full or nearly full to nearly empty, and the lack of the drop manhole would create severe splashing problems. All other details of the drop manhole are identical to those which have been described and discussed for the straight-through manhole in Fig. 7.3.

Figure 7.5 depicts a typical prefabricated concrete manhole. In this type of manhole one side remains vertical and the slope to the manhole frame and cover is taken up on the other side. Since this detail obviously came from a prefabricated manhole manufacturer's catalog, the question might fairly be raised as to why not simply specify the manhole and forget the detail. This can create installation problems. The detail, when used, must state an actual value for the vertical height which is noted as "DWN AS REQ'D" or depth as required. This value can easily be confused or inaccurately stated. The better solution for prefabricated manholes is to show at least a partially completed detail, as in Fig. 7.5, and then to provide a specification in sufficient detail to cover all the pertinent requirements that you feel the manufacturer you have selected, or any other alternative manufacturer, must supply. Further, as stated in note 4 on the detail, if a field built manhole is an alternative selection, the data contained in your detail and accompanying written specification must be sufficient to also cover this construction alternative.

Figure 7.6 is the companion figure for the manholes as detailed in Figures 7.3 and 7.4. Essentially, three types of manhole frames and covers are described. Patterns 1 and 4 are usually described as lightweight non-traffic patterns. Patterns 2 and 5 are called regular weight, or standard wheel loading, roadway and parking lot patterns, and pattern 3 is the one used for heavy wheel loading areas. These data are also contained in the catalogs of the foundry manufacturers who supply cast-iron frames and covers. It should be noted on your plans, as indicated in the detail notes, that the covers for storm manholes may be solid or perforated. The covers for sanitary manholes are always solid.

Wet Wells

When the plumbing designer is faced with the problem of elevating or lifting sanitary sewage flow within the building to meet a required sewer connection elevation, the problem is very similar to that which the civil-sanitary engineer must frequently solve in the design of municipal sewage systems. The first problem to be resolved is the probable flow rate of sanitary sewage at normal and peak rates. The second problem is how big should the sewage collection chamber be since any pump selection will require some liquid in its reservoir to enable the pump to operate with reasonable on and off cycles. The final problem is to size the pump.

Sewage pumps may be of the submersible type, which is always situated below the surface of the sewage effluent and within the sewage collecting tank. A dry pit pump is one that is located in a separate dry pit and draws the sewage effluent into the pump from a suction pipe connected into another pit that contains the affluent. Figure 7.7 represents a typical wet well for a dry pipe system.

The basic sizing of the wet well is the same whether it is for a dry pit pump or a submersible pump. The premise is that the pump may operate approximately two thirds of the time and that the sizing of the well is such that the operating period is at least 2 min and preferrably, 4 min.

In Figure 7.1 our typical sanitary system had a flow of 965 fixture units and a 6-in line. In terms of elevation the leaving line is 11.5 ft below the floor and as such would require a very deep pit. Thus, in actual practice, the piping and pumping would probably require rearrangement to lessen the pit depth.

However, the 965 fixture units and a 6-in line is a common value of units and pipe size that may require pumping. Presuming that the pitch is ⅛ in for this size, we find, using Table 7.2, that the flow with the sanitary line at half full is 111 gpm. Additionally applying a 50 percent safety factor, the peak flow is 166.5 gpm (1.5 × 111), which we round off as 165 gpm. This is our pump output value. To determine storage value the following formula can be used

$$T = \frac{V}{D - Q} + \frac{V}{Q}$$

where
T = time, min
V = storage volume, gal
D = pump capacity, gpm
Q = incoming flow, gpm
$V/D - Q$ = time to pump down wet well
V/Q = time to refill wet well

Using a trial time of 4 min, an incoming flow of 111 gpm, and a pump capacity of 165 gpm, the equation becomes

$$4 = \frac{V}{(165 - 111)} + \frac{V}{111}$$

Thus

$$2V = 4(54) + 4(111)$$

$$V = 326 \text{ gal}$$

Using the 4-ft 0-in wet well values, 4 ft of liquid would

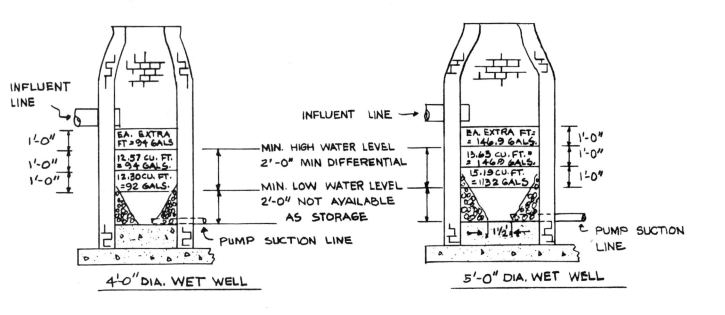

4'-0" DIA. WET WELL

INFLUENT LINE

1'-0"
1'-0"
1'-0"

EA. EXTRA FT = 94 GALS.

12.57 CU. FT. = 94 GALS.

12.30 CU. FT. = 92 GALS.

MIN. HIGH WATER LEVEL
2'-0" MIN DIFFERENTIAL

MIN. LOW WATER LEVEL
2'-0" NOT AVAILABLE
AS STORAGE

PUMP SUCTION LINE

5'-0" DIA. WET WELL

INFLUENT LINE

EA. EXTRA FT. = 146.9 GALS.

13.63 CU. FT. = 146.9 GALS.

15.19 CU. FT. = 113.2 GALS.

1'-0"
1'-0"
1'-0"

1½"

PUMP SUCTION LINE

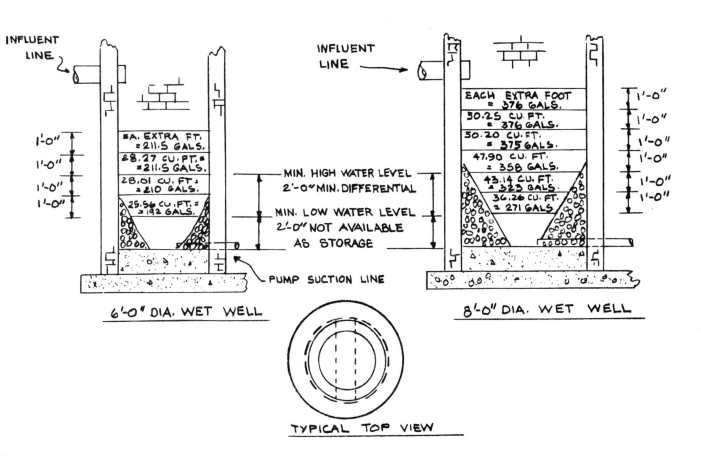

6'-0" DIA. WET WELL

INFLUENT LINE

1'-0"
1'-0"
1'-0"
1'-0"

EA. EXTRA FT. = 211.5 GALS.

28.27 CU. FT. = 211.5 GALS.

28.01 CU. FT. = 210 GALS.

25.56 CU. FT. = 192 GALS.

MIN. HIGH WATER LEVEL
2'-0" MIN. DIFFERENTIAL

MIN. LOW WATER LEVEL
2'-0" NOT AVAILABLE
AS STORAGE

PUMP SUCTION LINE

8'-0" DIA. WET WELL

INFLUENT LINE

EACH EXTRA FOOT = 376 GALS.

30.25 CU. FT. = 376 GALS.

50.20 CU. FT. = 375 GALS.

47.90 CU. FT. = 358 GALS.

43.14 CU. FT. = 323 GALS.

36.26 CU. FT. = 271 GALS.

1'-0"
1'-0"
1'-0"
1'-0"
1'-0"
1'-0"

TYPICAL TOP VIEW

TYPICAL WET WELL DESIGNS

FIGURE 7-7

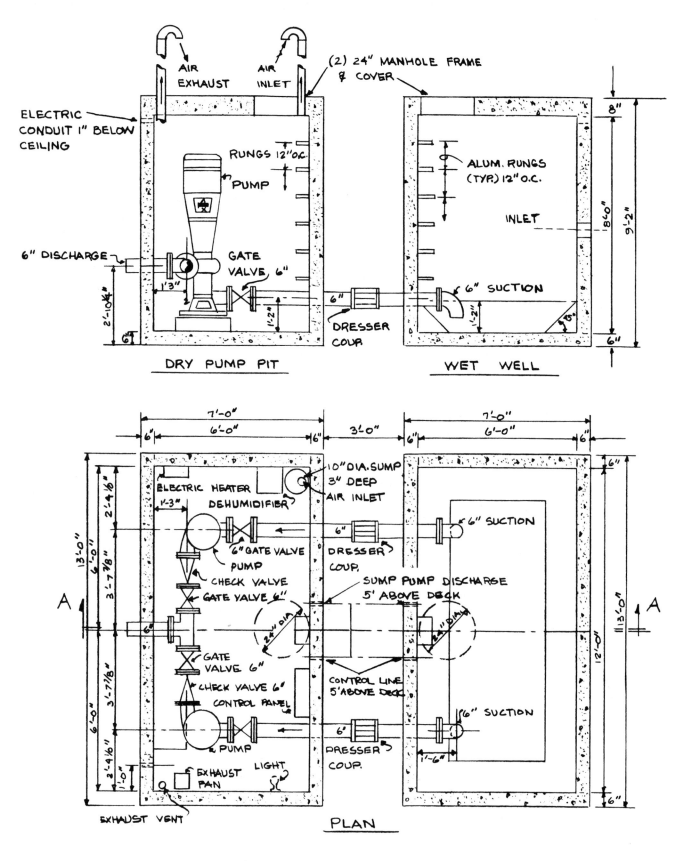

LIFT STATION WITH WET WELL
— NO SCALE —

FIGURE 7-8

equal 4 min and the well would hold 374 gal [(3 × 94) + 92]. The total depth would be 4 ft 0 in plus 2 ft plus 2 in to clear the incoming elevation, or 6 ft 2 in. Applying the same values to a 6-ft 0-in wet well, the depth would be 4 ft 2 in. The 326 gal equals 43.5 cu ft (326/7.5). If the chamber were square or rectangular to suit a submersible pump, the same 43.5 cu ft would have to be removed in 4 min.

The properly designed wet well is essential for a trouble-free lift pump installation. The well is not a septic tank and should not be used for storage, creating a situation in which the effluent could stagnate and decompose. Avoiding excessive starting and stopping of the pump is also required. Thus the "on" cycle has a range, in practice, of 2 to 5 min to completely empty the well. Well storage volume is directly related to effluent flow volume. The average flow, which we calculated at 111 gpm using half-full pipe values, could peak at a full flow of 222 gpm (2 × 211). While even higher values are at least theoretically possible in a tall structure that pressurizes the horizontal drain line, it is not too likely an occurrence except on a momentary basis.

Finally it may be noted that a further safeguard against septic action in the wet well is incorporated in the detail in Fig. 7.7 by sloping the bottom. This is to preclude a problem of sewage solids settling out near the end of the pumping cycle. The sewage wet well should be so sized that under no circumstances will the flow be contained for more than 10 min in the wet well. If there are long periods of very low flow, the pump cycle and related well capacity should be sized close to a 2-min cycle.

Lift Station with Wet Well

Figure 7.8, depicting a lift station with a wet well, may seem unrelated to the preceding Fig. 7.7, but in actuality it is merely a larger square pit which replaces the round one in Fig. 7.7; it is tied to a separate dry pit. Figure 7.8 simply is a large installation. Further, in large and sometimes in small installations the concrete wet well and pit construction details are shown and included in the building structural drawings.

The wet well as shown has a surface area of 72 sq ft which equals a volume of 72 cu ft/ft of vertical height. One of the basics of design is a flooded suction line. That is the pump-down lower limit point and is such that the suction line is always covered by liquid. In lieu of piping off the side of the pit, the turned down 6-in elbow creates the same allowable minimum level. No dimension is given for the well inlet in the detail, but, based on probable actual conditions, the well capacity available is 288 cu ft (4 ft high × 72 sq ft of area). This equates to 2,160 gal (288 × 7.5). This is sufficient to handle probable sewage flow from a 10-in sanitary sewer. This sort of situation could, using the formulas noted in the Fig. 7.7 discussion, handle a storm water flow of some 1 cfs or 450 gpm.

When the two pits are separated as shown in our detail, pipe joints are commonly Dresser couplings. The pits and ladders are similar to the wet well previously described. The pumps are of the vertical, centrifugal type. It should be noted that in any sewage lift pump installation there are two important requirements: The pump must commonly be able to handle 3-in solids, and the pump installation must be a duplex, not single, pump installation.

The pump discharge can be upward to any reasonable height so long as the pump is properly selected to discharge against the head imposed. The detail also shows a number of items in the pump pit that have been found, through actual experience, to be advantageous. Protecting the pump against freezing temperatures if the pit is outdoors is one item. In many areas hot humid summer whether can result in dripping that corrodes pipes, valves, and pump walls. Dehumidification or ventilation is a requirement. Some ventilation of the space is also desirable. Access is, of course, required through manholes as shown. Lights, and possibly exhaust fans, may be requirements. Finally, gate valves on both sides of the pumps for service work and check valves to preclude one pump pumping into the other are obvious requirements.

Normally operating control is based on a high-low float control in the wet well with each pump sized to take the full load and an alternator switch incorporated in the pump starting circuit so that at each new start the pump that did not complete the previous run will be the one to start. This ensures equal use of both pumps.

Submerged Wet Pit Lift Station

In the discussions of Figs. 7.7 and 7.8 the possible use of the details in these two figures to solve the problem of the sewer line in Fig. 7.1 was explored. Neither of these details seemed readily applicable. However, if Fig. 7.1 represented an actual case and a lift station was mandatory, a solution would have to be found. There are two possibilities and these are presented in the next two figures.

Figure 7.9 depicts a pumping arrangement with a pump very similar to that in Fig. 7.8 except for the extended shaft. This is also a standard manufactured item furnished by a number of pump manufacturers. The liquid surface area of 40 sq ft (8 × 5) can contain 300 gal/ft of depth (40 × 1 × 7.5). Using the pit dimensions as shown, the bottom of the pit receiving sewage effluent would be 12 ft 8 in or 12.67 ft, below the finished floor. If this finished floor corresponded to elevation 100.0 ft of Fig. 7.1, the inlet to the pit at 88.5 ft noted in Fig. 7.1 would change the inlet elevation of "variable" as noted in the Figure 7.9 detail to 11 ft 6 in. The difference between this elevation and the floor of Fig. 7.9 would be 1 ft 2 in, or 1.16 ft. Since the requirement of Fig. 7.1 as calculated in Fig. 7.7 is 326 gal and Fig. 7.9 has a capacity of 300 gal/ft, the available capacity of Figure 7.9 is 348 gal (300 × 1.16). In the actual setting of the float control some slight adjustment would be needed. But, in this case, the sewage flow shown in Fig. 7.1 could be lifted by using the detail in Fig. 7.9 with very minor

extension of the 13-ft 2-in dimension to perhaps 14 ft 0 in. Pumps of the type depicted in the detail are commonly available with ratings up to 400 gpm and 50 ft of total discharge head.

As in Fig. 7.8, the area containing the duplex pumps,

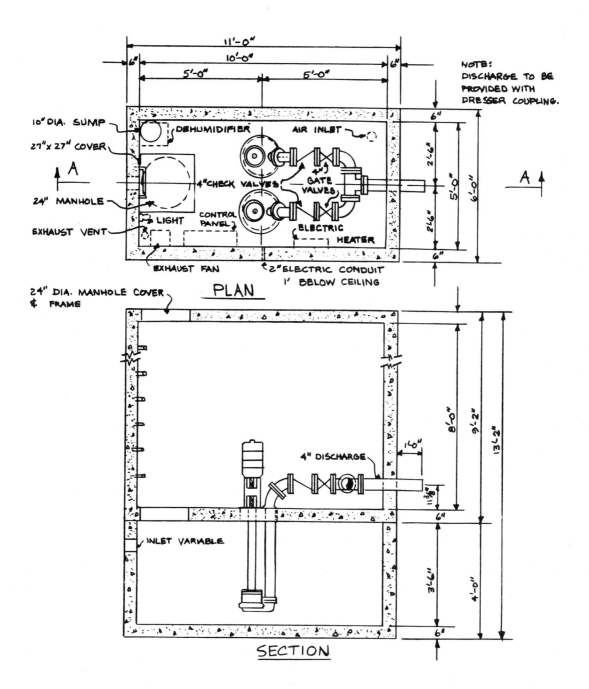

SUBMERGED TYPE·WET PIT SEWAGE LIFT STATION
— NO SCALE —

FIGURE 7-9

which are a requirement in all sewage pumping installations, depicts all of the items that practical experience dictates are required. If, as we've just noted, the detail is used to resolve a lift problem related to Figure 7.1, the actual location of the pit would very likely be just inside the exterior factory wall shown in Fig. 7.1. The electric heater in the interior pit could be omitted. However, the dehumidifier should remain even in an air-conditioned factory. And the inlet and exhaust pipes which are noted but not sized or shown would have to be extended through the factory roof with goosenecks similar to those shown for Fig. 7.8.

Submersible Wet Pit Pumping

Another solution to the lifting of sanitary sewage flow is shown in Fig. 7.10. It is handling the same wet well area that was shown in Fig. 7.9. The capacity of this type of standard manufactured product ranges up to 600 gpm and 300 ft of total discharge head.

The detail is slightly misleading in that the pump, which was compressed for the sake of detail clarity, is approximately 36 in high to the top of the motor. This solution could also be used in resolving any lifting of sanitary waste that might have been required in Fig. 7.1.

The submersible pump installation is by far the simplest method of sanitary sewage lifting. All that is required is a simple collecting chamber and two pumps which can be furnished with standard factory supplied operating controls. The common argument against this type of installation, when it is applicable for use, is the service problem when all parts are buried beneath the sewage liquid. While this is a valid argument, the pump in Fig. 7.9 is also buried in sewage liquid. In addition the liquid provides a natural cooling of the motor and a fairly steady state temperature environment for the motor's operation. Being in a totally sealed and enclosed environment, the motor operates in near ideal conditions of heat and temperature control.

As in all wet wells, no electrical services are shown or provided. Commonly the pit is sized so that the sewage level rises to approximately the top of the pump motor. In the detail no inlet was shown, but, as in Fig. 7.9, the capacity of the well is 400 cu ft ($8 \times 5 \times 1$), which equals 300 gal/ft of height. The legs supporting the pump create a required level in the bottom of the pit of 9 to 10 in to keep the pump suction intake covered at all times. Even at minimum sizes the pumps easily handle 3-in solids.

As we carefully note in the detail, the pipe spacing is necessary to contain gate and check valves. Some rearrangement of the pumps themselves could be done. It is possible to install these pumps in a pit that has a bottom dimension of 5 by 3 ft, but a practical minimum is more likely to be 6 by 4 ft.

When lowest first cost is a prime requirement, this submersible pumping arrangement is frequently a viable choice.

Sewage Ejector

Sewage ejectors use compressed air to discharge or eject sewage from the relatively small sewage receiver that is part of their system. Commonly they are described as pneumatic ejectors. The ejector is a standard manufactured device which may include the air compressor as an integral part of the package, as does the one depicted in Fig. 7.11. Two other alternative means of supplying compressed air are depicted and described in Fig. 7.12. A description of these systems follows this discussion. Standard plumbing specifications normally would describe the compressor's performance including the gallons per minute the unit can handle and the available discharge head. Ejectors are normally rated up to 300 gpm with discharge heads up to 50 ft.

Generally ejectors are used when flows are not great, but when the solids in the effluent are a considerable part of the volume of the effluent liquid. This problem occurs in certain domestic wastes, as well as in industrial wastes and heavy slurries. Inside the ejector shown in Fig. 7.11 are two open bells—one that can rise with the incoming sewage is at the top. The weight of the incoming liquid holds a check valve open while the weight of liquid on the lower bell and the lift line holds the unit discharge check valve closed. When the ejector finally fills the exhaust connection in the piston, the valve in the unit closes and compressed air is admitted to the receiver. The pressure of air on the surface of the liquid is greater than that of either the intake or discharge line. Internal check valves in both the intake and the discharge lines ensure that the liquid flows out of the discharge line and into the sewer.

To avoid complexity only one ejector is shown in Fig. 7.11 although as in the other sewage pumping details two are required. The second ejector is merely outlined as a dashed circle. A vent line outlet is depicted, but the actual piping is not shown. The vent line acts as a relief line off of the diaphragm valve. As described above, the air in the ejector must be exhausted so that the pump can fill. If the ejector is located exterior to the building and in an area permitted by local codes, the ejector pneumatic valve may be discharged into the pit and the pit itself may be vented, as the detail alludes, some 5 ft above the pit top, or deck, terminating in a gooseneck. In interior locations the vent would run separate from any other vent up through the building roof, terminating in a fashion similar to all roof vents.

Again, as in our other details and based on various experiences, we have included a humidifier, a sump pump, an exhaust fan, and a separate pit air intake and exhaust vent opening. In an indoor location it may be possible to discover that none of these items are required. But, depending on location, area temperatures, types of effluent handled, moisture problems, and groundwater conditions, some or all of these items may be required. They are all shown both to remind you to consider them and to depict their probable locations.

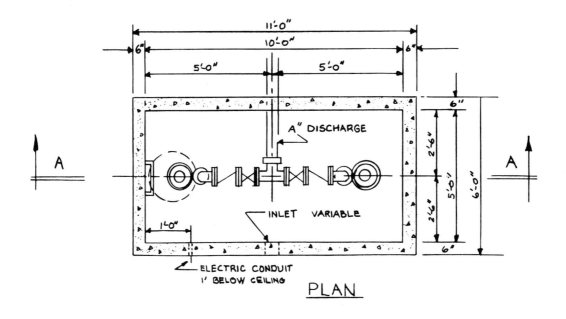

PLAN

NOTES:

1. ELECTRIC CONTROLS TO
   BE MOUNTED IN ENCLOSURE
   IN NEARBY BUILDING
   UNLESS OTHERWISE
   SPECIFIED.

2. DISCHARGE TO BE
   PROVIDED WITH DRESSER
   COUPLING.

A	3"	4"
B	2'-4 7/8"	2'-9 7/8"

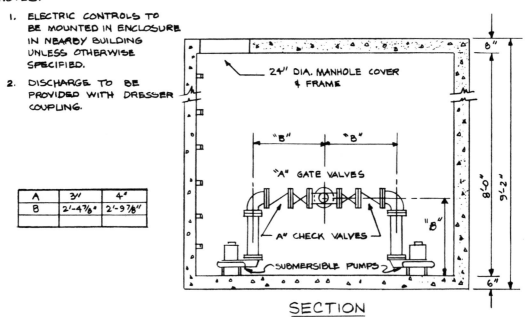

SECTION

SUBMERSIBLE WET PIT SEWAGE LIFT STATION
—NO SCALE—

FIGURE 7-10

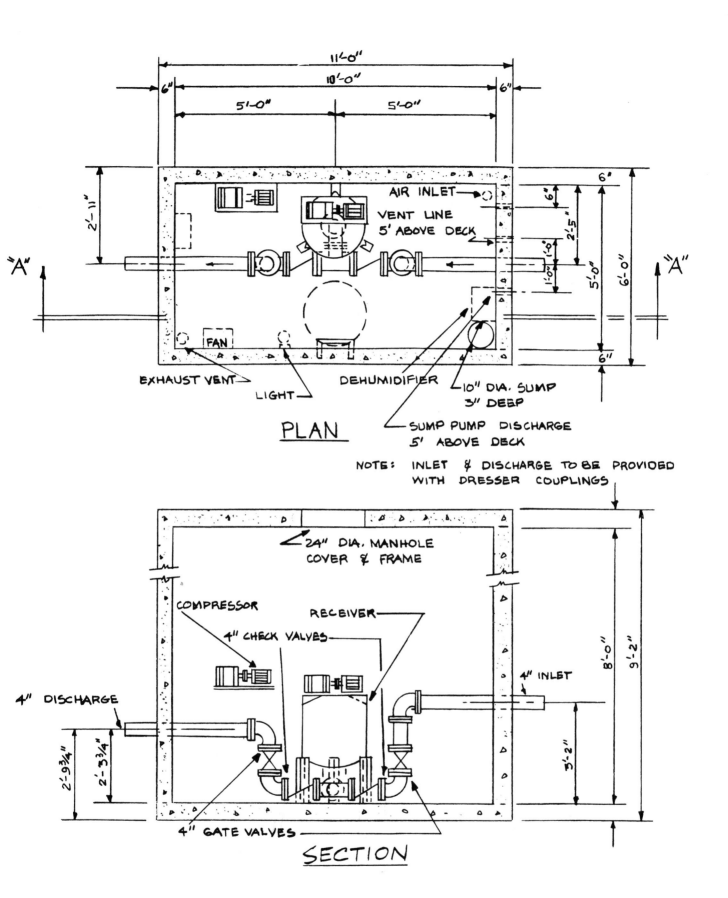

PNEUMATIC SEWAGE EJECTOR STATION

NO SCALE

FIGURE 7-11

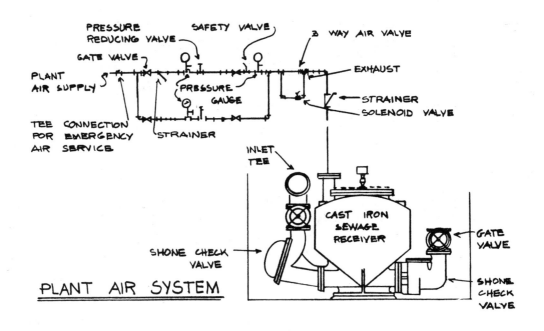

PLANT AIR SYSTEM

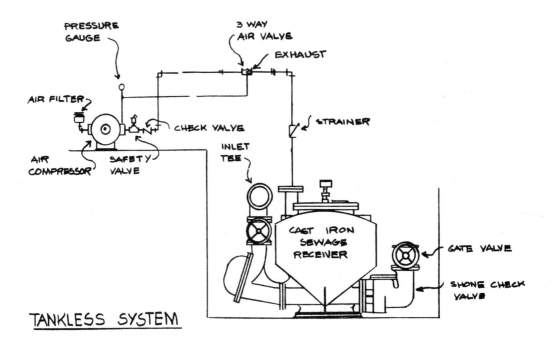

TANKLESS SYSTEM

PNEUMATIC EJECTOR

FIGURE 7-12

Within the 50-ft discharge head limit, the discharge could be above, below, or at the same level. In Fig. 7.11 the discharge is lower than the intake. Like most of the details in this book this detail came from an actual job. The discharge had to pass below an adjacent large storm sewer line that is not depicted and then rise to meet a sewer invert some distance away.

To keep the sewage lift details in a size format relative to each other, the pit is again the same size as it is in the previous pump figures. In actual practice your detail would have to show dimensions related to your particular problem.

Ejector Air Sources. Figure 7.12 depicts two alternative methods of supplying compressed air to the pneumatic ejector. In Fig. 7.11, the ejector system was of the type that is commonly described as a packaged system. The compressor and ejector are assembled and shipped as one package. The exhaust of the air control valve was noted as being possibly vented through the building roof or within a properly ventilated pump chamber.

In industrial installations it is not uncommon for the source of compressed air to be away from the pit and on the factory floor, which would be the case in the lower detail of Fig. 7.12. Industrial plants usually have their own 125 pounds per square inch gravity (psig) plant air system. Local compressors outside the pit also can be used as a source of compressed air, as depicted in the upper portion of Fig. 7.12. Not shown in Fig. 7.12 is a third arrangement which is identical to that as detailed for a tankless system except that the compressed air discharges into an air storage tank where it is stored under pressure and released as required by the action of the three-way valve downstream from the storage tank. The storage tank acts as a cushion and provides for more organized on and off periods of operation of the compressor. In the plant air system the extensive plant compressed air piping that usually exists plus a receiving tank at the source eliminates the need of a storage tank.

The details are not very complicated and clearly depict the pipe, gauges, filters, and other specialties that are required. Again the problem of what to do with the exhaust on the ejector fill cycle exists and is not resolved on the detail. Obviously as the ejector fills, the air being pushed out of the ejector through the exhaust relief in the air valve contains sewage odors. The exhaust may be piped to a separate vent through the building roof, or if a wet well or sewage manhole is nearby the exhaust air may be piped to the manhole *above* the maximum high-water level of the wet well or manhole.

There is some air noise in the air release process, and this may be resolved in the wet well or manhole application by the installation of commercially available, small noise silencers which are attached to the end of the discharge pipe.

Storm Drainage Detail. Figures 7.13 through 7.17 present details commonly used in a site storm drainage system. In the site drainage system discussion and plan shown at the beginning of this chapter the parking lot drainage system was arranged so that drainage was toward the center of the lot and the drains were located in the low points as flat surface drains. This is not always the case.

In a well-known architectural standards handbook a typical site development plan is depicted for a shopping center. The plan depicts land contours, building layout parking, and traffic flow patterns. The parking lot is pitched toward curbed areas adjacent to the main building, its site traffic entrance, and the egress roadway pattern. The drains depicted are curb inlet drains. Figure 7.13 depicts one version of the curb inlet drain. In essence the curb inlet drain is the same as the area drain in Fig. 7.14 except that there is a slotted opening in the curb above the drain. The water flow is normally caught on the grilled opening of the drain.

Catch basins depicted in Figs. 7.13 and 7.14 are very similar in construction to storm and sanitary manholes except that their overall top-to-bottom depth is usually around 5 ft. In the two figures there is an area below the leaving elevation of the drain line of 2 to 6 ft. This is an area that collects sand, gravel, bits of pavement, and other rubble that is washed into these drains with every rainfall of consequence. And this is the area that must be periodically cleaned out to avoid blockage of the storm drainage line. In many areas local and state codes very specifically describe how deep this area below the drain outlet must be; this should always be carefully investigated.

Figure 7.15 seemingly refutes the above discussion in that there is no collection space below the pipe leaving the basin. This type of installation is only recommended when the catch basin or catch basins (there could be two or three in near proximity to each other) deliver their flow to a storm manhole 25 ft or less from the basin and when this storm manhole has a refuse collection area below the lowest leaving point. Installations of this type are usually seen at street intersections and not in parking lot drainage of the type the plumbing designer would encounter.

Figure 7.16 illustrates the type of basin that, with adjustment to flow inlet and outlet, may be used to satisfy the storm drainage system described in the storm drainage discussion at the beginning of this chapter. To emphasize the point that the catch basin, with proper sizing of the basin and piping, can be used as a storm manhole, the detail depicts entering and leaving pipes at right angles to each other. In our previous storm drainage example the main storm drain line in the parking lot simply passed straight through the basins. However, whether the line passes straight through or leaves at some angle to the incoming line, this detail can be used, with proper sizing and proper depiction of enter-

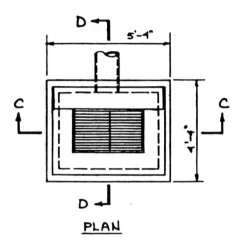

D

5'-7"

C ——— C

—— T —— (1' - 7")

D

PLAN

WITH CONCRETE MASONRY CORBELLING
WILL BE PERMITTED TO A MAX. OF 3"

CATCH BASIN WALL TO BE 12" THICK
WHEN DEPTH OF MANHOLE IS
GREATER THAN 10'(MASONRY ONLY)

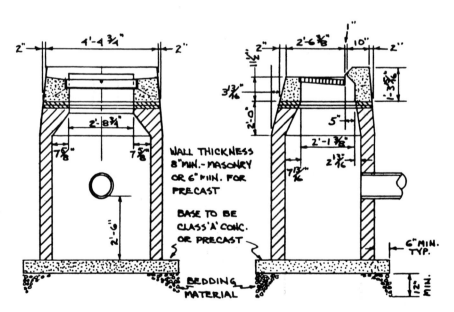

2"   4'-4¾"   2"

2'-8¾"

7⅝"   7⅝"

2'-6"

WALL THICKNESS
8" MIN. - MASONRY
OR 6" MIN. FOR
PRECAST

BASE TO BE
CLASS 'A' CONC.
OR PRECAST

BEDDING
MATERIAL

SECTION C-C

2"   2'-6⅜"   10"   2"   1"

1½"

3¹³⁄₁₆"

2'-0"

5"

5"   3⁵⁄₈"   1'-3⅜"

2'-1⅜"

2¹³⁄₁₆"

7¹³⁄₁₆"

6" MIN.
TYP.

12"
MIN.

SECTION D-D

CURB TYPE CATCH BASIN

NOT TO SCALE

FIGURE 7-13

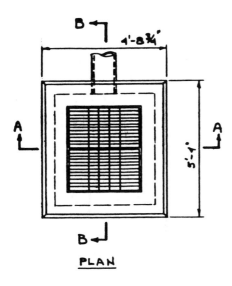

WITH CONCRETE MASONRY CORBELLING
WILL BE PERMITTED TO A MAX. OF 3"

CATCH BASIN WALLS TO BE 12" THICK
WHEN DEPTH OF MANHOLE IS GREATER
THAN 10'. (MASONRY ONLY)

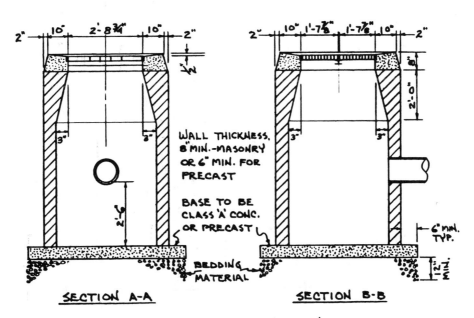

SECTION A-A   SECTION B-B

DOUBLE GRATE CATCH BASIN

NOT TO SCALE

WALL THICKNESS,
8"MIN.-MASONRY
OR 6" MIN. FOR
PRECAST

BASE TO BE
CLASS 'A' CONC.
OR PRECAST

FIGURE 7-14

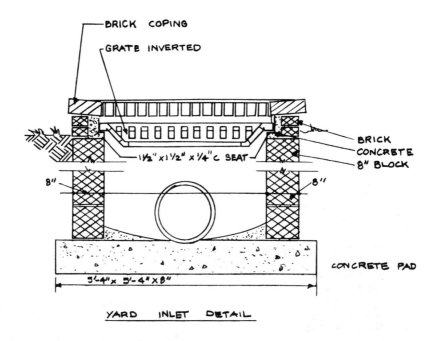

BRICK COPING

GRATE INVERTED

BRICK

CONCRETE

8" BLOCK

1½" x 1½" x ¼" C SEAT

8"

8"

CONCRETE PAD

5'-4" x 5'-4" x 8"

YARD INLET DETAIL

TYPICAL SECTION THRU CATCH BASIN

FIGURE 7-15

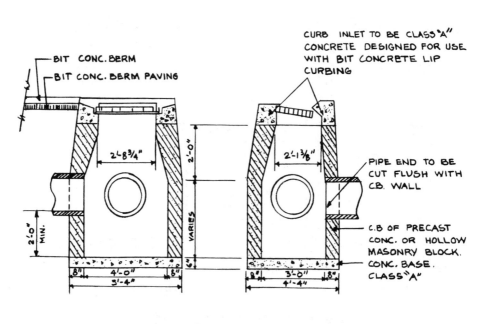

BIT CONC. BERM

BIT CONC. BERM PAVING

CURB INLET TO BE CLASS "A" CONCRETE DESIGNED FOR USE WITH BIT CONCRETE LIP CURBING

2'-8¾"

2'-0"

2'-1⅛"

2'-0" MIN.

VARIES

PIPE END TO BE CUT FLUSH WITH C.B. WALL

6"

8" 4'-0" 8"

5'-4"

8" 3'-0" 8"

4'-4"

C.B OF PRECAST CONC. OR HOLLOW MASONRY BLOCK.

CONC. BASE. CLASS "A"

TYPICAL SECTION THRU PRECAST CATCH BASIN

FIGURE 7-16

ing and leaving lines that are based on your actual design problem.

Finally, as alluded to in our discussion of lift pumps, we depict in Fig. 7.17 the storm drainage system that requires a lift pump to meet the storm line invert. Although you cannot see both of the lift pumps, there are two and the operation is a stepped system. In this system one pump operates unless the incoming flow exceeds its capacity. If that occurs, the second pump with the higher level control will be brought into operation.

This system as depicted is one that requires more site drainage knowledge than is usually possessed by the average plumbing designer. Since the catch basin is used not only for the local area drainage but also, as implied, for one or more additional drains, the sizing of the catch basin is the same, in this case, as the sizing of the wet well we previously covered. We would strongly recommend that the drainage requirements be at least reviewed by a competent licensed civil-sanitary engineer because the drainage volume that Fig. 7.17 can handle is definitely limited in scope.

Again, the plumbing designer could be faced with the problem of lifting roof storm water drainage within the building. The detail would still be valid, and the only alteration required would be to change the catch basin into a wet well. In essence, in this case Fig. 7.17 would be very similar to Fig. 7.8, and the calculations of flow would be similar to those described in our previous discussion of storm drainage and wet wells. The premise of this discussion is to illustrate that plumbing design problems may frequently be readily extrapolated to the resolution of site drainage design problems. The concern of this book, and especially of the discussion in this chapter, is to try to cover building storm drainage, which is definitely a plumbing design problem, showing the proper disposition of storm water on its way to the site disposal area or street utility line without either ignoring or getting the plumbing designer overly involved in what happens to the line between the building and the street.

Roof Drains. In all the discussion of the storm drainage system design and calculation, the roof drain, which, as the code specifies, must consist of at least two drains in a roof area of less than 10,000 sq ft, has not been depicted. Figure 7.18 depicts a typical roof drain, and Fig. 7.19 depicts the offset which was mentioned in our discussion of Fig. 7.2. The plumbing designer usually has at his or her disposal several catalogs from roof drain manufacturers. There are pros and cons for the various types, and Fig. 7.18 is not to be construed as expressing favoritism for any particular type.

The installation of each roof drain should be detailed since there are a number of items related to the drain installation that directly affect the integrity of the roof. Commonly, though not always, an attempt is made, as depicted in Fig. 7.18, to create a drainage basin around the drain by depressing the roof material so that there is positive slope toward the drain. The roof drain usually is specified with a deck clamping device. In addition the plumbing specification usually requires the plumbing contractor to supply the drain flashing material.

The roof, affected by changes in temperature, does expand and contract, and the piping must be arranged to accommodate this movement. The drain, or sump receiver, is commonly furnished to the general contractor, who normally has the roofing contractor set (install) the drain.

In Fig. 7.19 we illustrate what we meant in the Fig. 7.2 discussion concerning rain leaders rising adjacent to side of a column and being offset to the natural low point of the roof. For example in a typical 24-ft by 32-ft bay, the building has vertical columns on 24-ft centers in one direction and on 32-ft centers in the direction that is at a right angle to the 24-ft direction. The four columns are rigid building frame members. The framing usually runs in one direction (sometimes both directions) on the column line. The low point is either on the center line between the 24-ft-on-center columns or the 32-ft-on-center columns depending on the framing system. If the framing runs in both directions, the low point is exactly in the center of the 24 by 32 rectangle. The high point is located by each of the columns.

It is not always possible to locate the drain at the exact low point, but every effort should be made to get the drain as far away as possible from the column.

Frequently the drain receives rain water that cools the rain leader below the dew point of the surrounding air. The pipe will definitely sweat. While messy the sweat, or condensation, on the vertical pipe will simply run down the side of the pipe, and when the space between the pipe and its sleeve is large enough, the condensation will end up in the earth. But the condensation will frequently drip off the horizontal offset portion of the pipe and the interior exposed bowl portion of the roof drain. The simple addition of a 1-in fiberglass insulation to the bowl and horizontal portion of the rain leader line will solve this problem. Invariably the humidity buildup of dropped ceiling portions of the building make this condensation drip problem much more likely and much more severe.

Summary

This chapter has covered a phase of plumbing design and detailing which usually is the simplest of all plumbing tasks. For the majority of projects the excellent precise data contained in the BOCA *Basic Plumbing Code* are more than sufficient to resolve all design problems that are part of storm and sanitary drainage piping. In most cases piping sized from the code tables, which we have reproduced in this chapter, and pitched at minimum allowable pitches will result in storm and sanitary lines leaving the building at invert elevations sufficiently high to properly continue to street and site connections. Further, most design projects that include site drainage

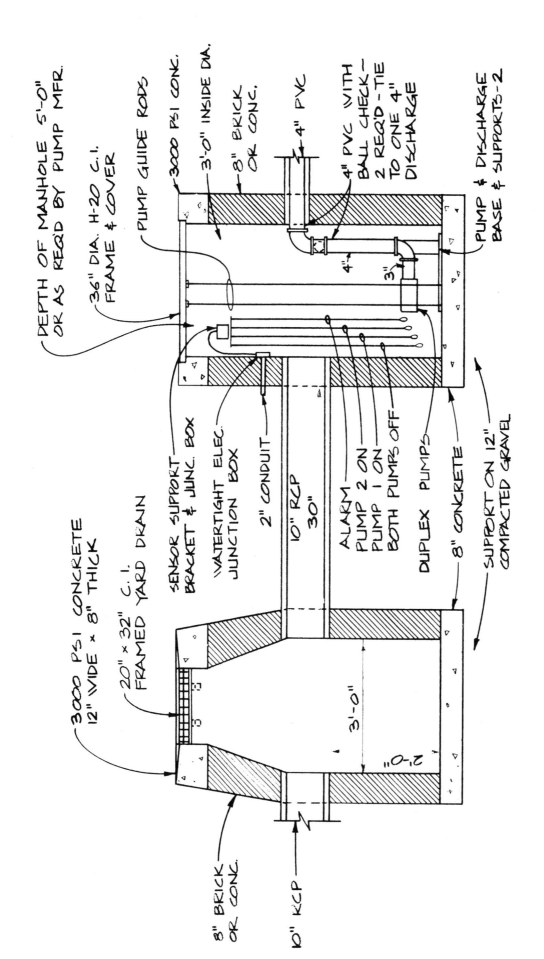

DEPTH OF MANHOLE 5'-0"
OR REQ'D BY PUMP MFR.

36" DIA. H-20 C.I.
FRAME & COVER

PUMP GUIDE RODS

3000 PSI CONC.

3'-0" INSIDE DIA.

8" BRICK
OR CONC.

4" PVC

4" PVC WITH
BALL CHECK—
2 REQ'D-TIE
TO ONE 4"
DISCHARGE

4"

3"

PUMP & DISCHARGE
BASE & SUPPORTS-2

3000 PSI CONCRETE
12" WIDE × 8" THICK

20" × 32" C.I.
FRAMED YARD DRAIN

SENSOR SUPPORT
BRACKET & JUNC. BOX

WATERTIGHT ELEC.
JUNCTION BOX

2" CONDUIT

10" RCP
30"

ALARM
PUMP 2 ON
PUMP 1 ON
BOTH PUMPS OFF

DUPLEX PUMPS

8" CONCRETE

SUPPORT ON 12"
COMPACTED GRAVEL

8" BRICK
OR CONC.

10" RCP

3'-0"

2'-0"

CATCH BASIN & LIFT PUMP STATION

NO SCALE

FIGURE 7-17

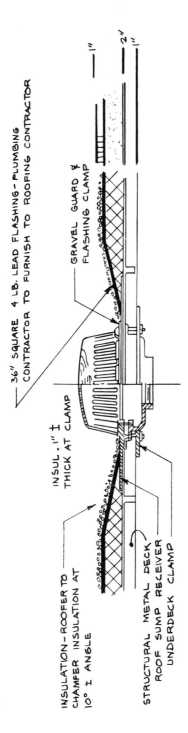

36" SQUARE 4 LB. LEAD FLASHING-PLUMBING
CONTRACTOR TO FURNISH TO ROOFING CONTRACTOR

GRAVEL GUARD &
FLASHING CLAMP

INSUL. 1" ±
THICK AT CLAMP

INSULATION-ROOFER TO
CHAMFER INSULATION AT
10° ± ANGLE

STRUCTURAL METAL DECK
ROOF SUMP RECEIVER
UNDERDECK CLAMP

1". LOCATE AT LOW POINTS OF ROOF.

2. PIPE WITH EXPANSION JOINTS OR
   OFFSETS TO ALLOW FOR EXPANSION.

3. FURNISH SUMP RECEIVER TO GENERAL
   CONTRACTOR SO RECEIVER CAN BE
   ATTACHED TO THE DECK .

ROOF DRAIN DETAIL

FIGURE 7-18

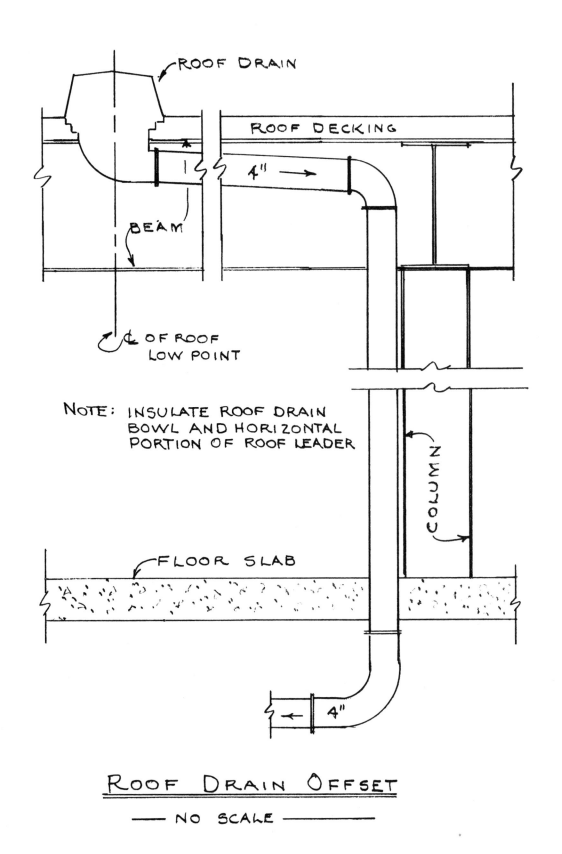

ROOF DRAIN

ROOF DECKING

BEAM

4"

℄ OF ROOF
LOW POINT

NOTE: INSULATE ROOF DRAIN
BOWL AND HORIZONTAL
PORTION OF ROOF LEADER

COLUMN

FLOOR SLAB

4"

ROOF DRAIN OFFSET

— NO SCALE —

FIGURE 7-19

work will have a separate design specialist handling that phase of the work. Since this design specialist is usually a civil-sanitary engineer, or one skilled and competent in that area, he or she is aware of the plumber's leaving invert problems and usually can resolve them in the site work. The plumbing drainage design will end 6 to 10 ft outside the building, where it will be picked up by the drainage specialist.

But this is not always the actual situation. Professional pride, client pressure, and a number of other reasons may impel or compel the plumbing designer to include in the plumbing design that "simple little bit of site drainage."

The additional site drainage may well be a simple additional amount of exterior drainage. Having had a number of simple exposures, the competent experienced plumbing designer may well slowly acquire at least a rudimentary knowledge of site drainage design. In 17 of the 19 details we have presented in this chapter we have constantly delved into this twilight zone of plumbing design. In Fig. 7.2 the twilight zone has suddenly enlarged to a serious question of professional, legal, and ethical capability on the part of the plumbing designer.

The details presented are correct. While their application does not seem unusually difficult, it should be clearly understood that the specification for the work beyond the building is a separate specification called site drainage or site utilities and is definitely not part of the plumbing specification. While the purpose of this book is to provide workable, usable knowledge of plumbing design, the mere fact that the material has been presented should definitely not be construed to mean that anyone can design this exterior work or that this book contains all you really need to know. If that were true, this book would supplant all other engineering texts on sanitary engineering. It, of course, is not true and is a very dangerous assumption. So long as the exterior sanitary line to the street and some simple added paving drainage is the limit of the work accepted by the *skilled*, experienced plumbing engineer or designer, there is no reason for undue concern. With care the design should be perfectly satisfactory, and the use of the details and data in this chapter can readily be a part of the design effort.

EIGHT

Sewage Disposal

The subject of sewage disposal design may readily be debated as to whether or not it is within the area of the plumbing designer's expertise. So long as the limit of the area of sewage disposal design is confined to septic tank and field solutions, all of the available evidence indicates that plumbing designers can, and have, acquired the expertise to design septic tank and field sewage disposal systems and have been very successful in their efforts in this area. This chapter limits itself to septic tank and field sewage disposal.

Those readers who live in unsewered areas are usually familiar, at least in usage terms, with the residential septic tank and effluent disposal field. Local requirements can and do vary. One of the common requirements is for a 1,000-gallon (gal), concrete, buried septic tank with one or more removal cleanout manholes and 150 feet (ft) of drainage tile that is usually arranged as two 75-ft or three 50-ft lines on 6-ft to 8-ft centers. While this chapter will not specifically cover the simple residential system, all of the discussion in this chapter, if reduced to minimal values, would at least apply to the residential leaching field which, like all leaching fields, is a key element in the system. Regardless of whether the field lines are used for a small or a large installation, the same rules of selection, design, and installation apply.

Commonly, laws affecting sewage disposal are enforced by the local or state Department (or Board) of Health. The officials of the Department of Health may or may not have some direct input in the approval of the sewage disposal system designer. They definitely will have an input into what is and is not an acceptable design. Thus, in whatever state you practice or intend

to practice, your first step is to get a copy of the local and state rules and regulations and be certain that you fully understand all of the system requirements as stated in that document. One of the common rules is that the system design is your design and approval by the local authorities does not in any way enable you to shift the blame for system failure to said code or authority. As far as system operation is concerned, it is your design that succeeds or fails.

System Design Premise

The basic premise of any septic tank and field sewage system is that the liquids and solids flow into a holding chamber or tank where they are detained long enough to be dissolved into a liquid which can then leave the tank and enter the soil via a series of distribution lines. The decomposition of the mixed solid and liquid in the tank creates odors which can flow back through the incoming effluent lines and be vented out through the building vent system. This venting of septic tank gases is not a perfect system. In densely populated suburban subdivisions there can be odor problems requiring special vent treatment.

Septic Tank Design

Smaller residential and commercial septic tanks are commonly supplied by companies that specialize in the construction of prefabricated concrete tanks. In general the rule is that the size of the tank shall be sufficient to hold the total amount of sewage effluent it receives for a period of 18 to 24 hr. During the septic reduction process, some of the solids entering the tank will settle to the bottom and form a sludge. Bacteria which thrive in a septic tank digest much of the solids that settle out of the sewage and convert some of the solid matter to gases and liquids which pass out of the tank with the other liquids. These gases also help keep away the scum that floats on the top of the tank. The accumulated tank solids thus consist of settled sludge and floating scum which, from time to time, must be removed. A good indication of a poorly designed tank is lack of scum created over long periods of time. In this instance the solids are actually being passed into the disposal field and will eventually clog this field. Even when properly designed tanks are installed, large flows of disinfectants and chemicals can destroy the tank bacterial action and create clogged field lines.

The rectangular septic tank should be designed so that the length in the fluid flow direction is twice the width and the width is approximately equal to the depth of liquid. For example, the tank could be 9 ft 0 inches (in) long by 4 ft 6 in wide with 5 ft 0 in of liquid depth and perhaps 6 ft 0 in of overall top to bottom depth. All dimensions stated are inside dimensions. Further discussion will be provided when we illustrate specific details. In general the tanks are concrete and either prefabricated or site constructed.

Dosing Tank Design

Both government rules and regulations and practical engineering design frequently demonstrate the need for a dosing tank. This dosing tank receives the liquid from the septic tank. In the dosing tank a patented device creates a seal that is similar to a water closet trap seal which at a given liquid level enables the chamber to rapidly discharge its contents to the leaching field.

As can be readily visualized, the normal flow of liquid is fairly slow out of the septic tank to the field. Long lengths of leaching tile tend to get uneven usage with the portion of the lines nearer the tank getting more than their designed share of the flow. Some states have fixed rules such as any field containing more than 500 ft of tile shall be supplied by a dosing chamber or a properly selected and approved pump. From a logic standpoint the leaching field should be arranged for good, even distribution and should be able to provide some oxygen to be drawn through the pores of the soil and into the leaching trenches.

Intermittent application of sewage through dosing can accomplish both even distribution and oxygen absorption. The working capacity of the siphon chamber should be sufficient to half fill the volume of leaching tile in one application. The rates of siphon discharge can be secured from the siphon manufacturer. Too long a dosing period may not create good distribution since the sewage may seep away too rapidly in one portion of the field. The rate of discharge at minimum head should be at least 25 percent in excess of the flow to the dosing tank so as to overbalance the flow of incoming sewage

and vent the siphon. Otherwise the siphon might discharge continuously.

Distribution System Design

Once the sewage leaves the septic tank, the line receiving the sewage flow usually enters a diverting or distribution box which allows the flow to be split up into a series of leaching lines. A very small residential field consisting of two to four lines of leaching piping can have two lines connected by wye fittings, and these lines feeding two sublines are also connected with a wye fitting which originates at the sewage tank outlet. More piping than those four lines is safely and properly connected with each of two lines tied to one wye and the feeder for the wye connected to a distribution box outlet.

Leaching Field Design

The ability of the soil to absorb moisture must be demonstrated by a test performed by a properly certified testing consultant. Many engineering firms that are frequently involved in septic tank and field design will arrange to have one or more of the members of the firm certified. In some states field verification of the field testing is done by a local health officer. Regardless, the test results produce a value described as the percolation rate. The percolation rate is the rate, in minutes per inch, that the water in the test hole drops as it is absorbed in the surrounding earth.

There are various rules and requirements to properly describe design data based on percolation rates. Allowable maximum percolation rates (slowest value of soil absorption) commonly vary from 40 to 60 minutes per inch (min/in). The following is a typical presentation of allowable design values:

Percolation rate (min/in)	Disposal trenches (maximum rate of gal/sq ft/day)	Disposal beds, seepage pits, cesspools (max-application) (gal/sq ft/day)
2	3.5	2.5
3	2.9	2.0
4	2.5	1.8
5	2.2	1.6
10	1.6	1.1
15	1.3	0.9
20	1.1	0.8
25	1.0	
30	0.9	
40	0.8	

In viewing the above data the first question to be answered is what is a square foot as noted in the above table. A square foot is based on the total bottom area of the disposal trench or the entire bottom area of the disposal bed. Since a trench is commonly 2 ft in width, its area is 2 sq ft/lineal foot. Using the above table and a 4-min percolation rate, the gallons per day per lineal foot equals 5 gal/day/ft (2 × 2.5) of trench. And a flow of 4,000 gal/day requires 800 ft (2000/5) of trench.

Again the various local regulations have somewhat differing rules on the construction of beds and trenches. The following are general rules that would apply to the 800-ft of trench we just calculated. It should also be noted that a dosing tank may also be required.

Minimum lines per field or bed	2
Maximum length per line	100 ft
Minimum diameter of distribution lines	4 in
Grade of distribution lines	2 to 4 in/100 ft (no gradient needed if dosed by siphon or pumps)
Maximum width of disposal trench bottom	3 ft
Minimum distance between walls of adjacent trenches	5 ft
Minimum cover over distribution lines	12 in
Minimum distance between distribution lines	6 ft
Maximum depth of invert of distribution pipe	2.5 ft (below finished grade)
Minimum distance between adjacent beds	10 ft
Length of bell and spigot clay pipe lines	2 ft
Openings at joints of bell and spigot	0.5 in (clay pipe lines)
Distance between distribution lines and edge of bed shall not be less than	2 ft
Termination of distribution lines from end of trench	2 ft

The distribution lines may consist of clay or tile bell and spigot pipes, perforated pipe, or other suitable pipe.

Site Testing

While the testing is usually done by a separate consultant and/or Board of Health representative, the septic system designer may, on occasion, be presented with data which, based on his or her experience, seems questionable. In the paragraphs that follow a typical test procedure is outlined. The supplier of doubtful data should be asked how the data were obtained to determine if the procedures generally compare to those in the following paragraphs.

Dig two or more test holes within the area of the proposed seepage system, not less than 10 ft apart. One of the holes should be at the depth of the bottom elevation of the proposed seepage system, and the second hole should be at a depth of about 18 in below the bottom elevation of the proposed seepage system. This is to evaluate the consistency of the soil with depth of the seepage qualities of the soil. The size of the seepage system must be based on the highest percolation rate obtained. The holes shall be not less than 6 in in diameter to 6 in square, nor should they be greater than 8 in in diameter or 8 in square. Scarify the bottom and sides of the test holes and remove all loose material. Place about 2 in of coarse sand or fine gravel in the holes to prevent bottom scouring.

Fill the holes with clear water to a minimum depth of 12 in above the coarse sand or fine gravel. Keep water in each hole for at least 4 hr and preferably overnight by refilling. If necessary to maintain water in each hole for this period, provide a reservoir of water and an automatic siphon to deliver it to the holes intermittently, or the percolation test holes should be soaked and maintained full to the top for not less than 4 hr before the percolation test is made. In uncompacted sandy soils containing no clay or silt, the above saturation procedure is not necessary and the test can be made as soon as the water from one filling has seeped away.

When the saturation process is complete, the water depth should be adjusted to 6 in over the coarse sand or fine gravel before the test is begun. The drop in water level should be measured from a fixed reference place, such as a board laid across the hole, over 30-min intervals, refilling the holes to a depth of 6 in as necessary.

When three consecutive readings at 30-min intervals read the same rate, the test may be considered complete. If no stability is reached between three 30-min readings, then not less than 4 hr of readings must be taken. The drop in water level which occurs during the final 30-min period is then used to calculate the percolation rate. This rate shall be expressed in minutes per inch.

In soils in which the first 6 in of water seeps away in less than 30 min, after the saturation period, the time interval between measurements should be reduced to 10 min and the test run over a period of 1 hr. The drop in water level which occurs during the final 10-min period is used to calculate the percolation rate. This rate shall be expressed in minutes per inch. If an unanticipated cut is made, the results of any percolation test made prior to the cut is invalid. A new percolation test shall be made under the changed conditions.

Finally there may be a concern over groundwater. Usually this is a winter and spring problem and should be determined between December and May. In making this determination it is necessary to bore or dig an adequate number of holes of convenient size in the proposed leaching area to a depth of at least 5 ft below the lowest point of the proposed subsurface seepage system or to the water table if encountered sooner. An open perforated pipe at least 4 in in diameter shall be installed. Such pipe should be installed at the beginning of the wet season and remain in place for a period of time as defined in the local regulations. This pipe shall be capped at the top and mounded to prevent the collection of surface water. All water table test holes shall be properly witnessed by the certified engineer and/or the local health officer.

Septic Tanks. As previously noted, septic tanks are constructed of concrete. They are prefabricated and delivered to the site, or they are properly formed and poured in place at the site. The normal minimum strength value of the concrete is 5,000 psi after 28 days. Steel reinforcement is usually ASTM-A615-68, grade 40 at 1-

ft minimum cover. The design loading is normally AASHO-H10-44. These data are general and commonly apply to the tank depicted in Fig. 8.1.

There are three important points in any septic tank. Figure 8.1 was deliberately selected to illustrate these specifics. First, the tank as depicted is deliberately large enough to illustrate that, in this instance and in many others, a single manhole in the center of the tank is very impractical. If the tank is going to clog for any reason, it most likely will clog at either the inlet or the outlet. Items accidentally flushed into the tank such as cloths and brushes commonly will stick at one of these two locations. Thus servicing of the tank is far easier if there is a manhole at the influent and effluent locations.

Second, to create a sort of primary-secondary detention solution, especially when relatively high peak to normal flows exist, the tank can be compartmentalized as we have indicated with the divider partition shown as a pair of dashed lines in the sectional elevation. The partition effectively divides the tank into two segments of one-third and two-thirds of the total tank volume. The pass-through opening is approximately one-third

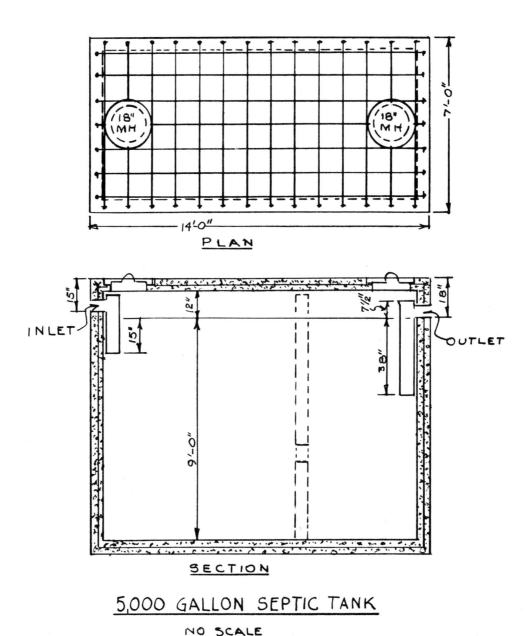

FIGURE 8-1

from the bottom of the tank. Generally, and required in some state codes, partitioning is used or required only in tanks whose capacity is 5,000 gal or greater.

Third, the rules governing inlet and outlet tees vary in the various state regulations. In general the tees which are open at the top must end at least 12 in below the liquid line for the inlet tee. Our detail shows 15 in. The outlet tee must terminate 18 in below to one-third of the liquid depth. Ours shows a depth of 38 in which is slightly more than one-third (36 in) of the 9-ft 0-in liquid depth.

The top of the open tee must also, commonly, be at least 6 in above the liquid level. Ours shows 7½ in, and just to be certain that the terminating point is really in the air space, we depict a 12-in space above the top of the liquid. Finally we depict minimum size manholes of 18-in diameter.

These various values noted above must be checked against local and state rules in your area. Our tank is twice as long as it is wide. However, the depth of liquid is greater than the tank width. This may not be permitted in your area. Some local rules require depth and width to be equal to each other. Further there frequently are requirements that tanks of 2,000-gal capacity and larger must have manholes brought up to finished grade.

Septic tanks have a considerable odor that is very

offensive and which is created by sewage material decomposition occurring in the tank. As such septic tanks must be watertight and reasonably airtight. In Fig. 8.1 the dimensions and the flow line of the inlet and outlet indicate a 3-in differential. Commonly a 2-in differential is the minimum allowable value. The tank must be placed on a firm foundation that is not subject to settlement which could adversely alter this differential.

Because the primary concern is to avoid locating the leaching lines too far below the surface of the leaching field and because the ideal field is one that is level or nearly level, the top of the buried septic tank is frequently 12 to 18 in below finished grade. The controlling factor in the septic tank elevation is the depth of a leaching line termination which is at the farthest point from the septic tank. It is this elevation, properly pitched upward and back to the tank, that is the critical determinant of septic tank size, depth, and amount of earth cover.

Dosing Chambers. In the previous discussion of septic tank design, information and basic design data were provided for sizing dosing chambers. There is no rule that the dosing chamber is directly attached to the septic tank as we have depicted in Fig. 8.2. If the septic tank and the dosing chamber are supplied by a prefabricated

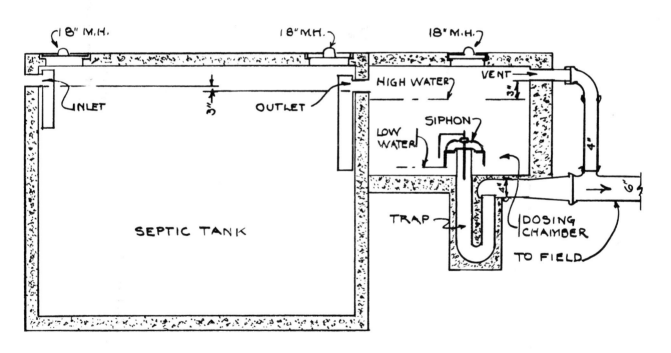

SEPTIC TANK WITH DOSING CHAMBER

NO SCALE

FIGURE 8-2

tank manufacturer, usually the two tanks will be supplied as two separate items and a short piece of piping will be required to connect the effluent of the septic tank to the inlet of the dosing chamber.

As depicted in the detail, the dosing apparatus consists of a trap seal that is similar to a water closet seal. The upper, patented device allows the effluent to the dosing chamber to slowly fill the chamber until, at a fixed height noted as the liquid upper level or slightly below it, the pressure of the column of liquid breaks the seal and flushes the trap which discharges the liquid to the leaching field or drainage area. Manufacturers of the patented bell and/or of the prefabricated tank and bell system can provide detailed design data. If standard 4-in drainage tile is used in the leaching field, the capacity of the tile is 0.58 gal/lineal foot ($0.33 \times 0.33 \times 3.14/4 \times 7.5$). A field containing 600 lineal feet of 4-in leaching tile holds 350 gal (600×0.58). Since the rule previously stated was that the dosing chamber must be capable of filling the field at least half full, the cham-

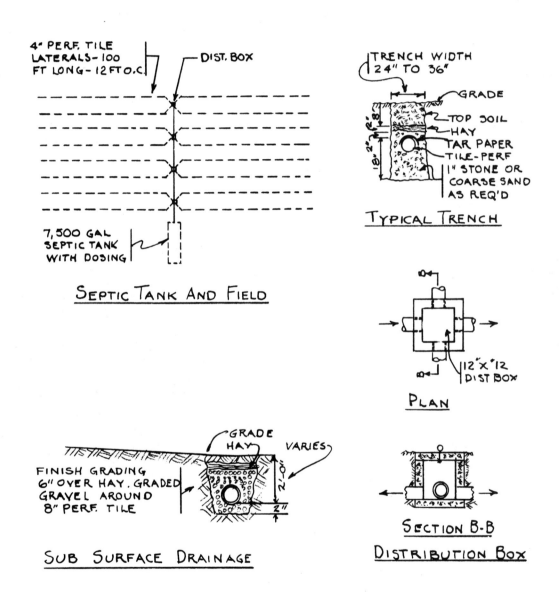

SEWAGE DISPOSAL SYSTEM

NO SCALE

FIGURE 8-3

ber volume would be 175 gal (0.50 × 350). At 7.5 gal/cubic foot (cu ft) the volume is 23.3 cu ft (175/7.5).

Figure 8.2 gives no size values of the dosing chamber. Presuming the septic tank was the same as the one shown in Fig. 8.1, the width of the chamber would be 7 ft, and as the detail somewhat implies, the length would be 4 ft. Thus, with a fairly common 12 in, or 1 ft, of liquid drawdown, the dosing chamber in Fig. 8.2 equals 28 cu ft (7 × 4 × 1) which is 210 gal (28 × 7.5).

When large volumes of effluent are applied at timed intervals, the liquid distribution is fairly equal over the entire field and the absorption bed has a chance to dry out between dosings. This type of effluent application frequently adds to the life of the field.

The construction requirements for the concrete tank are the same as for the septic tank. The detail depicts a 4-in to 6-in pipe commonly called an increaser which is presumed to be needed to handle the discharge volume. The trap is vented into the air space of the dosing chamber. In the roof of the dosing chamber an 18-in manhole is centered above the siphon.

Proper design of the siphon-operated dosing system does involve some hydraulic problems and head losses. In general, the premise of the dosing design is to allow for 3 to 4 hr between dosing. This may result in the need for more than one dosing apparatus. Two siphons may also be installed in one chamber. In this type of installation the siphons will dose alternately and each siphon should be piped to one-half of the disposal field.

Leaching Fields

Using the design data we previously presented, the length of the tile in the leaching field can be readily calculated. Frequently one of the major limitations of the leaching field is the amount of reasonably level surface available for the field. It is very important to keep firmly in mind that the basic premise of any field design is to distribute the liquid from the septic tank evenly over the entire area. Thus the ideal leaching field is or should be nearly level. If there is a slope to the field, all of the leaching lines should flow in the same direction as the slope of the land and not at right angles to this slope.

Figure 8.3 is a composite detail depicting the various elements of a typical leaching field. The details are not related to each other. The field as depicted is on 12-ft 0-in centers. One of the advantages of this wide spacing of trenches is that, assuming the soil tests were satisfactory, additional rows of tile could be installed between the 12-ft 0-in rows on 6-ft 0-in centers. Since sewage disposal space tends to get used for other purposes, this wide spacing tends to "lock in" the space and, if the load increases, provides land for the expansion of space loading.

On the right side of the figure are typical details of leaching trenches and distribution boxes. The construction of the trench as depicted is common for a soil that is not tight and has no unusual problems. This is the type of detail that should always appear in the septic field drawings. While it can be described in words, the detail eliminates any confusion. Below the detail on the trench are two details that cover the distribution box. They are related to the distribution box shown on the typical septic tank and field drawing. The detail does not show invert elevations, which should be part of the drawings and are based on starting inverts at the septic tank. The boxes are "flow through" boxes with side outlets which connect to double wyes as shown on the field drawing.

Finally we added on the detail a typical subsurface drainage line. If the field is surrounded by land at higher elevation, the tendency is for both surface and subsurface water to flow over and through the septic field. If this is the case, the flow must be directed around and away from the disposal field by surface trenches and subsurface drainage lines. The subsurface drain, as shown, is sometimes called a "curtain" drain. Not shown on the detail is the final discharge of this drain. Commonly the drain system is arranged to trap the water on the high side of the field and to pipe this water away from the field to a termination point which may be a lower surface area, a pond, or some exterior drainage collection area.

Special Situations. In leaching field design as in other areas of design there are, on occasion, marginal situations. Generally these are not situations that involve privately owned buildings. In almost all cases they involve publicly owned buildings, and, by far, the most common of these situations is a public school in a district in which there is not a municipal sewer system and no possibility of a municipal sewage system in the future.

The two details depicted in Fig. 8.4 are a solution to the problem that has on occasion been accepted by local health authorities. The upper detail depicts a 3-ft 0-in-wide trench with a 30-in bed of sand under the leaching pipe. The soil is presumed to be reasonably impervious. No soil is truly impervious. However, this detail presumes that the absorption rate is so slow that by soil test standards it is relatively impervious. Ultimately the leachate will fill the sand and slowly trickle through the gravel and be carried off to some acceptable location.

The lower half of Fig. 8.4 is a solution that uses much the same premise. The entire field is a subsurface sand filter with sand to the depth of 36 in covering the entire field. This sort of field is best served by a siphon, and generally the allowable dosage rate is 1 gal/sq ft/day. A thousand-pupil school based on 15 gal per pupil per day would require a field of 15,000 sq ft (1000 × 5), which is slightly more than one-third of an acre.

As can be seen in both of these details, the sewage effluent has to drain off somewhere. Somewhere is usually a stream or a pond that has been carefully selected and approved by the local health authority. The solution is definitely not ideal and cannot be compared to an open sand filter. While the effluent that has passed through

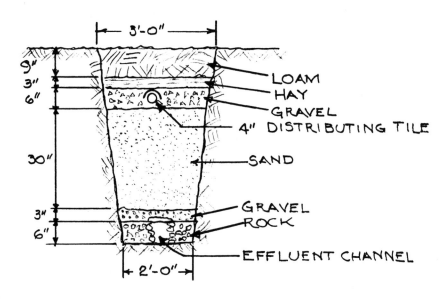

LEACHING TRENCH FOR TIGHT SOILS

——— NOT TO SCALE ———

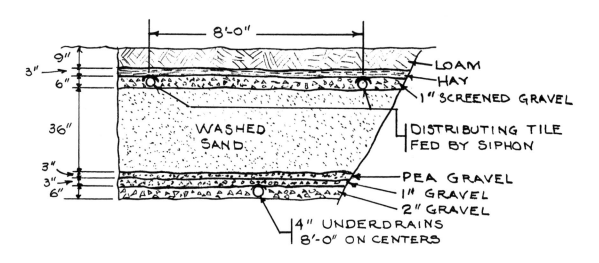

SUB SURFACE SAND FILTER

——— NOT TO SCALE ———

FIGURE 8-4

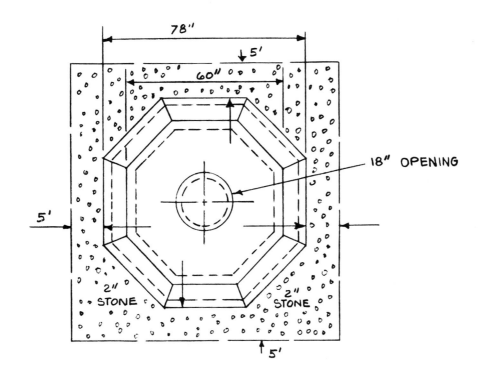

18" OPENING

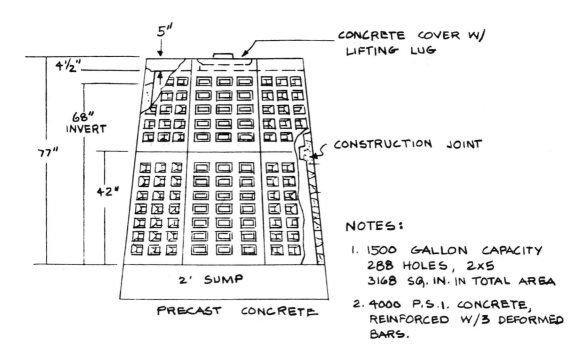

CONCRETE COVER W/ LIFTING LUG

CONSTRUCTION JOINT

2' SUMP

PRECAST CONCRETE

NOTES:

1. 1500 GALLON CAPACITY
   288 HOLES, 2×5
   3168 SQ. IN. IN TOTAL AREA

2. 4000 P.S.I. CONCRETE,
   REINFORCED W/3 DEFORMED
   BARS.

TYPICAL LEACHING GALLEY

NO SCALE

FIGURE 8-5

the sand is usually clear, it still is putrescible and as such has an odor. Over time the subsurface sand filtered effluent may show the presence of iron and the precipitation of iron may result in the discoloration of the receiving stream or pond. In any situation, except the above noted public school site problem, it is unlikely that this solution would be acceptable.

Leaching Cesspools. Generally leaching cesspools are not an acceptable means of sewage effluent disposal. These are also commonly described as seepage pits. Most codes restrict the use of seepage pits unless some special condition justifies such a solution. The designer of a proposed seepage pit should always meet with the local health authority and be prepared to explain in complete detail why the seepage pit solution is being proposed before proceeding very far with this design.

All seepage pits, or as our detail terms them "leaching galleries," are lined with stone, brick, prefabricated perforated walls, or cement block. Figure 8.5 delineates a precast concrete wall. There are various rules and regulations covering the installation of these pits or leaching galleys. Our detail merely shows an 18-in opening in the top. Sometimes, if the depth of burial is 12 in from the top to the surface, a removal cover with no access manhole may be allowed. More commonly the access to the removable cover must be via a masonry manhole extended to grade with a cast-iron frame and cover for access to the pit cover. Again, the number of pits allowed is subject to local rules. Commonly a separation rule that is effective is one that requires the distance between side walls of adjacent pits to be at least 20 ft and the distance from any well to be 200 ft.

Usually the application of leaching pits is limited to pits for residential systems or very small buildings containing 15 or fewer occupants. While data on sizing pits vary, a pit such as the one depicted in Fig. 8.5 would be sufficient for five people in a residence if the soil is an average condition of sand or light, sandy loam. For fifteen people three pits, as depicted in Fig. 8.5, would be required although some codes would permit two $8 \times 8 \times 8$ pits.

Summary

A review of the details and data presented in this chapter should readily convince the plumbing designer that sew-age disposal, on site, using a septic tank and field design is well within his or her ability to master. The actual problem in most cases is the failed system created by shipshod testing and design.

Testing is the essence of the design solution. There are many areas of the nation in which by one means or another the testing results may be forced into a conclusion that simply will not stand the test of time. While this problem is not prevalent, the more insidious problem is the marginal design coupled with the marginal installation. There have been many tests of actual sewage flow in a given situation, after the system has been installed and operating, which clearly demonstrate that actual flows are far less than calculated design flows. Invariably these sorts of results affect the more experienced designer much more than they affect the designer with a limited amount of completed and operating systems. The net result, to the experienced designer, is the tendency to not be a stickler for installation details. Usually this results in leaching trenches with a little less area, perhaps in some cases almost zero line pitch, or some of the fill materials being not quite what was specified and detailed.

The above paragraph should be taken as a serious warning to all new and experienced septic disposal system designers. If the test is to be done, it should be done properly and, if anything, the results viewed conservatively. The small amount of effort required to properly supervise and insist on proper quantities and qualities of materials is minor both to the designer and to the installer. The difference very frequently is a septic tank and field system with problems 5 years (yr) after installation versus one that 15 yr later is still trouble free.

Simply applying the data requirements as listed in this chapter and designing accordingly will produce a satisfactory result in all but the most unusual case. The one clear problem has always been and still remains. That problem is proper estimation of sewage flow. In general the data as given are conservative. Areas that should always be carefully investigated are those such as sports centers and restaurants. These invariably have wide flow variations and a successful design in one case may prove to be a problem in another. Every available means of verifying actual water use should be explored.

NINE

Fire Protection

The essence of fire protection is the preservation of human life in the case of an outbreak of fire regardless of the fire's cause or source. Two of the elements of this or any fire protection system are part of electrical, not plumbing, design and are not covered in this chapter. These two elements are the fire alarm system and emergency lighting. While many people equate emergency lighting and emergency power with the resolution of the problem of power failure, they are overlooking the fact that frequently fires and power failures are related, sometimes directly related. Many fires have their origin in a failure or short circuit in the electrical system.

The plumbing designer is frequently involved in all types of sprinkler and standpipe fire protection systems within the building. Logically he or she can also be involved in an exterior protection system provided by a system of strategically located fire hydrants. As in other areas of plumbing design, we approach a gray area of water supply for the fire protection system when the water supply is not available from a public or private utility source. In this instance we do not believe that the design of large storage tanks and large-volume deep well pumps associated with large fire protection systems is logically an area that should be under the design con-, trol of the plumbing designer. This type of water source, its solution, and design are quite properly in the realm of a professionally trained civil-sanitary engineer; therefore, this chapter and this book will not cover this area of the fire protection water supply system.

Every new building requires some specific type of fire protection. Fire alarms and emergency lighting afford a measure of protection to the building occupants but not to the building itself. The building fire protection system may consist of sprinklers, fire standpipe and hose systems, or other systems as required by the insurance underwriters, local fire departments, or the building's owners. Of all of the available systems the most common is the fire standpipe system.

In all systems in which water in some form is the source of fire protection, a basic design concept is that the water in any such system must be available both in sufficient quantity and in sufficient pressure to adequately service the intended fire suppression source. It is seemingly self-evident that no fire protection system can be termed safe and adequate when the source of water supply is questionable. Thus, an absolutely mandatory design requirement is an adequate volume and pressure of water at the source of the fire system supply.

For all of the areas of design the final source of information is the National Fire Protection Association (NFPA). While there are local rating bureaus that frequently must give their stamp of approval before any design is considered acceptable, it is the NFPA that is the sole final arbitrator of what is and is not an acceptable fire protection design solution. All of the data given in this chapter are based on data taken directly from the various NFPA publications. The address of the NFPA is Batterymarch Park, Quincy, Massachusetts 02269. The NFPA is also the publisher of the *National Electric Code.*

Standpipe and Hose Systems

A standpipe and hose system is a method of providing a reasonable degree of protection for a building and its occupants in the event of a fire. The basic elements of the system consist of strategically located pipe risers, valves, hose outlets, and hoses arranged in such a fashion that water can be discharged in streams or spray patterns through the hoses to extinguish a fire and protect the occupants. The supply for this system depends on available water supply and pressure and may require pumps or storage tanks to meet the supply requirements.

While not a substitute for a sprinkler system, the standpipe system may be a necessary complement since it can furnish a reliable means of supplying effective water supply streams to a fire in the shortest possible time. This feature of the standpipe and hose system is especially valuable in the upper floors of high-rise buildings, in a large floor area in low-rise buildings, and in structures in which the construction makes fighting the fire difficult to solve with exterior hose steam application.

The NFPA Standard 14 in Volume 2 of the NFPA code books contains specific requirements for standpipe and hose systems. The essence of these requirements is a required pressure of 65 pounds per square inch gravity (psig) at the topmost outlet, a minimum flow of 500 gallons per minute (gpm) for the first standpipe, and a limit of 275 feet (ft) above grade for the highest hose outlet. A further restriction is that pressure at any hose outlet in excess of 100 psig must be reduced by an approved device which cannot be adjusted.

In Figs. 9.1 to 9.3 we present typical details of standpipe systems for three fairly typical sizes of buildings. These are diagrammatic illustrations and not every valve and specialty item is shown since, at the small scale of this book, the details would lose their clarity. In order of presentation Fig. 9.1 is a fairly low-rise four-story building whose uppermost hose cabinet is 40 ft above the incoming street line level. Figure 9.2 is a 13-story building (fairly common to many communities). The penthouse hose cabinet is approximately 150 ft above incoming street line level. Figure 9.3 is a 40-story building that represents the common tallest structure in all but a few very large cities in the United States.

Height may well be a problem even in the four-story building application. The NFPA defines three classes of service. Classes 1 and 3 require 500 gpm for 30 minutes (min). Class 2 is a more limited service requirement of 100 gpm for local fire assistance. The residual pressure requirement of this class is 65 psig with the flow. There are many locations at which the street pressure is below 65 psig. Since the lowest structure, Fig. 9.1, is 40 ft to the highest outlet, the static head equals 17.5 psig. Thus the street pressure, rounded, would have to be 83 psig. Recognizing this fact, the NFPA code does allow a lower pressure provided the standpipe riser is sized as per their requirements, which are stated in NFPA 14.

As the building increases in height, the pressure in the standpipe necessary to reach the hose cabinet increases. Thus when the pressure is above 100 psig, either a nonadjustable pressure-reducing valve (PRV) or an orifice plate is required. The PRV device is noted in our diagrammatic details. We have shown them where the pressure is in excess of 50 psig. Technically in Fig. 9.1 the PRV could have been shown on the hose cabinets of the first two floors.

Figure 9.1 has an *uninterrupted* line from the street that goes to the hose cabinet via the standpipe riser. If this line, as shown, is the only building supply, the domestic service and its meter would be tapped off this line. The standpipe line flow is not restricted by passing through a meter. Figure 9.2 is the typical medium-sized building that is served by a house fire tank. The volume of this tank to match class 1 or class 3 requirements would be 15,000 gallons (gal) (500 × 30) plus a reserve of 10 to 15 percent or 1,500 to 2,250 gal. For the local fire assistance requirement of class 2 the tank would be 2,000 gal (100 × 30) plus 15 percent spare, or 3,450 gal. The difference in elevation from the tank to the first hose cabinet is some 25 ft, and the pressure is only some 11 psig. This is one of the problems of the roof tank installation. The tank is filled by a lower rated discharge pump with the proper discharge head capacity.

Typical of the system depicted in Fig. 9.2 is the dual use of the storage tank as a supply of both domestic and fire service water requirements. In this application, as noted on the detail, the tank must be large enough to always contain a minimum of 3,500 gal for fire assistance purposes as required by a class 2 application. Note that this tank is also the source for the siamese hydrant connection at the street level.

Here again there is a pressure problem. As in all roof tank systems, the pressure of the water column is from the top down. Here the detail denotes the use of a pressure-reducing valve when the head is some 60 psig (135/2.3). While the NFPA requirement dictates the use of a pressure-limiting device at 100 psig, there can be problems in using a hose at 100 psig, especially by an unskilled user in panic situations. The 60 psig value is simply a more practical, easier to handle value.

Pressure-Reducing Devices. Pressure-reducing devices, as noted in Fig. 9.2 and as will be noted in upcoming Fig. 9.3, can be pressure-reducing valves with fixed nonadjustable springs as indicated, or they can be properly calculated and sized orifice plates at the nozzle inlet. The orifice plates, when properly sized and installed, work equally well and, being contained in the piping, are less subject both to failure and to unauthorized "adjustments."

In American units the formula for calculating orifice plates is

$$A = 0.0425 \, Q \sqrt{P_1 - P_2}$$

where A = orifice area, sq in
$\quad Q$ = flow, gpm
$\quad P_1$ = standpipe pressure at hose inlet, psig
$\quad P_2$ = hose nozzle outlet pressure, psig

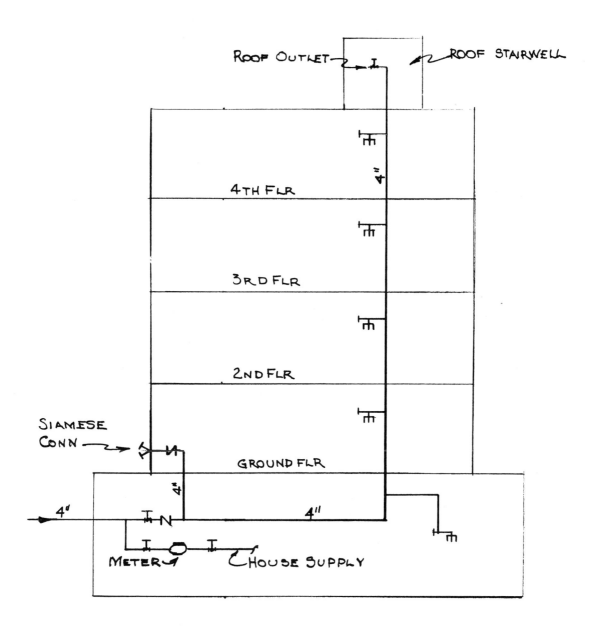

STANDPIPE SYSTEM
LOW RISE BUILDING
— NO SCALE —

FIGURE 9-1

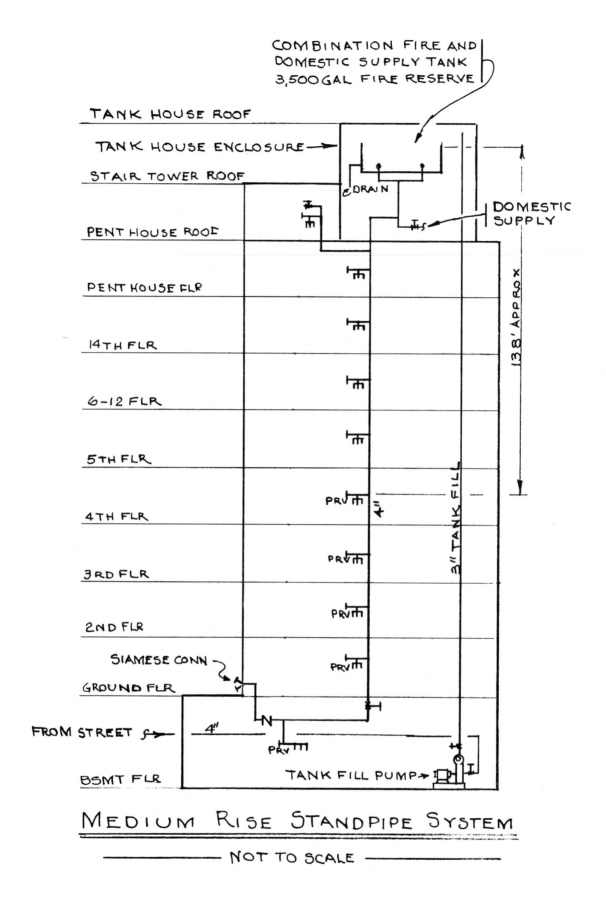

COMBINATION FIRE AND
DOMESTIC SUPPLY TANK
3,500 GAL FIRE RESERVE

TANK HOUSE ROOF

TANK HOUSE ENCLOSURE →

STAIR TOWER ROOF

DRAIN

DOMESTIC
SUPPLY

PENT HOUSE ROOF

PENT HOUSE FLR

14TH FLR

6-12 FLR

5TH FLR

4TH FLR

PRV

4"

3RD FLR

PRV

2ND FLR

PRV

SIAMESE CONN

PRV

GROUND FLR

FROM STREET

4"

PRV

BSMT FLR

TANK FILL PUMP →

138' APPROX

3" TANK FILL

MEDIUM RISE STANDPIPE SYSTEM

— NOT TO SCALE —

FIGURE 9-2

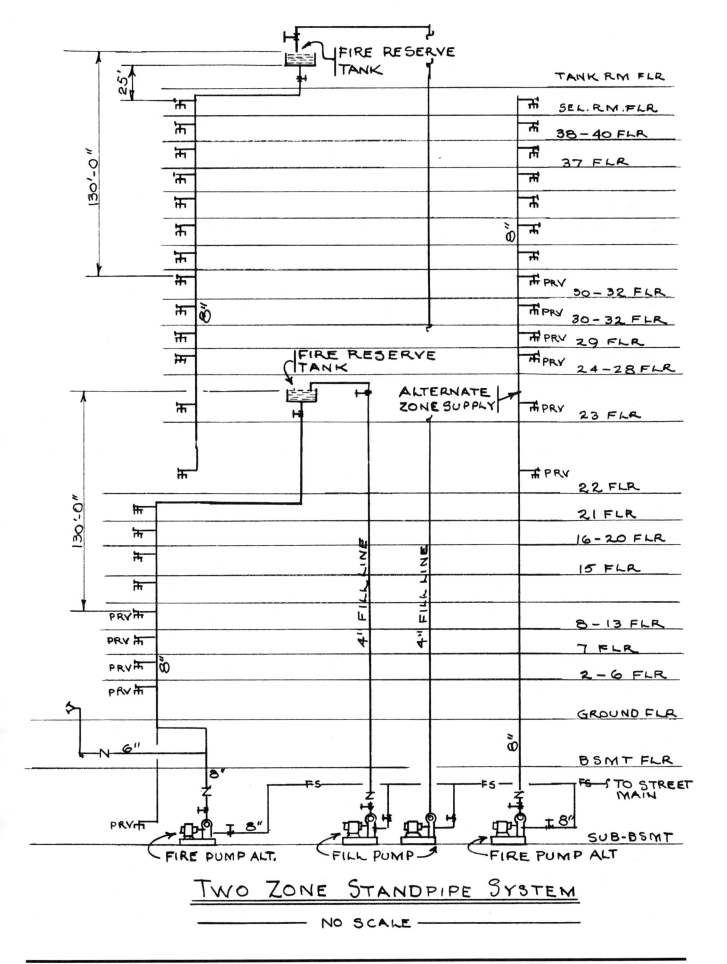

Two Zone Standpipe System

—— No Scale ——

FIGURE 9-3

In standard international units the formula for calculating orifice plates is

$$A = 1.902 \, Q \, \sqrt{P_1 - P_2}$$

where A = orifice area, square millimeters (sq mm)
Q = flow, liters per minute (lpm)
P_1 = standpipe pressure at hose inlet, bars
P_2 = hose nozzle outlet pressure, bars

High-Rise Structures

Figure 9.3 is a composite detail developed to portray more than one possible solution to a 40-story building. To avoid clutter and to keep the detail sufficiently large so as to be both usable and readable, a taller building was not depicted. If the building were 60, 80, or more stories high, the most likely solution would be to greatly enlarge or duplicate the tank installation on the fortieth floor. Fill pumps equal to the 20-story and 40-story fill pumps could draw from the 40-story supply tank and fill two additional fire supply tanks on the sixtieth and eightieth floors which would downfeed 20 stories in each case as is depicted for tanks shown at the 20-story and 40-story levels. The floor-to-floor height is presumed to be approximately 12 ft 0 inch (in). With a lower floor-to-floor value a few more stories (two to five) could be added to each section.

The scheme on the left side of the detail depicts the supply to the standpipe risers from properly located storage tanks. The capacity of the fire reserve tanks for a class 2 application is 3,500 gal. Again, not shown to avoid clutter is the domestic water supply. In situations such as the one depicted in Fig. 9.3 an economical, logical use of the fill pump would be for it to fill a separate domestic storage tank located next to and slightly higher than the fire tank. A controlled gravity flow arrangement would allow the domestic tank supply to be the source of fill to the fire tank.

Somewhat incongruous is the fire pump feeding up into the supply and noted as "alt." This pump is related to the pump on the right side of the detail that is also labeled "alt." What is meant by this designation is that in lieu of the tank system both the upper and lower sections could omit the use of fire tanks and use fire pumps that are properly designed for the pressures involved and located in the basement. As far as pumps in lieu of tanks are concerned this is as high as you can go. The pressure required to travel up 40 floors, or usually some 480 ft, is near the limit of pressure (250 psig) the pipe and fittings can withstand. Above this height the only practical solution is to utilize tanks and upper floor fill pumps. Note also that the siamese connection is off the lower level pump or tank.

As in Fig. 9.2, there is a requirement for a pressure-reducing valve or an orifice. In this detail, as in the preceding two details, the standpipe system is really not the source of water to put out the fire. It is a source, in general, to contain and control small fires before they become serious. In a real fire fighting situation the fire department may try to further pressurize the standpipe system and connect its hoses to special outlets in the hose cabinets. Fire fighting apparatus parked on the street is usually limited to fighting fires in buildings having 12 stories or less.

Hydrants and Hose Cabinets

Figure 9.4 depicts two common details of a fire protection system. The upper portion of the detail covers the typical fire hydrant detail. This detail is part of the exterior fire protection which is normally quite limited in scope and application. When the building is located in built-up commercial areas, the local municipality may already have in place all, or all except one or two, fire hydrants that will be required. In totally rural or large-site applications the exterior fire hydrant system should be designed by a group or individual with special expertise in exterior fire protection. This is an important requirement when the water supply system also requires on-site fire storage tank facilities. The exterior hydrant system in this instance is definitely not part of plumbing design nor is it an area to be covered by a book on plumbing design. The upper part of Fig. 9.4 is intended for use when the sprinkler and/or standpipe situation requires one or two hydrants tapped off the incoming fire line service. It definitely is not a unit to be supplied by a building field reserve tank.

In the lower part of Fig. 9.4 is a typical fire hose cabinet that is a standard part of the standpipe system. Not noted in the detail is the pressure-reducing valve or orifice plate which has previously been covered. These devices could be noted on the detail as a requirement. A separate connection in the hose cabinet is provided for fire department use. Commonly the fire department much prefers to use its own hoses and may elect to make other uses of this connection. The object is to provide the local department with the freedom to choose exactly how it will integrate this outlet in its overall fire fighting scheme. These cabinets are standard items that may simply be specified and no specific detail be provided. As in many other cases of plumbing design solutions, details, however simple in nature, do clarify the intent of the specifications.

Automatic Sprinkler Systems

The design of the automatic sprinkler system in today's building plans is, in one sense, much more stringent than that which was once considered acceptable. In the not too distant past the designer could simply show the correct number of heads and the sprinkler riser main location and let the sprinkler system contractor make the actual, required piping layout. Many old-time sprinkler contractors will still say that is all that is necessary and that frequently the engineer's design creates more con-

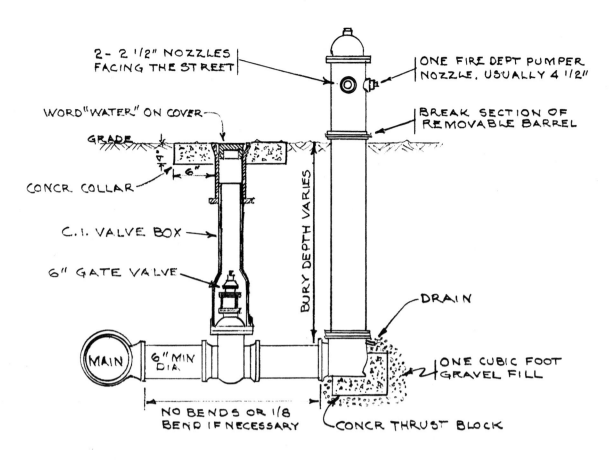

2 - 2 1/2" NOZZLES FACING THE STREET

ONE FIRE DEPT PUMPER NOZZLE. USUALLY 4 1/2"

WORD "WATER" ON COVER

GRADE

BREAK SECTION OF REMOVABLE BARREL

CONCR. COLLAR

C.I. VALVE BOX

6" GATE VALVE

BURY DEPTH VARIES

DRAIN

MAIN

6" MIN DIA.

ONE CUBIC FOOT GRAVEL FILL

NO BENDS OR 1/8 BEND IF NECESSARY

CONCR THRUST BLOCK

HYDRANT DETAIL
— NO SCALE —

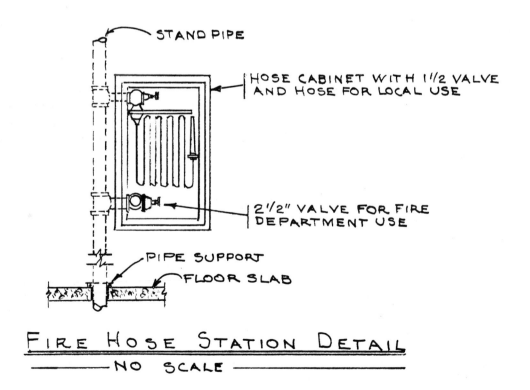

STAND PIPE

HOSE CABINET WITH 1 1/2 VALVE AND HOSE FOR LOCAL USE

2 1/2" VALVE FOR FIRE DEPARTMENT USE

PIPE SUPPORT

FLOOR SLAB

FIRE HOSE STATION DETAIL
— NO SCALE —

FIGURE 9-4

fusion and problems than it resolves. In one sense this argument still has real merit since the "authority having jurisdiction" will insist on a completely dimensioned plan covering every line and branch before approval is given. As the engineer's plan is seldom, if ever, dimensioned to the detail required by the authorities, the sprinkler contractor will always have to redesign the plan to meet this requirement.

The "authority having jurisdiction" is a very broad term when used by the National Fire Protection Association (NFPA). This authority can be federal, state, local, regional, or individual such as fire chief, fire marshall, chief of fire prevention, etc. In every installation the insurance aspect of an automatic fire protection system is very important. For insurance purposes an insurance inspection department, rating bureau, or other insurance representative may be the authority having jurisdiction. Also, in every situation, this authority must be determined and all design questions cleared through it. Many engineers and designers make the mistake of presuming that this authority is literally a free source of design information and the answer to all design questions. This is far from true. The authority commonly will advise that the authority's role is to interpret special code requirements and not to design the system.

A sprinkler system is any automatic fire protection system that is fed from a single source. When the number of heads exceeds this requirement, the operative word is systems, not system. The source may be a utility source, with or without pumps, tanks, and other items and includes piping above and below grade. The design may be in accordance with predetermined pipe sizes based on tables provided by NFPA or on hydraulic calculations made in accordance with NFPA permitted procedures.

The NFPA has five classifications of sprinkler systems. These are wet pipe systems, dry pipe systems, preaction systems, deluge systems, and combined dry pipe and preaction systems. While this portion of the book will cover applicable details and some limited application of design parameters, it will not repeat or cover in detail the information contained in the NFPA publication, *Automatic Sprinkler Systems Handbook*. A copy of this very valuable book should be in every designer's office and may be obtained by writing to the National Fire Protection Association, Batterymarch Park, Quincy, Massachusetts 02269.

Water Supplies. In Chap. 3 and in the beginning of this chapter the subject of water supplies for domestic and standpipe systems was discussed. The sprinkler system is simply another water user with certain basic requirements. These are spelled out in detail in the NFPA publications. The building sprinkler system commonly is adjudged to be one of light or ordinary hazard. For light hazard situations the NFPA requires 500-gpm to 750-gpm capability at the base of the riser for a period of 30 to 60 min at the residual pressure of 15 psig or higher. If this is to be supplied by a storage tank, the net usable capacity ranges from 15,000 to 45,000 gal (30 × 500 or 750 × 60). The values for ordinary hazard are 700 to 1500 gpm for 60 to 90 min, or 42,000 to 135,000 gal. This does not include a minimum allowance of 3,500 gal for the standpipe riser. Sprinkler system pressures are limited to 175 psig from any source.

No details are presented in this chapter for pumps and storage tanks for sprinkler systems since, except for pump and tank size, the general arrangement is as has been presented for standpipe systems. While there are rules in some areas that meters must be installed on any and all water-using systems, the NFPA very strenuously objects to these rules since both flow and pressure can be adversely affected by the passage of water through any type of meter.

In many industrial areas both the quantity and the pressure of water are low. Frequently the source of water is a well. And also this situation has requirements for exterior fire hydrants, interior sprinkler systems, and domestic water use. The situation may seem to the plumbing designer to be relatively uncomplicated. There are manufacturers of standard water supply tanks, the exterior piping to the hydrants is quite likely 8 in, and the fire and the domestic demand is reasonably easy to calculate. As in the previous chapter on storm damage, this is a gray area. However, it is not part of plumbing or engineering design. It is a civil-sanitary engineering task and should be designed by an experienced specialist in that area of water supply systems. The plumbing designer would be well advised to resist any attempt to extrapolate plumbing and interior piping expertise into water supply system engineering especially when it is related to fire protection systems.

Sprinkler System Sizing

The basic sprinkler pipe materials acceptable to the NFPA are steel and copper. While other materials may be acceptable to the NFPA, the circumstances surrounding their acceptance must be based on NFPA rules and conditions. The NFPA hydrostatic test requirements state that all new systems including yard piping must be tested at 200 psi for 2 hours (hr) or at 50 psi *above* the maximum maintained pressure when the yard pressure exceeds 150 psi.

There are two other requirements that are standard design elements. The first of these requires that no more than eight sprinklers be installed on a branch line. There are certain conditions under which nine or ten are permitted, but eight is the general rule. The area of sprinkler head coverage is the second item. For light hazard this is 15 ft both for distance between branch lines and resistance between heads. The area of coverage per head is allowed 200 sq ft maximum. For any project the first step is to determine the type of occupancy such as light

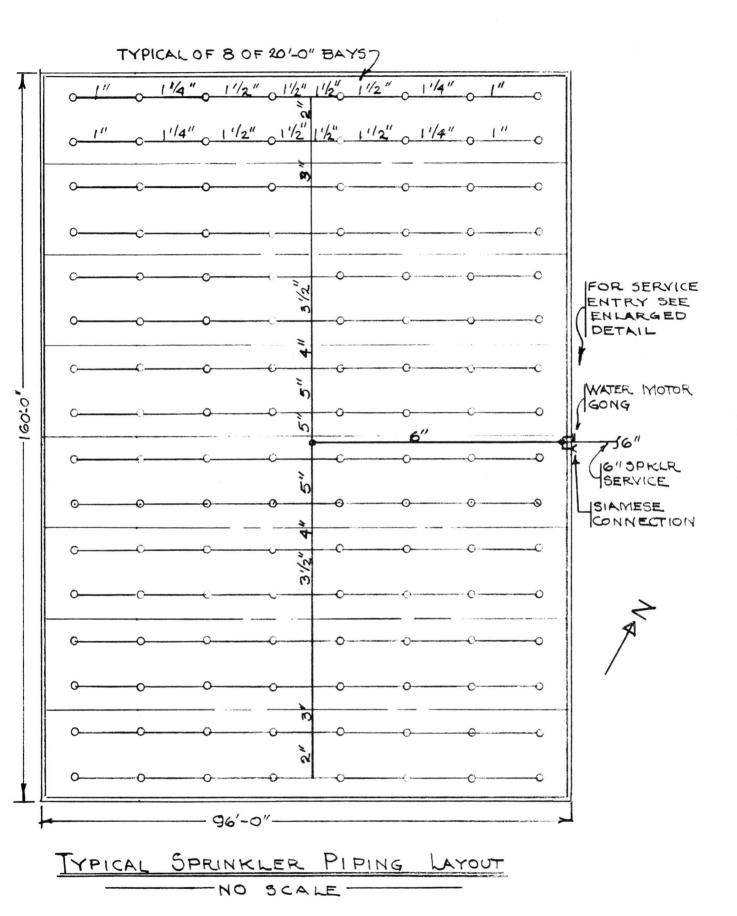

TYPICAL OF 8 OF 20'-0" BAYS

1" 1¼" 1½" 1½" 1½" 1½" 1¼" 1"

1" 1¼" 1½" 1½" 1½" 1½" 1¼" 1"

2"

3"

3½"

4"

5"

5"

5"

6"

6"

6" SPKLR SERVICE

FOR SERVICE ENTRY SEE ENLARGED DETAIL

WATER MOTOR GONG

SIAMESE CONNECTION

5"

4"

3½"

3"

2"

160'-0"

96'-0"

N

TYPICAL SPRINKLER PIPING LAYOUT
NO SCALE

FIGURE 9-5

hazard, ordinary hazard, etc. A further restriction on spacing is the type of ceiling such as a flat slab, open joist, etc. From a knowledge of the surface a layout of heads based on NFPA rules and regulations can be prepared and a piping layout designed.

Figure 9.5 depicts a typical plan for an ordinary hazard, structure bay application. The plan is not to scale. The coverage per head is 120 sq ft and in this case the code limitation is 120 sq ft. The sizing of the piping in each bay and for the main is based on piping tables for ordinary hazard systems.

Not all the information that is required on a sprinkler plan is shown on Fig. 9.5 because of the constraints of clarity required to fit this book. The actual plan, acceptable to the rating bureau, would be a totally dimensioned plan showing pipe lengths, elevation, and other data. This plan will be prepared by the installing sprinkler contractor. The designer's plan, while complete and accurate as to pipe size and number of heads and other items, is not the code authority's acceptable plan.

The required items on the designer's sprinkler plan are:

1. Name of owner and occupant
2. Location, including street address
3. North arrow
4. Construction and occupancy of building
5. Building height in feet
6. Street main size and pressure
7. Distance from nearest pumping station or reservoir
8. A water flow test of city main, if required
9. In small towns data on water works system, as required
10. Fire, walls, fire doors, unprotected window openings, large unprotected floor openings and blind spaces
11. Distance to and construction and occupancy of adjacent buildings
12. Spacing of sprinklers, number per floor, total number on each riser and each system per floor, type of system
13. Capacity of dry pipe system, if used
14. Whether property is in floor plan area
15. Name and address of party submitting the layout

The authority having jurisdiction may not want the designer's plans, but others involved in the building approval such as the fire department, the building department, and others commonly will. They require the plans prior to approval of the project, especially one using city, state, or federal funds.

The NFPA code has been slowly evolving since 1978. Properly prepared pipe calculations using the hydraulic method of computation are now acceptable. This is a specialized form of calculation which is not covered in this text. In 1983 the NFPA published the *Automatic*

Sprinkler Systems Handbook wherein the hydraulic method of pipe calculation was outlined in detail. Copies of this book may be obtained by writing to the National Fire Protection Association, Batterymarch Park, Quincy, Massachusetts, 02269.

There are many categories and special situations related to the number and location of sprinkler heads and the proper, legal definition of the space to be protected. What has been depicted in this book is an example of a typical system. The designer must have at least the NFPA 13, *Standard for the Installation of Sprinkler Systems*, in his or her possession to properly select a system that will conform to all requirements. In addition two other standards are also required. These are NFPA 13A, *Recommended Practice for Inspection, Testing and Maintenance of Sprinkler Systems* and NFPA 13D, *Standard for the Installation of Sprinkler Systems in One-and-Two-Family Dwellings and Mobile Homes.*

The area served by a single sprinkler system is a space on one floor which for ordinary or light hazard installation equals 52,000 sq ft or less. The number of stories in each building is not a factor. Buildings whose floor area does not exceed 52,000 sq ft but are 100 stories high are still one system since each floor area constitutes a separate fire area.

To achieve the layout as depicted the first step is to know the type of ceiling (open, flat slab, etc.). There are a number of fairly obvious conclusions which are further explained in the NFPA code. In Fig. 9.5 the design was restricted by beam spacing as well as the 130 sq ft per head rule. The 20-ft bay spacing dictated two rows of sprinklers on 10-ft 0-in centers for even distribution across the bay. The 96-ft 0-in width of the building is divided into spacing of heads on 12-ft 0-in centers. While the heads could have legally been on 13-ft 0-in centers, eight heads would still have been required since 7×13 is 91 not 96. And on 13-ft centers the coverage would not have been evenly divided.

Once the heads are located, the branch and main piping is laid out. The rule of eight heads on a branch line would have allowed the main to run along one wall and feed eight sprinklers. This would have required 2-in piping based on the NFPA tables for ordinary hazard pipe sizing to the first three heads and 1½-in pipe to the next three heads. Since there was no reason not to center the main as shown, this was the arrangement selected as being more economical and simpler to install. The sizing of the pipe was simply and easily accomplished by using the appropriate NFPA table of pipe sizes for heads connected.

Commonly the scale of the drawing, even ¼-scale drawings, is too small to depict all the items at the incoming sprinkler service. Usual practice is to do what is shown on Fig. 9.5. That is, to indicate the incoming service, the water motor gong, and the siamese connection and to refer the contractor to a separate, larger detail depicting these items.

Sprinkler Service. Figure 9.6 is a typical sprinkler service detail. As depicted, it is a composite detail that has more elements than are required by the standard wet pipe installation shown in Fig. 9.5. To relate Fig. 9.6 to the note on Fig. 9.5 simply change the 6 by 4 tee to straight pipe and delete the 4-in line and the dry pipe portion of the detail. With the deletion Fig. 9.6 relates directly to Figure 9.5.

The purpose of the composite detail is to provide a detail that fits almost all the common sprinkler service

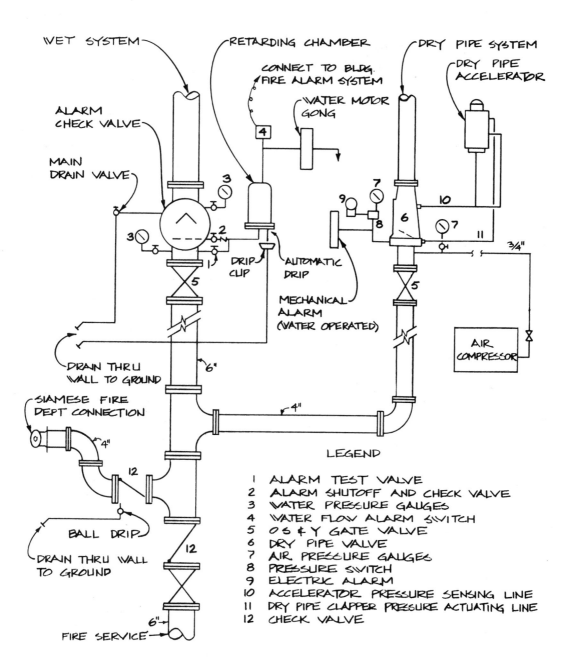

SPRINKLER SERVICE

NO SCALE

FIGURE 9-6

situations and can, by deleting the wet or dry portion, be used for either a wet or dry situation. In many sprinkler piping and system installations there frequently is an unheated area such as an attic or storage space that is a part of the design problem. Thus while a majority of the sprinkler system is suitable for a standard, wet pipe system, a portion (sometimes a substantial portion) requires a dry pipe system because of its location in an unheated area in which a water filled pipe would readily freeze and burst.

The detail depicts all of the items required for the wet pipe system including the alarm valve, the water-actuated motor alarm gong, drain piping, siamese connections, and the gate and check valve. All of these items are part of the wet pipe system. The dry pipe system is under pressure provided by an air compressor. If a head

in the dry pipe system opens in a fire and the pressure drops, the dry pipe valve opens to allow water to be delivered to the system to control the fire. The dry pipe system has an electric alarm which operates anytime the pressure falls, and it has a water operated alarm which only operates when water is flowing. Thus the system has a built-in alarm in the case of pressure failure.

The sizing of the piping for a dry system is the same as that for the wet system previously described, as is the location and spacing of heads. At a point farthest from the service the wet pipe system has a drain line on the end of the most remote branch line with a test valve and drain connection to the outside of the building or an acceptable drain receptacle. This connection must take the flow of one sprinkler receptacle and is called the inspector's test tee. Its purpose is to assure that

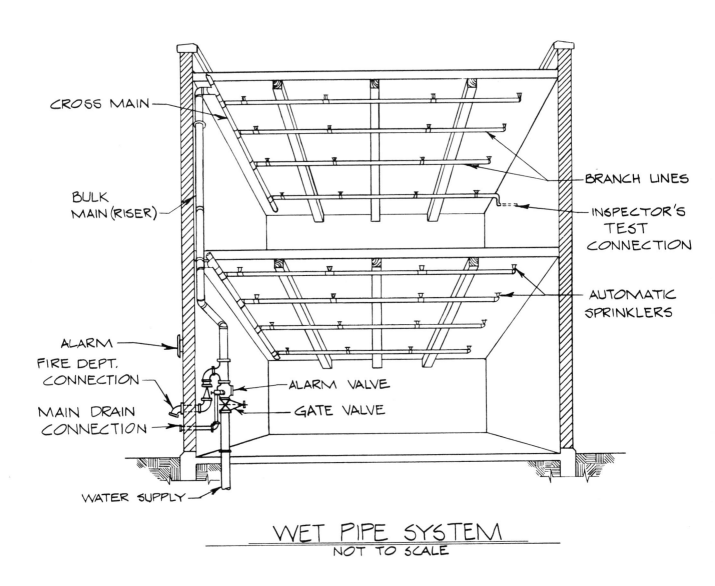

CROSS MAIN

BULK
MAIN (RISER)

ALARM
FIRE DEPT.
CONNECTION

MAIN DRAIN
CONNECTION

WATER SUPPLY

BRANCH LINES

INSPECTOR'S
TEST
CONNECTION

AUTOMATIC
SPRINKLERS

ALARM VALVE

GATE VALVE

WET PIPE SYSTEM
NOT TO SCALE

FIGURE 9-7

proper flow is available to the farthest point in the system. A similar arrangement is required in the dry pipe system. The purpose of the dry system test is to measure the approximate time required for water to flow from the control source to the most distant location in the dry pipe system. The dry pipe test valve must be sealed with a plug or a nipple when not in use to avoid leakage of air and to avoid accidental tripping of the dry pipe valve.

Wet Pipe System Overview. Figures 9.7 and 9.8 are composite three-dimensional drawings whose purpose is to assist the reader in grasping the essentials of Figs. 9.5 and 9.6, which were previously discussed, as well as to illustrate some of the points of typical sprinkler installation.

At the end of the discussion of Figure 9.6 the requirement of the inspector's test tee was covered. In Figs. 9.7 and 9.8 the location of the inspector's test tee is depicted. As can be seen, the test tee is located at the end of the most remote run on the *second* floor. A two-story pictorial representation was used to emphasize this point. If the building under design is a one-story structure, careful measurement should be made to ascertain which is the most remote point. Because of the symmetrical arrangement of piping in Fig. 9.5, either the northwest or northeast corner sprinkler could be selected.

The pipe sizing tables previously discussed are used to size all branch mains, cross mains, and bulk mains. There are two pictorials used instead of one because there are systems in which the total installation is, for

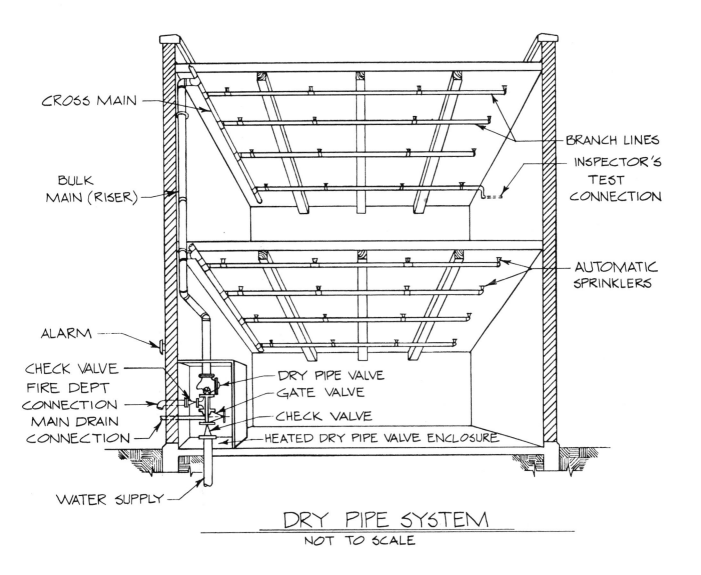

DRY PIPE SYSTEM
NOT TO SCALE

FIGURE 9-8

example, used to sprinkle an unheated warehouse. In this case, as is depicted in Fig. 9.8, water is present up to the dry pipe valve. Thus some form of heated enclosure is required as shown.

Finally, although it may sound redundant, the standard method of installing the sprinkler heads is on top of the main in open ceiling spaces as depicted in Figs. 9.7 and 9.8.

Dry Pipe Service. As pointed out, Fig. 9.6 is a composite wet and dry sprinkler service detail. There may be some chance for error in trying to use Fig. 9.6 for a dry pipe system. Figure 9.9 not only depicts and repeats the dry pipe detail, but it illustrates the remaining elements in a dry pipe sprinkler and fire protection system. Again, this figure is somewhat of a composite drawing in its depiction of exterior partitions of sprinkler and fire service piping. Previously we depicted an exterior fire hydrant detail. This detail somewhat elaborates, at least for location and piping, the detail used in Fig. 9.4.

What is shown in Fig. 9.9 is, discussing the interior, building portion of the building system first, a complete dry pipe sprinkler system. As was previously noted in the discussion of Figs. 9.6 and 9.8, a dry pipe system is not totally "dry." There is water up to the dry pipe valve and to the exterior siamese connections. The dry pipe system has a water motor gong which is water flow actuated and an electric alarm. Since water exists up to the dry pipe valve, there is a drain, similar to a wet pipe system, of the dry pipe valve's "wet" side. Among the standard devices, again depicted in this detail, are the dry pipe accelerator, the accelerator or sensing line, the air pressure actuating line, and the air compressor.

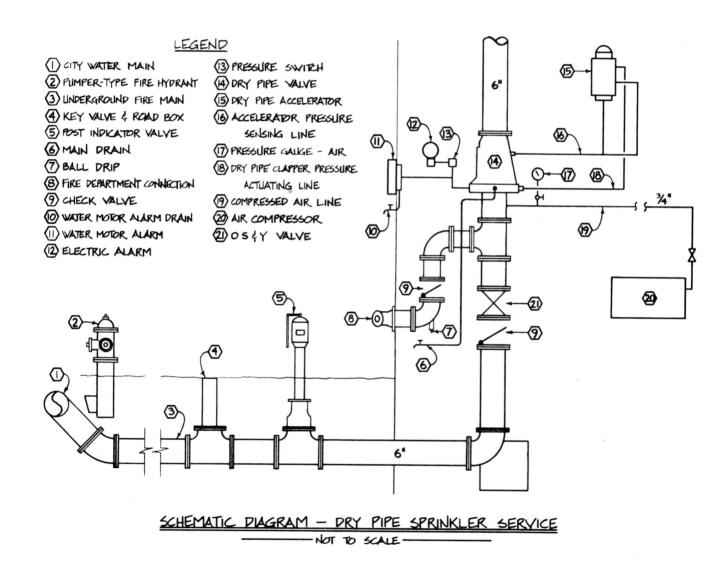

LEGEND

① CITY WATER MAIN
② PUMPER-TYPE FIRE HYDRANT
③ UNDERGROUND FIRE MAIN
④ KEY VALVE & ROAD BOX
⑤ POST INDICATOR VALVE
⑥ MAIN DRAIN
⑦ BALL DRIP
⑧ FIRE DEPARTMENT CONNECTION
⑨ CHECK VALVE
⑩ WATER MOTOR ALARM DRAIN
⑪ WATER MOTOR ALARM
⑫ ELECTRIC ALARM
⑬ PRESSURE SWITCH
⑭ DRY PIPE VALVE
⑮ DRY PIPE ACCELERATOR
⑯ ACCELERATOR PRESSURE SENSING LINE
⑰ PRESSURE GAUGE - AIR
⑱ DRY PIPE CLAPPER PRESSURE ACTUATING LINE
⑲ COMPRESSED AIR LINE
⑳ AIR COMPRESSOR
㉑ OS&Y VALVE

SCHEMATIC DIAGRAM — DRY PIPE SPRINKLER SERVICE
———— NOT TO SCALE ————

FIGURE 9-9

The additional items, exterior to the building, are the post-indicating valve, the key valve and road box, and the hydrant or hydrants, if more than one is required. Of the three items the key valve and road box are commonly not used if avoidance is at all possible.

The post-indicating valve (PIV) is literally a valve that is precisely what its name implies. In the NFPA code the basic requirement is that all valves to sprinkler systems shall be approved and listed indicating valves, unless a nonindicating valve, such as an underground gate valve with an approved roadway box, complete with T-wrench, is accepted by the authority having jurisdiction. The post-indicating valve has a handle, as indicated on the detail, with a movable sign on the face that indicates whether the valve is open or closed. When the valve absolutely must be installed in a roadway situation and when nothing can protrude above the surface, the au-

thority will accept this valve in lieu of the post-indicating valve. In general the presumption is that the contract plan will show the post-indicating valve and *not* the roadway valve.

In Fig. 9.9 the fire hydrant depicted illustrates that it is somewhere out near the curb line although the actual location, or locations, will be determined by the particular site requirements.

Special Systems. While the two most common sprinkler systems are the wet pipe and the dry pipe, there are certain special and high-hazard situations that require a different sort of treatment. These include highly flammable situations and very cold situations in which there is a danger of serious water damage resulting from damaged automatic sprinklers or broken piping.

Figure 9.10 depicts a deluge system which is very sim-

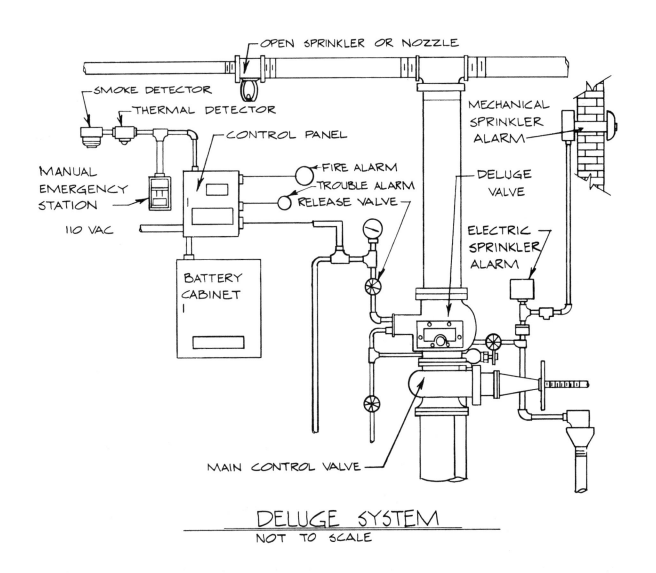

DELUGE SYSTEM
NOT TO SCALE

FIGURE 9-10

ilar to a preaction system. The primary difference between a deluge and a preaction system is that in the deluge system the heads are always open while in the preaction system the heads are closed. In the deluge system the supply of water to the open heads is controlled by a detector in the same area as the heads. When the detector actuates the valve, a fast total application of water covers the extra hazardous area. The deluge system is limited to 225 heads which may be installed upright or pendant as depicted. Commonly the system includes interconnecting fire emergency stations such as control panels, fire alarms, direct current battery backup,

and other safety precautions. If Fig. 9.10 were a preaction system, it would be very similar in appearance with the major difference being that the detector may sound an alarm but the heads will not release water until they are fused as in a normal wet or dry system. The preaction system operates faster and reduces fire and water damage when compared to a conventional dry pipe system. Preaction systems are limited to 1,000 heads. Pendant sprinklers must be for an approved dry type only if they are in an area subject to freezing.

Figure 9.11 is an on-off system which is also described as a fire cycle system. Though it uses heat detectors

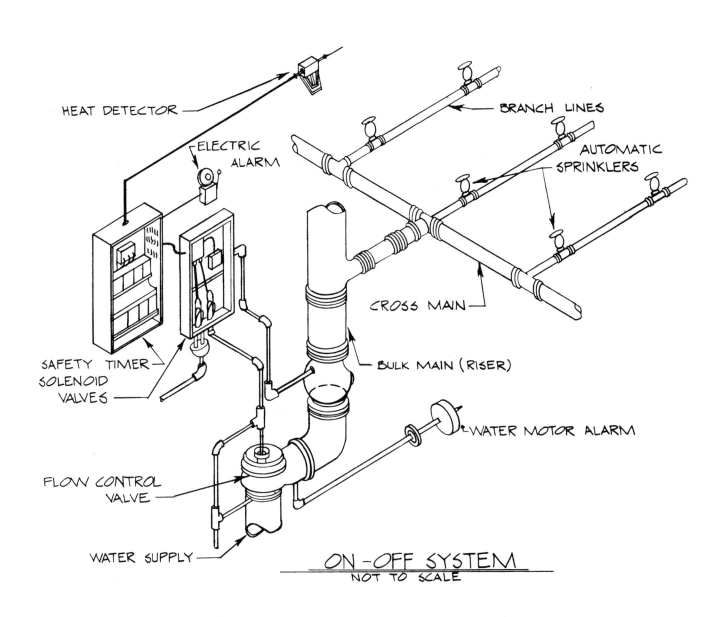

FIGURE 9-11

and controls, this system has the capability of continued on-and-off cycling while controlling the fire and of shutting off the water when the fire is extinguished. It is widely used when the minimizing of water damage is a prime requirement. Since flow is started by the detection system, accidental water damage is eliminated. In addition water turn-off is not required for maintenance or after a fire is out. As such the chance of water not being available because of accidental reasons is also eliminated.

In operation the flow control valve is supervised by water pressure. Solenoid valves control upper chamber pressure and the valves are in turn controlled by heat detectors whose signal passes through a control cabinet containing a safety timer. On a fire call from the heat detector the solenoid actuates the flow control valve and sends water to the sprinklers. At this point electric and water motor alarms sound, sprinkler heads fuse, and water is applied to the fire. When the temperature of the involved area drops below a fixed value, the detectors signal the safety timer which, after a preset period, closes the flow control valve. Water is shut off. Should a fire recur, the system cycle as described is repeated.

TEN

Special Items

In a general sense all hydraulically involved systems are related to plumbing design. However, hydraulics is not part of plumbing. On the contrary plumbing is part of the overall flow of fluids design called hydraulics. Those who are involved in plumbing design know that this area of design is more than simply supplying water, waste, and vent piping to plumbing fixtures. With certain limits, as we have noted in past chapters, plumbing design commonly includes exterior drainage and water piping. In addition, as a special item, sprinkler and fire protection systems within the building are part of the overall plumbing design.

In view of the expanding limits of plumbing design the question the author of this book and the plumbing designer both have to face is what are the limits of plumbing design. The special systems covered in this chapter together with the fire protection and exterior work previously mentioned do, in the author's judgment and experience, constitute work that commonly falls to the plumbing designer. The special systems covered in this book could be expanded to many other types of fluid piping including oil, gasoline, waste water, and the like. Almost all of these are areas of very special expertise. Within reasonable limits the following systems are considered part of plumbing design and are covered in this chapter:

Swimming pool filter systems

Waste interceptors

Gas systems

Compressed air systems

Vacuum systems

Fire extinguishing systems

Water treatment systems

The approach taken in handling the discussion and detail presentation of the above systems is that they occur as an adjunct to the plumbing system. For example, compressed air and vacuum systems, water treatment systems, and gas systems have each been the subject of more than one book that is totally dedicated to the particular subject. This book does not delve into that degree of engineering and related design data. While none of the subjects listed above are treated superficially, they are treated in a limited sense. The premise of this chapter is to supply the necessary information to resolve the design problem contained in a building such as a hospital, industrial plant, laboratory building, hotel kitchen, and the like when, in addition to the basic plumbing and sprinkler system design, you are confronted with one or more areas that require the design of one of the seven special listed systems.

Swimming Pool Filter System

While there are exceptions, the most common fact of swimming pool filter design is that the pool design and its gutter collection system are part of the architectural planning and design effort. In addition there are state and federal regulations, most commonly state Board of Health regulations, that govern the overall required filtration result. Thus the plumbing designer who becomes the pool filter system designer must provide a solution that satisfies the architect and the state Board of Health and also the designer's own design engineering judgment.

In essence a pool is drained back to the filter system by main pool drains (one or more) in the deep water end of the pool and by the gutter drains. The drainage rate, called pool turnover, is the first item of concern to the health officer. All of the water in the pool must pass through the filters at least once in 6 to 8 hours (hr). This requirement at your site must be ascertained by checking local health regulations.

Taking a typical indoor publicly used pool as an example, we can assume a total, probable pool capacity of 175,000 gallons (gal). (Even when you are given a specific value of capacity, it is your responsibility as designer to check the pool plans and verify the capacity of the pool). We can further assume a 6-hr turnover requirement. Thus the pool filter circulating pump capacity is 486 gallons per minute (gpm) [175,000 gallons/6 hr × 60 min/hr)]. This is a fairly common value.

In Chap. 4 there is a chart for fairly smooth pipe. To handle a flow of 486 gpm the typical pump would have at least a 5-inch (in) suction and 4-in discharge. Using Table 4.8 in Chap. 4, and not only wanting to relate pipe size to pump size but also to keep the piping flow below 10 feet/second (fps) velocity, we can readily find that a 6-in line at a 1 pound (lb)/100-ft drop will handle 500 gpm. It may be noted that a 5-in line could handle the flow at a 3 lb/100-ft drop and a fluid velocity of 8 fps. The 6-in line velocity value is 6 fps and is in the preferred velocity range.

Filter manufacturers rate their filters on a gallons-

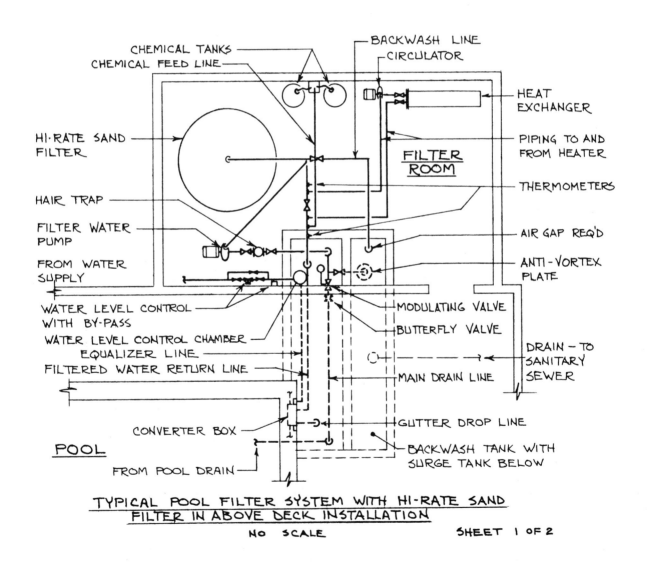

TYPICAL POOL FILTER SYSTEM WITH HI-RATE SAND
FILTER IN ABOVE DECK INSTALLATION

NO SCALE   SHEET 1 OF 2

FIGURE 10-1

per-minute-per-square-foot-of-filter-capacity basis. Presuming a fairly common value of 4 gpm/square foot (sq ft), the required filter area is your pump capacity, 500 gpm, divided by the filter capacity, 4 gpm/sq ft. Thus the required pool filter area is 125 sq ft (500/4).

With the above data the pump and filters can be selected using available manufacturers' catalogs. The pump head capacity must be sufficient to overcome all piping and filter head losses. The basic design of the pool filter piping system is a straightforward calculation and the distribution lines from the pump to the filter to the pool and back again is not a difficult task.

Even the selection of the filters themselves is not overly difficult. Frequently the owner or architect has very definite ideas about the type of filter material and either a sand or diatomaceous earth unit is preselected for the plumbing designer. The real design effort is correctly arranging all the components so that they can be installed, operated, and serviced within the space allocated in the pool filter room.

Filter Room Piping. Figure 10.1 represents a plan view and Fig. 10.2 represents an elevation view of a filter system taken from an actual project in which the pool capacity was 183,000 gal. A corner of the pool area can be seen in Fig. 10.1. The problem was to find space for the backwash and surge tank and the filters and related equipment in a constricted equipment space requirement. To resolve this problem the capacity of the backwash and surge tanks had to be calculated and a tentative layout of the filter equipment made.

In Fig. 10.1 the pool building line is partially represented by a second broken angle line that is above and to the right of the pool line. These lines are below deck level; building and pool lines and the space between the pool and the building are the pool deck area. The section depicted is a portion at the deep end of the pool.

The purpose of the surge tank is to take the overflow from the pool gutter. A pool is normally filled to within an inch or two of the lip of the gutter drain. The displacement of water is created by the bodies of swimmers. This causes some water to overflow and drain, through gutter drains and piping, back to the surge tank. The filter system pump circulates water via supply openings below the pool water level into the pool and returns water via a main pool drain or drains in the lowest point of the pool. As seen in Fig. 10.2, it also picks up gutter drain water as the return line passes through the surge. The surface of the pool in the deep end is sloped toward this drain or drains to facilitate the return drainage system.

The amount of water overflowing from the pool to the gutters and their drain piping cannot be calculated. This is an empirical value which is usually a stated value in the health department standards. A reasonable presumption is 0.5 to 1 gal/sq ft of pool water surface area. The quantity of the backwash water is calculated from the data provided by the filter manufacturer. If we used the hypothetical value of 500 gpm for the pump capacity and a filter manufacturer's backwash time of 3 min, the total gallons to be emptied into the backwash tank is 1,500 gal (3 × 500). Since the tank has a gravity drain, usually 6 in, very little of this water is drained in 3 min. Thus the backwash tank must be large enough to hold 1,500 gal.

In Fig. 10.1 it can be seen that part of the backwash and surge tanks extend into the filter room. This was done, in this instance, to provide some access to these tanks for cleaning and maintenance purposes since the majority of each tank is, as seen in the detail, permanently covered by the pool deck. Further, all of the equipment—filter, chemical treatment, pump, circulator, heat exchanger, and piping—was required to be arranged as seen in Figs. 10.1 and 10.2. The spaces and arrangement were located as shown after a number of meetings with the architectural designer.

As may be surmised, this was an unusual installation, and it is presented to illustrate not only the filter system but also the importance of knowing the size of each item, the clearances required for service and maintenance, and exactly what space and arrangement are required for the piping.

The peculiar arrangement of the filter room, its piping, and related backwash and surge tanks was required because of the presence of ledge rock on the site. The space for the pool itself required a large amount of very expensive blasting. Obviously no one wanted any more blasting than absolutely necessary. This resulted in a filter room that was outside the basic building and below grade. Normally a swimming pool excavation extends to the pool building walls, and the designer will have corridors under the deck on three sides and at the deep end will locate the diving board on a reinforced concrete slab that extends from the pool to the building wall. The filter equipment room will be on the floor below this deck and at the same level as the deep end of the pool. If, for example, the pool is 45 ft wide and has a 14-ft deck at the deep end and a 10-ft pool depth, the result would be a room 630 sq ft (45 × 14) with a clear ceiling height of probably 9 ft 0 in for pool filter equipment. In many pool details, including the ones that follow in this book, the impression is given that there is always ample room for filters, surge tanks, backwash tanks, and the like. This is commonly true. But, as shown in Figs. 10.1 and 10.2, which were specifically selected to illustrate the designer's problem, it is not necessarily true every time.

These two details are not details in the typical sense in that they do not show all of the piping and connections that are depicted in details that follow. However, the details are superior to other details both in this book and in other books in terms of depicting actual size relationships. There is no scale depicted, but all items shown have been proportionally reduced from scaled drawings. As a result, the size of the filter room relative to the filter and the backwash and surge tanks is in proper relationship. The two drawings are not schematic. The installation was completed as shown and has been satisfactorily operated, serviced, and maintained since the winter of 1980–81.

Filter Types. There are four types of filters that have been used in a number for swimming pool installations. These filters and their filtering rates, gallons per minute per square foot, are as follows: high rate sand at 15 to 20 gpm, rapid-sand at 3 to 4 gpm, pressure diatomite at 2 to 3 gpm, and vacuum diatomite at 1½ to 3 gpm. In each case the lowest of the two stated rates will produce the best overall results. Both 8-hr and 6-hr rates are used in pool designs. The 8-hr rate will provide 95 percent water purification and requires 3 days to reach equilibrium. The 6-hr rate will provide 98 percent water purification and requires 2 days to reach equilibrium.

This book does not take a stand in the never ending argument over sand and gravel filters versus diatomaceous earth filters. There are numerous studies and reports comparing initial costs, operating costs, and performance results for each system. Each system has its good and bad features. Each can be designed and installed to produce a satisfactory result. Selection should be made by the architect, engineer, and owner, based on the given requirements of the individual installation and the preferences of each of the parties to this decision. In the details that follow a high rate sand filter system, a conventional rapid-sand system, and a vacuum diatomite system will be presented.

Pool Supply and Drainage Systems. In Figs. 10.1 and 10.2 the emphasis was on the need, in some cases, to prepare plans that clearly define the spaces required for the various elements of the filter system. In Fig. 10.3 and those that follow the presumption behind each presentation is that there is ample space to install the filters and related systems and that the need is to clearly rep-

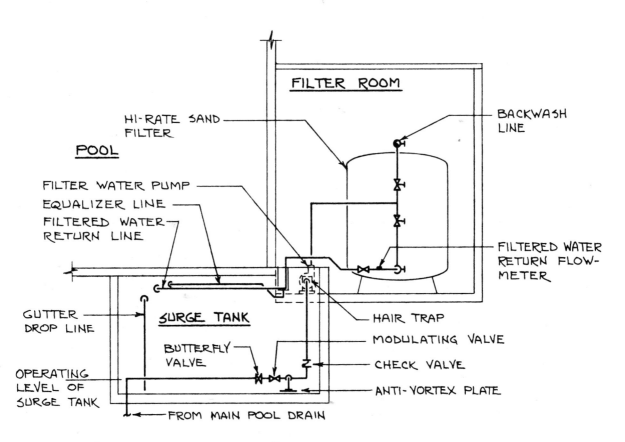

TYPICAL POOL FILTER SYSTEM WITH HI-RATE SAND FILTER IN ABOVE DECK INSTALLATION

NO SCALE   SHEET 2 OF 2

FIGURE 10-2

resent the many related items and all of the associated piping.

Not shown in any of these details is the water supply to the pool and the drainage from the pool. In many pool installation situations the building water supply is a 3-in line and the building sanitary drain is not more than an 8-in line. As in Chap. 4 and Fig. 4.17, it is reasonable to presume the capacity of the 3-in line to be 100 to 150 gpm. While this line normally must serve other plumbing fixtures in the building, the extension of this 3-in service at full size to the pool is required. To fill the typical 180,000-gal pool the 3-in line would require 30 hr [(180,000/100) × 60]. Since the line is used for normal plumbing requirements during the typical day, it quite commonly takes a week of partial operation at night to fill the pool.

The 8-in sanitary sewer discussed in the Fig. 10.1 explanation is yet another problem. Based on Table 7.2 in Chap. 7, an 8-in line can handle 340 gpm flowing half full and 680 gpm when flowing full at a ¼-in pitch and with a velocity of 4.34 fps. Obviously the line is large enough, but again draining 180,000 gal of water takes time.

The plumbing code requires that water from swimming or wading pools, backwash from pool filters, and deck drains be discharged into an indirect waste by means of an air gap. It requires the same indirect waste system when the recirculating pump is used to discharge waste pool water to the drainage system.

Based on the above, it is fairly obvious that the draining of the pool through a backwash tank arrangement as indicated in Fig. 10.1 will apply to all details that follow. The drainage strainers in the backwash tank should be removed during the pool drainage operation to facilitate flow. Further, the valving should be carefully adjusted so that the recirculating pump does not overfill the backwash tank.

Conventional Sand Filter. Figure 10.3 is the first of three filter details. The filter area of the system as de-

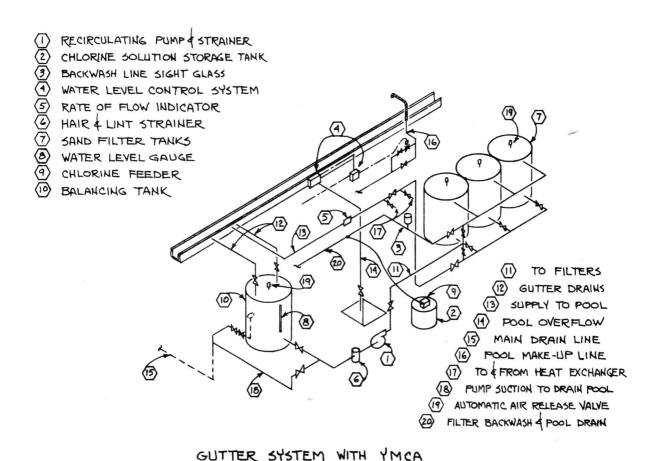

① RECIRCULATING PUMP & STRAINER
② CHLORINE SOLUTION STORAGE TANK
③ BACKWASH LINE SIGHT GLASS
④ WATER LEVEL CONTROL SYSTEM
⑤ RATE OF FLOW INDICATOR
⑥ HAIR & LINT STRAINER
⑦ SAND FILTER TANKS
⑧ WATER LEVEL GAUGE
⑨ CHLORINE FEEDER
⑩ BALANCING TANK

⑪ TO FILTERS
⑫ GUTTER DRAINS
⑬ SUPPLY TO POOL
⑭ POOL OVERFLOW
⑮ MAIN DRAIN LINE
⑯ POOL MAKE-UP LINE
⑰ TO & FROM HEAT EXCHANGER
⑱ PUMP SUCTION TO DRAIN POOL
⑲ AUTOMATIC AIR RELEASE VALVE
⑳ FILTER BACKWASH & POOL DRAIN

GUTTER SYSTEM WITH YMCA
3 TANK SAND FILTER SYSTEM
NO SCALE

FIGURE 10-3

tailed corresponds to the conventional or rapid-sand filter rate of 3 to 4 gpm of flow per square foot of filter area. At 3 gpm the probable filter area of a common indoor pool is 160 to 170 sq ft. Each of the three filters shown in the detail would have a filter area of approximately 55 sq ft.

In this detail the previously discussed gutter overflow goes into a manufactured balancing tank, although a concrete tank could have been utilized as was shown in Figs. 10.1 and 10.2. The sizing premise remains the same.

The premise behind the schematic presentation shown in Fig. 10.3 is that there is sufficient room in the available pool filter area to properly show the filters, tanks, pumps, and piping and that the basic need is not only to show all the related component parts but all the piping arrangement and system flow.

If there is sufficient space for the filter system installation, the schematic detail in Fig. 10.3 is ideal. The detail was condensed to fit the confines of this book. It can readily be enlarged. Arrows showing flow direction

and all pipe sizes could readily be shown on the detail. In addition certain key piping elements such as the lines in and out of the filters, to and from the pool, and from the balancing tank through the pump to the filters could have their sizing repeated (with a careful note that it is being repeated) on the basic ⅛-scale or ¼-scale floor plans. This procedure would clearly identify the relationship of the detail to the plan.

Among the items that are easily depicted on a detail like Fig. 10.3 are the small specialty items such as sight glasses, hair strainers, level controllers, and air release valves. By comparing what was shown on Figs. 10.1 and 10.2 with what is shown in Fig. 10.3, you can readily appreciate the value of this type of presentation.

Frequently there are minor problems with pool overflow, and in the closed and vented balancing tank system an overflow resource is required. In addition indoor pools are commonly heated pools. Heating of water for the pool is no different from heating hot water for domestic use. The pool water is normally maintained

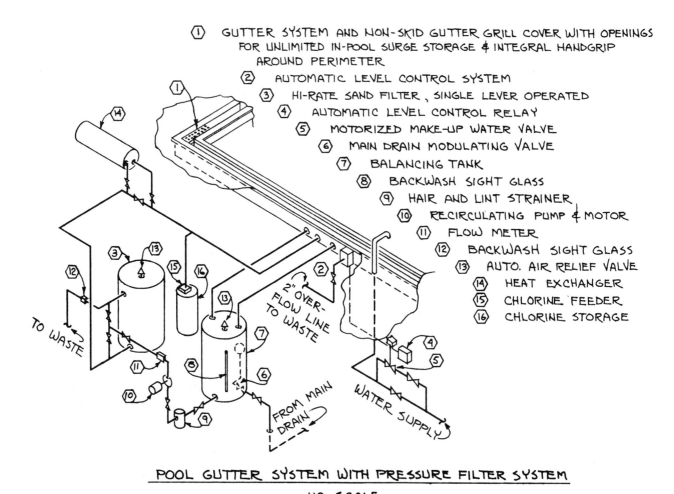

① GUTTER SYSTEM AND NON-SKID GUTTER GRILL COVER WITH OPENINGS FOR UNLIMITED IN-POOL SURGE STORAGE & INTEGRAL HANDGRIP AROUND PERIMETER
② AUTOMATIC LEVEL CONTROL SYSTEM
③ HI-RATE SAND FILTER, SINGLE LEVER OPERATED
④ AUTOMATIC LEVEL CONTROL RELAY
⑤ MOTORIZED MAKE-UP WATER VALVE
⑥ MAIN DRAIN MODULATING VALVE
⑦ BALANCING TANK
⑧ BACKWASH SIGHT GLASS
⑨ HAIR AND LINT STRAINER
⑩ RECIRCULATING PUMP & MOTOR
⑪ FLOW METER
⑫ BACKWASH SIGHT GLASS
⑬ AUTO. AIR RELIEF VALVE
⑭ HEAT EXCHANGER
⑮ CHLORINE FEEDER
⑯ CHLORINE STORAGE

POOL GUTTER SYSTEM WITH PRESSURE FILTER SYSTEM

NO SCALE

FIGURE 10-4

at 80°F. The calculations for sizing the pool heater were covered in detail in Chap. 5.

Gutter details have not been presented as they are a part of the basic pool construction and are the architect's design responsibility.

Pressure Filter System. As previously noted, the pressure high rate sand filter system is a system specifically designed and guaranteed to permit much higher rates of flow per square foot of filter area. When clean, the head loss through the filter is usually 3 pounds per square inch (psi) and when the pressure differential between inlet and outlet line gauges is 15 psi, the filter must be backwashed. When sizing the circulating pump, the head loss for dirt load through the filter and face piping is normally presumed to be between 15 and 20 psig.

The pressure high rate sand filter system shown in Fig. 10.4 is very similar to the detail shown in Fig. 10.3. And the design premise behind the depiction of Fig. 10.4 is also the same as in Fig. 10.3. The filter system is located in an area that is sufficiently large to preclude any real installation problems, and the basic equipment and piping with key sizes marked for detail correlation are shown on the filter room ⅛-scale or ¼-scale plans. As in Fig. 10.3, this detail could be enlarged and all of the pipe sizes could be shown on the detail.

Diatomaceous Earth Filter System. Figure 10.5 depicts the detailing for a vacuum diatomite filter system with a rectangular diatomaceous earth fiberglass tank filter. Using a 2-gpm/sq ft filter design, these filters, on a 6-hr turnover, will handle pools having capacities of 100,000 to 604,800 gal.

Again the premise for this system is the same as for the two previous details. There is no problem with space and the detail presents the easiest and clearest way to present the required piping.

In this detail the draining of the pool is via a gate valve and drain line to a sump, and the pool has a gravity, not pumped, drainage system.

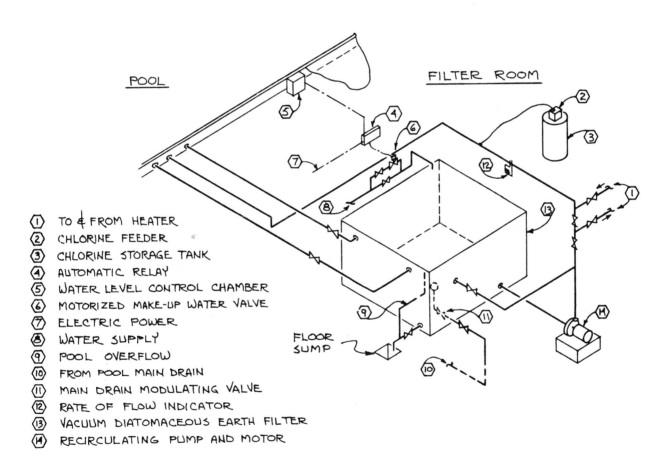

① TO & FROM HEATER
② CHLORINE FEEDER
③ CHLORINE STORAGE TANK
④ AUTOMATIC RELAY
⑤ WATER LEVEL CONTROL CHAMBER
⑥ MOTORIZED MAKE-UP WATER VALVE
⑦ ELECTRIC POWER
⑧ WATER SUPPLY
⑨ POOL OVERFLOW
⑩ FROM POOL MAIN DRAIN
⑪ MAIN DRAIN MODULATING VALVE
⑫ RATE OF FLOW INDICATOR
⑬ VACUUM DIATOMACEOUS EARTH FILTER
⑭ RECIRCULATING PUMP AND MOTOR

POOL SYSTEM WITH DIATOMITE FILTER

NO SCALE

FIGURE 10-5

Filter Systems Summary. The details have been represented with a minimum of discussion to preclude any assumption of favoritism. All pool filter system designers discover very quickly that each system's proponent is both rabid and vocal. Certainly, judgment as to the quality of results plays an important role. For the designer of the system, space not only to install the system but to remove and replace particular components should also be given serious consideration. For the individual or firm that designs pool filter systems infre-quently, it would be well worthwhile to visit representative installations of each of the three types and to confer with the operators of these systems before making a final system selection.

Waste Interceptors

Nearly all of the problems of handling problem fluids in the waste water stream are handled by specialists in the field, and the plumbing designer's problem is limited to merely making connections as per the special waste

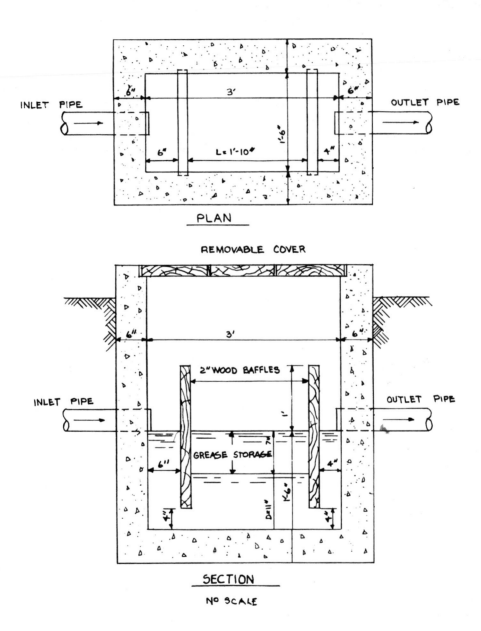

DETAILS OF GREASE INTERCEPTOR

FIGURE 10-6

disposal system engineering requirements. The plumbing code primarily concerns itself with indirect waste connections. Here the primary concern is not the contamination *of* the sanitary drainage system but rather the contamination *by* the sanitary drainage system. In Chap. 6 piping connections to acid and flammable waste receptors which were standard products were detailed. In the discussion that follows two other devices to intercept oil and grease which may or may not be standard items are covered.

Figure 10.6 is a typical exterior grease interceptor that may be used for the separate interior drainage system that connects, outside the building, to the sanitary line going to the septic tank. As in many other details in this book, this one was taken from an actual job. While the arrangement can be used as is, the dimensions must be checked and sized to fit your particular project requirements.

As is well-known, there are standard commercial grease interceptors available to be piped and installed in kitchen areas. However, there are frequently a number of different sources of grease in the average food preparation area. In addition the head of the food preparation staff frequently objects to grease traps that have to be opened and cleaned and that have the cleanout cover located in the kitchen area. Since the principles of separation of water and grease are fairly simple to master and basically consist of a tank capable of holding the water long enough to enable the grease to rise to the top and a set of baffles to contain that grease, it logically follows that a field fabricated tank solves the system designer's and the user's problems.

Figure 10.6 is simply the tank solution. The tank is large enough to hold the greasy water long enough to allow the grease to separate. The baffles simply provide a containment area for the separated grease. The wood baffles rapidly absorb some grease and thus become effectively sealed from all but the very slow decomposition. The trap was contemplated, in this particular application, to be cleaned at least once a week. This particular design has been installed and the septic system, not shown, has operated reasonably grease free for a number of years.

Oil Interceptors

Automotive service areas of all types are a very common area of plumbing design. And, in every instance, there is a problem of drainage that contains spent, spilled, or wasted oil. Figure 10.7 is somewhat similar to the previous presentation of pool filter systems. In this case there is always room to show the items involved in this figure on the plans. The advantage of using a detail such as Fig. 10.7 is that it is a nondimensional, schematic composite that depicts all the related items, shows how they interconnect, and provides a scheme with which all related sizes and control devices can, if desired, be clearly shown.

All oil contains a certain amount of heavy solids (sludge) which, given the proper detention time, will fall to the bottom of the sludge pit as shown. When the solids build up, they may be vacuumed or scraped out of each pit and disposed of off site in an environmentally legal manner. The liquid that remains is a mixture of oil and water which is piped to a commercially available standard oil interceptor. Here oil and water are separated and water is discharged through a backflow valve to the sanitary sewer. The oil trapped in the interceptor is drained into a holding tank. A line, as shown, terminates on the floor in a properly tight fitting covered connection. This line allows a testing of the quantity of oil in the storage tank and a means of connecting a suction pump to empty the tank. This waste oil may be sold or otherwise disposed of off site, also in an environmentally legal manner. The related vent piping is also much more readily depicted in a detail such as Fig. 10.7.

Gas Systems

While there are special situations in which manufactured gas is used, for all practical purposes and for the purposes of this book the discussion and details that follow will be limited to natural and propane gas applications. Plumbing and gas piping are commonly associated together in plumbing design, and gas piping is commonly part of the plumbing contract work.

Reduced to essentials the basic design premise of the gas piping system is to provide a system that delivers gas at the point of use at a pressure, and thus not only must the pipe supplying the gas be large enough to supply the proper volume of gas, but it must also be large enought to preclude the drop in pressure between the point of use and the gas meter, however distant, from exceeding the allowable design criteria. If the gas at the outlet of the meter is 3 psi and the point of use required pressure is 2 psi, the *total* allowable pressure drop is 1 psi (3 − 2) regardless of length of run. This can be a problem that should be seriously investigated when the point of use is 500 or 1,000 ft or more from the gas meter.

The size of the gas piping depends upon the following factors:

1. Allowable loss in pressure from the meter to the point of use
2. Maximum gas consumption to be provided
3. Length of pipe and number of fittings
4. Specific gravity of the gas
5. Diversity factor

Gas Pipe Sizing

To enable the user of this book to design a small gas system or a larger system with short runs of piping, data on gas systems piping information in the McGraw-Hill book by Crocker and King, *Piping Handbook*, have been abstracted in part in the table and example that follow. Figure 10.8 depicts a simple residential all-gas installation. The load values are taken from the involved appliance manufacturer's published data of gas input requirements. Pipe sizing Table 10.1, taken from the Crocker handbook, is applicable to this and similar situations. Gas systems up to a specific gravity of 0.70 may be sized directly from this table. For other values of specific gravity utilize the multipliers given in Table 10.2. Table 10.1 is based on straight lengths of pipe.

In Fig. 10.8 the upper portion of the figure gives distances and the British-thermal-unit-per-hour (BTUH) rating of connected appliances. The first step in the piping solution is to convert all BTUH values into cubic feet per hour (cfh) of gas by dividing the BTUH by

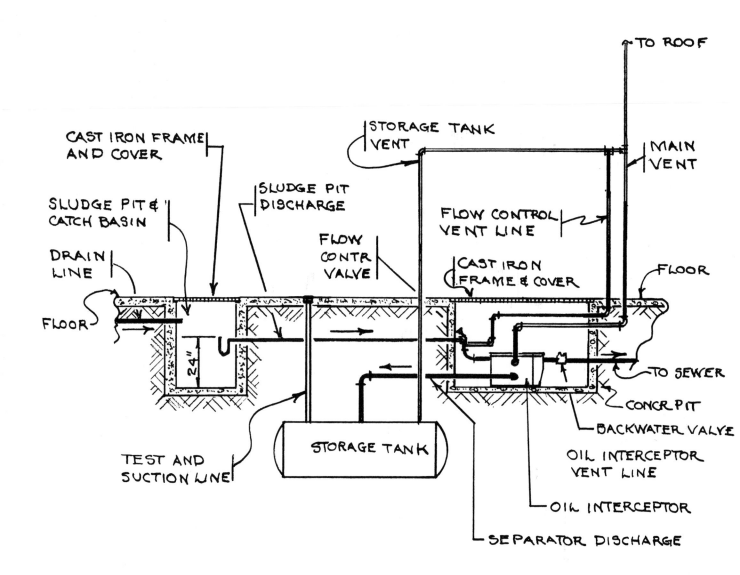

OIL INTERCEPTOR

— NO SCALE —

FIGURE 10-7

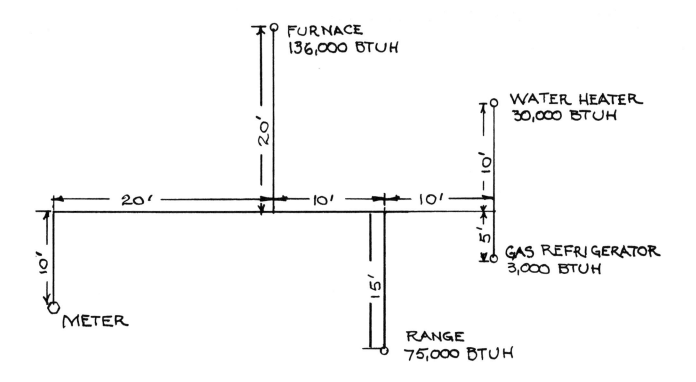

FURNACE
136,000 BTUH

WATER HEATER
30,000 BTUH

20'

20'

10'

10'

10'

5'

GAS REFRIGERATOR
3,000 BTUH

10'

15'

METER

RANGE
75,000 BTUH

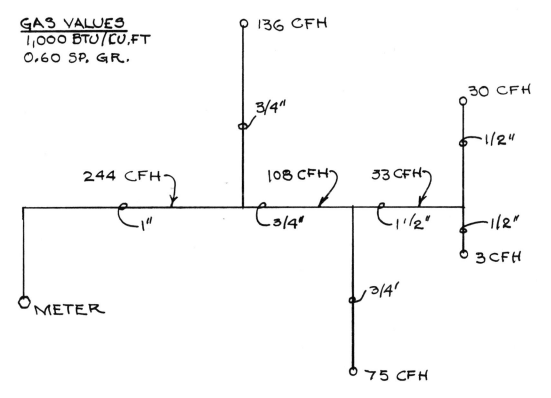

GAS VALUES
1,000 BTU/CU.FT
0.60 SP. GR.

136 CFH

30 CFH

3/4"

1/2"

244 CFH

108 CFH

33 CFH

1"

3/4"

1 1/2"

1/2"

3 CFH

3/4'

METER

75 CFH

GAS PIPE SIZING SCHEMATIC

— NO SCALE —

FIGURE 10-8

Table 10.1 Maximum Capacity of Pipe in Cubic Feet of Gas per Hour[a]

Nominal iron pipe size (in)	Length (ft)													
	10	20	30	40	50	60	70	80	90	100	125	150	175	200
½	175	120	97	82	73	66	61	57	53	50	44	40	37	35
¾	360	250	200	170	151	138	125	118	110	103	93	84	77	72
1	680	465	375	320	285	260	240	220	205	195	175	160	145	135
1¼	1,400	950	770	660	580	530	490	460	430	400	360	325	300	280
1½	2,100	1,460	1,180	990	900	810	750	690	650	620	550	500	460	430
2	3,950	2,750	2,200	1,900	1,680	1,520	1,400	1,300	1,220	1,150	1,020	950	850	800
2½	6,300	4,350	3,520	3,000	2,650	2,400	2,250	2,050	1,950	1,850	1,650	1,500	1,370	1,280
3	11,000	7,700	6,250	5,300	4,750	4,300	3,900	3,700	3,450	3,250	2,950	2,650	2,450	2,280
4	23,000	15,800	12,800	10,900	9,700	8,800	8,100	7,500	7,200	6,700	6,000	5,500	5,000	4,600

[a]Based upon a pressure drop of 0.5 inch water column and 0.6 specific gravity gas. For other pressure drops, multiply by $\sqrt{\Delta P/0.5}$, where ΔP is the pressure drop, inches water column.

Table 10.2 Multipliers to Be Used Only with Table 10.1 When Applying the Gravity Factor

Specific gravity	Multiplier	Specific gravity	Multiplier
0.35	1.31	1.00	0.78
0.40	1.23	1.10	0.74
0.45	1.16	1.20	0.71
0.50	1.10	1.30	0.68
0.55	1.04	1.40	0.66
0.60	1.00	1.50	0.63
0.65	0.96	1.60	0.61
0.70	0.93	1.70	0.59
0.75	0.90	1.80	0.58
0.80	0.87	1.90	0.56
0.85	0.84	2.00	0.55
0.90	0.82	2.10	0.54

1,000, which is the Btu-per-cubic-feet-per-hour value for natural gas. As in all other types of pipe sizing the load, in cubic feet per hour, is cumulatively added to the gas main. As can be seen in the detail, the line load is 33 cfh at the farthest end and 244 cfh is the total load.

The procedure to size the pipe using Table 10.1 first requires the calculation of the length of pipe to the farthest outlet. In Fig. 10.8 this is the water heater whose distance from the gas meter is 60 ft (10 + 20 + 10 + 10 + 10). This value of 60 ft becomes the *controlling* value. All other pipe sizes *must* be taken from the 60-ft column heading. Since the minimum value is 60 cfh for a ½-in pipe, the lines to the water heater (30 cfh) and the gas refrigerator are each ½ in, as is the main that feeds these two branches that supplies a total of 33 cfh. Using the same column, the range and furnace each require a ¾-in line, and the main, when the load passes 151 cfh and is less than 285 cfh, becomes a 1-in line. This is depicted in the lower half of Fig. 10.8.

While this may seem to be a very simplified situation, it is not uncommon to find a low-pressure system of considerable volume of gas (3,000 to 4,000 cfh or 3,000,000 to 4,000,000 BTUH) with all using burners and appliances located within 200 ft of the meter.

Low-pressure gas is commonly defined as gas delivered at a pressure between 5 and 14 in of water column pressure.

Larger gas systems in industrial applications generally utilize gas delivered at higher pressures. In general the 10 percent pressure loss rule is the standard of design. When gas pressures are 0.5 psig or less, the allowable pressure loss should not exceed 0.5 in of water. When higher pressures are used, the allowable pressure loss should not exceed 10 percent of the initial gauge pressure unless the point of use appliance has its own pressure regulator.

Thus a 5-psig line would have a total drop of 0.5 psig, a 20-psig line drop would have 2 psig, and a 50-psig line would have 5 psig. Tables 10.3 to 10.5 utilize these stated values.

In using Tables 10.3 to 10.5 the same procedure is followed as was previously described for a simple resi-

Table 10.3 Pipe-Sizing Table for 5-lb Pressure[a]

Pipe size of Schedule 40 standard pipe (in)	Total equivalent length of pipe (ft)									
	50	100	150	200	300	400	500	1,000	1,500	2,000
1	1,860	1,320	1,070	930	760	660	590	410	340	290
1¼	3,870	2,740	2,240	1,930	1,580	1,370	1,220	860	700	610
1½	5,860	4,140	3,390	2,930	2,390	2,080	1,850	1,310	1,060	930
2	11,420	8,070	6,600	5,710	4,660	4,050	3,610	2,550	2,080	1,810
2½	18,400	13,010	10,640	9,200	7,510	6,530	5,820	4,110	3,350	2,920
3	32,860	23,240	19,000	16,430	13,410	11,660	10,390	7,340	5,990	5,220
3½	48,480	34,280	28,030	24,240	19,780	17,200	15,330	10,820	8,840	7,690
4	67,700	47,880	39,140	33,850	27,630	24,020	21,410	15,120	12,340	10,750
5	123,790	87,540	71,570	61,890	50,530	43,920	39,160	27,640	22,570	19,660
6	202,138	142,950	116,870	101,060	82,500	71,720	63,940	45,140	36,860	32,100

[a]Capacity of pipes of different diameters and lengths in cubic feet per hour. For an initial pressure of 5 psig with a 10 percent pressure drop and a gas of 0.6 specific gravity.

Table 10.4 Pipe-Sizing Table for 20-lb Pressure[a]

Pipe size of Schedule 40 standard pipe (in)	Total equivalent length of pipe (ft)									
	50	100	150	200	300	400	500	1,000	1,500	2,000
1	4,900	3,470	2,810	2,450	2,000	1,730	1,550	1,070	890	770
1¼	10,190	7,210	5,840	5,090	4,160	3,600	3,220	2,230	1,860	1,610
1½	15,420	10,900	8,830	7,710	6,290	5,450	4,870	3,370	2,810	2,440
2	30,030	21,230	17,190	15,010	12,260	10,610	9,490	6,570	5,480	4,760
2½	48,390	34,220	27,710	24,190	19,750	17,110	15,290	10,590	8,830	7,670
3	86,420	61,110	49,490	43,190	35,280	30,550	27,310	18,910	15,770	13,690
3½	127,480	90,130	73,000	63,710	52,040	45,070	40,280	27,900	23,270	20,200
4	178,040	125,880	101,950	88,980	72,680	62,940	56,260	38,500	32,500	28,210
5	325,530	230,170	186,410	162,700	132,890	115,080	102,870	71,240	59,420	51,590
6	531,530	375,820	304,370	265,660	216,990	187,910	167,980	116,330	93,030	84,240

[a]Capacity of pipes of different diameters and lengths in cubic feet per hour. For an initial pressure of 20 psig with a 10 percent pressure drop and a gas of 0.6 specific gravity.

Table 10.5 Pipe-Sizing Table for 50-lb Pressure[a]

Pipe size of Schedule 40 standard pipe (in)	Total equivalent length of pipe (ft)									
	50	100	150	200	300	400	500	1,000	1,500	2,000
1	10,530	7,450	6,090	5,150	4,350	3,790	3,330	2,350	1,920	1,650
1¼	21,880	15,490	12,650	10,700	9,050	7,870	6,920	4,890	3,990	3,430
1½	33,110	23,430	19,130	16,190	13,690	11,910	10,470	7,410	6,040	5,190
2	64,450	45,610	37,250	31,530	26,660	23,190	20,400	14,420	11,770	10,110
2½	103,870	73,510	60,040	50,820	42,960	37,370	32,870	23,240	18,970	16,300
3	185,490	131,270	107,220	90,750	76,720	66,730	58,700	41,510	33,870	29,100
3½	273,600	193,620	158,140	133,850	113,170	98,430	86,590	61,230	49,970	42,930
4	382,110	270,420	220,870	186,940	158,050	137,480	120,930	85,510	69,780	59,960
5	698,660	494,430	403,840	341,800	288,980	251,360	221,110	156,360	127,600	109,630
6	1,140,780	807,320	659,400	558,110	471,860	410,430	361,040	255,310	208,340	179,010

[a]Capacity of pipes of different diameters and lengths in cubic feet per hour. For an initial pressure of 50 psig with a 10 percent pressure drop and a gas of 0.6 specific gravity.

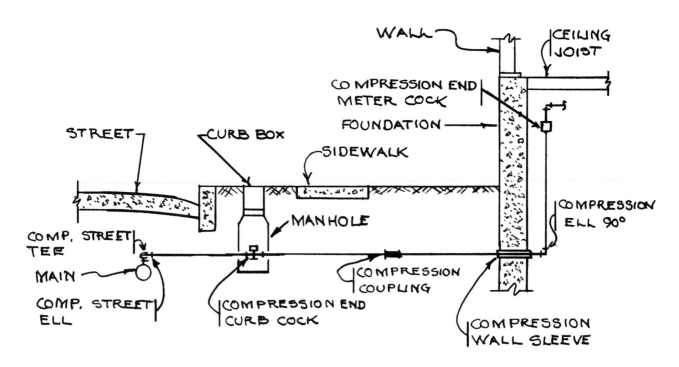

Gas Service with Compression Couplings
and Threaded Packing Nuts
— No Scale —

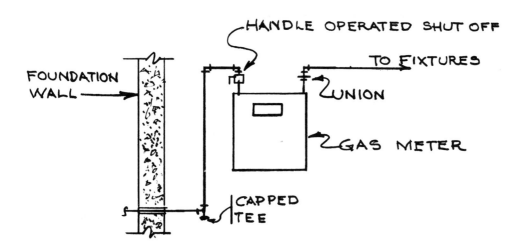

Conventional Gas Meter Piping
—————— No Scale ——————

FIGURE 10-9

dential system. Regardless of complexity, always make a schematic layout as was done in Fig. 10.8. Note lengths of runs, cubic-feet-per-hour values, and find the distance to the farthest user. As before that distance becomes the controlling pipe column selection. If it is 1,000 ft, for example, then all sizes are taken from the 1,000-ft column as was done in the example of the simple house with its 60-ft longest run.

In the majority of installation situations the gas utility company will bring the gas service to the gas meter, which is commonly located inside the building. If, in your particular project, the gas service is not brought to the meter, the requirements relative to who specifically owns what part of the service and who owns the meter must be carefully investigated. In addition the utility company must advise as to what is the incoming pressure and what pressure drop pipe sizing method shall be used for the incoming service piping. Consideration must be given to connected load, length of service, and main pressure. Generally the sizing is based on an allowable pressure drop related to the utility system main pressure, and the pipe size is usually based on the length of the service line and the anticipated load. A low-pressure main at a 6-in to 8-in water column would probably require a 0.3-in to 0.5-in drop. Systems at 1 to 15 psig would usually allow a 0.5-psig drop. Higher-pressure systems usually allow 0.5-psig to 3-psig drop. These are typical average values. The exact value should be secured, in writing, from the utility company.

Gas Services. There are a variety of piping requirements for gas services and seemingly no specific standard that is applicable throughout the United States. In many cases this fact does not affect the gas piping system designer since the gas company brings the service into the building and *terminates* it in a meter supplied and installed by the utility company.

Figure 10.9 depicts, in the upper half of the detail, a typical sketch of a simple low-pressure gas system service from the street using compression couplings with threaded packing nuts, and it depicts a street tee and a street ell at the main, an outside stop cork with a curb box, an inside stop cock, and a line to the gas meter. Piping is generally pitched back from the building meter to the street main using a pitch of 1 in/10 ft.

No one standard exists for the various shutoffs. In general it is common practice to install outside shutoffs (which may be curb cocks or, in the case of outside meters, meter inlet cocks) on all services where a pressure regulator exists and on all services to densely populated buildings. Whether the meter is located indoors or outdoors, the piping connections are very similar. The lower half of Fig. 10.9 depicts a typical low-pressure indoor meter installation. While Fig. 10.9 is reasonably representative of the specialties involved in the design of the low-pressure gas service, the designer cannot simply use this detail as is. A sketch of it could be presented

to the local gas supplier with a letter of transmittal requesting the supplier's specific written comments and approval.

Gas Heating System Piping. One of the more common design problems for the plumbing and gas piping system designer is the piping to gas or combination gas-oil boilers and furnaces. The oil piping, if required, is part of the heating contract. The gas piping up to and sometimes including the gas burner train is part of the plumbing contract.

Figure 10.10 addresses the standard situation in which the gas contract work terminates at the burner train. As noted on this detail, the gas service is, in each equipment connection, brought to the gas train with the piping on the train specified to be done from the connection point onward by another contractor. The basic gas service that the heating equipment includes is the service valve, the service regulator, and a gas meter. The individual feeders to each item of heating equipment include a plug cock and a pressure regulator.

The regulator has some specific requirements. It must have a pressure rating at least equal to that of the distribution system. The capacity of the regulator must be 15 percent larger than the maximum burning rate of the burner served. The regulator spring should be capable of an adjustment of 50 percent under to 50 percent over the desired pressure. A straight run of pipe should be used on both sides of the regulator to ensure proper operation, especially when pilot-operated regulators are used. The regulator can be installed close to the gas train connection, but 2 to 3 ft of straight run piping should always be specified on the upstream side of the regulator.

The valves and controls and the meter (items 1, 2, and 3) may be supplied by the utility company. This must be verified in writing by the utility company. The sizing of the gas line from the meter to the gas pressure regulator can be very important if the available gas pressure is marginal for proper burner operation. The required gas pressure is presumed to be available at the contractor's connection point. The factors affecting line sizing are the gas pressure available at the meter (item 3), the rate of flow required, the length of run, and the required pressure at the connection point.

While smaller boilers and furnaces can utilize low-pressure gas systems, the pressure requirement for larger boilers increases considerably. A typical boiler manufacturer would require 5 in at the gas train for maximum operation of a 200-horsepower (hp) boiler with this value rising to 42 in for a 600-hp boiler. The standard conversion factor to change inches to pounds per square inch gravity is 0.0361. This 15 in is 0.5415 psig (15 × 0.0361) and 42 in is 1.5162 psig (42 × 0.0361). If only low-pressure gas is available in sufficient quantity, a gas pressure booster pump would be required for boilers larger than 150 hp since the 150-hp boiler requires 11 in or 0.3971 psig of pressure.

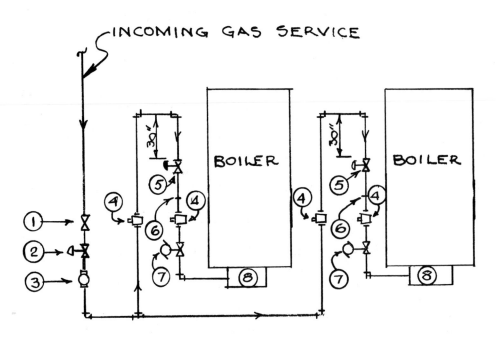

INCOMING GAS SERVICE

BOILER

BOILER

CIRCLED NOTES

① SERVICE VALVE

② SERVICE REGULATOR

③ GAS METER

④ PLUG COCK

⑤ PRESSURE REGULATOR AT BURNER

⑥ GAS PIPING CONTRACTOR CONNECTION POINT

⑦ PRIMARY GAS SHUT OFF VALVE

⑧ GAS BURNER

GAS PIPING TO BOILER PLAN

——————NO SCALE——————

FIGURE 10-10

Burner Gas Trains. Figures 10.11 and 10.12 depict slightly different versions of gas trains. As noted in the discussion of gas heating system piping, these gas trains are usually supplied by the heating contractor. There are occasions when all of the work on large gas systems that heat large volumes of water is part of the plumbing contract. For this reason work that is usually part of the heating contract has been shown in these two details.

In all cases, the line to the burner continues from the connection stopping point shown on Fig. 10.10 through the line to the burner shown on Figs. 10.11 and 10.12. All of the devices that are part of the train are depicted in Fig. 10.11 and the upper part of Fig. 10.12. The lower part of Fig. 10.12 depicts a system with high- and low-pressure switches that some burning systems may require.

In all three situations the pilot line is ¼-in pipe or ⅜-in tubing. If the line is exceptionally long, which would

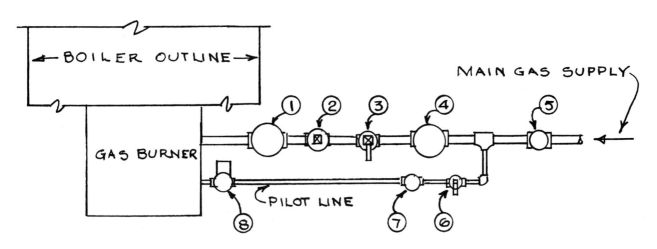

CIRCLED NOTES

① ELECTRIC GAS VALVE

② GAS COCK WITH SEPARATE HANDLE. OMIT IF GAS FLOW SET BY OTHER DEVICE.

③ MANUAL SHUTOFF WITH ATTACHED HANDLE

④ SHUT OFF TYPE GAS PRESSURE REGULATOR

⑤ DUST TRAP / STRAINER

⑥ PILOT GAS COCK WITH ATTACHED HANDLE

⑦ PILOT PRESSURE REGULATOR

⑧ PILOT SOLENOID VALVE

GAS BURNER INSTALLATION PIPING

——— NO SCALE ———

FIGURE 10-11

be somewhat unusual, the sizes may be increased one or possibly two pipe sizes.

Gas trains may not be required or the pilot operator portion of each gas train may not be required in a water heating situation for smaller sized systems. However, the larger water heating systems will normally require the system depicted in Fig. 10.11 with or without the pilot-operated feature.

Compressed Air Systems

Compressed air piping is sometimes included in the plumbing specification and as such becomes part of the design effort of the plumbing system designer. The selection of the air compressor, unless it is a fairly small, standard 125-psig unit, is not part of a plumbing specification and frequently is the subject of a special section of the specification, especially when a complete new compressor and piping system is specified.

The premise of the details presented and the discussion that follows is that an existing system is being altered or the owner has purchased a compressor or compressors and the piping is included in the plumbing contract. Local union and nonunion standard procedures may also determine whether plumbers or steam filters do the actual installation.

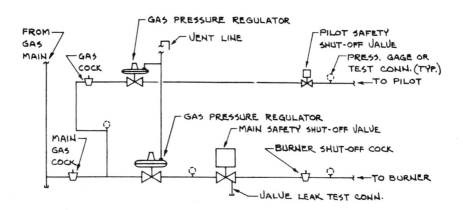

GENERAL PURPOSE

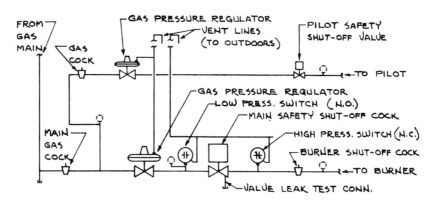

WITH GAS PRESSURE SWITCHES

TYPICAL GAS BURNER
GAS LINE TRAINS

FIGURE 10-12

Compressed Air Pipe Sizing

There are a number of formulas, tables, and charts that are available to size the compressed air piping. Frequently the designer does not have an applicable chart to fit the design requirements. For just such a situation we are presenting below the Harris formula which has stood the test of time and is a commonly used formula. Our source of this formula is the McGraw-Hill *Piping Handbook* by Crocker and King. The formula is as follows:

$$\text{Pressure drop} = \frac{LQ^2}{2,390 P_c d^{5.31}}$$

where L = length of pipe, ft
$\quad Q$ = cubic feet per minute (cfm) of air handled
$\quad 2,390$ = constant
$\quad P_c$ = (initial pressure + final pressure)/2
$\quad d$ = pipe diameter in inches

A sample problem, based on 3,00 cfm of air, 1,000 ft of pipe, a 120-psig initial pressure, and a final pressure of 118 psig, is to find the required pipe size. To solve the problem the pressure drop becomes 2 (120 − 118). The term P_c becomes 119 (120 + 118)/2. The formula would be rearranged to

$$d^{5.31} = \frac{LQ^2}{2,390 P_c}$$
$$= \frac{1,000 \times (3,000)^2}{2,390 \times 119}$$
$$= 31,644$$

Therefore

$$d = 31,644^{1/5.31}$$
$$= 7.04 \text{ in}$$

Another very common use of the formula is to find out how much air an existing pipe can handle, with the desired value being 3,000 cfm. Presume that a 6-in pipe is available. The result, using the length of 1,000 ft, the initial pressure of 120 psig and the final pressure of 118 psig as the values, and rearranging the formula, is

$$Q^2 = \frac{(120 - 118) \times 2,390 \times 119 \times 6^{5.31}}{1,000}$$
$$= \frac{2 \times 2,390 \times 119 \times 13,551}{1,000}$$
$$= 7,708,106$$

Therefore

$$Q = 7,708,106^{1/2}, \text{ or } 2,776 \text{ cfm}$$

This is less than the desired 3,000 cfm. Since the result is within some 10 percent of the desired value, one could increase the pressure drop to 2.5 psig. In the above formula, this slight change would result in a flow of some 3,100 cfm.

The importance of the above formula is that, as demonstrated, the designer can specifically state under what conditions an existing line can deliver the desired cubic feet per minute. All guesswork and table or chart interpretation is eliminated.

Basic Compressed Air System. Stripped of all essentials, Fig. 10.13 is a simplified detail for a compressed air system with a limited-use capability. It is typical of a small, integral tank mounted compressor with piping as noted on the detail for oil burners and plant use. In this instance plant use would most likely be various uses in a boiler room or adjacent repair and maintenance area.

While the detail may appear to be very simple, it is essentially complete. Except for a possible air silencer on the compressor air intake, there is no other piping. As we progress to more complex details, the air distribution system will not be repeated since that part of the compressed air design is nearly always identical to that shown in Fig. 10.13. The piping, similar to gas, water, or vacuum piping, is simply pipe from the compressor to items of equipment with a gate valve at the point of final connection.

Air Cooled Compressor Piping. Presuming that compressors and compressed air systems will have the same distribution design as noted in Fig. 10.13 even if the system becomes larger and much more complex than Fig. 10.13, one can compare the simple compressor mounted on a storage tank with the system shown in Fig. 10.14. Because the system being served is much larger, the compressor is much too large to sit on the tank and the tank is also much larger. In Fig. 10.14 the air cooled compressor with its own built-in intercooler creates a supply of hot compressed air. In the detail the hot compressed air is passed through a water-to-air heat exchanger and cooled before it is stored in the air receiver. The air delivered to the process in the plant passes through an air filter before going to the plant area to avoid equipment damage.

As pointed out in our opening discussion, this larger system and its components are commonly prepurchased by the owner and, if the compressed air piping becomes a part of the plumbing work, the detailer's assignment is to show, as in Fig. 10.14, all of these items properly arranged and connected.

In terms of work division the water piping most likely would be done by the plumbing contractor. Quite possibly the air piping, in this case, would be done by a steam fitter. However, the detail cannot logically be shown partly on the plumbing drawings and partly on piping drawings.

The drain line from the moisture eliminator is connected to the waste line from the one-pass air cooler. This should be carefully investigated since local codes may preclude other than limited amounts of water to

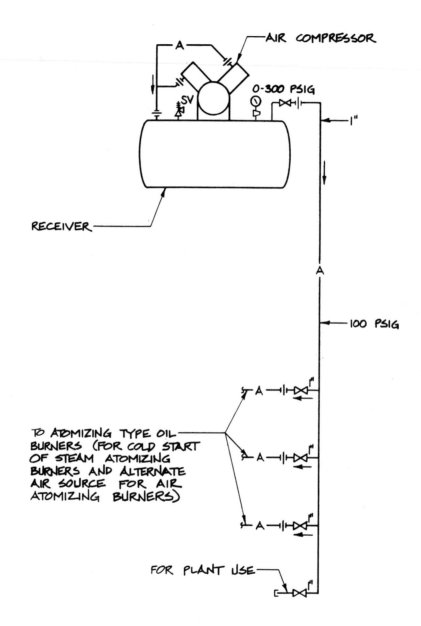

COMPRESSED AIR PIPING DIAGRAM
— NOT TO SCALE —

FIGURE 10-13

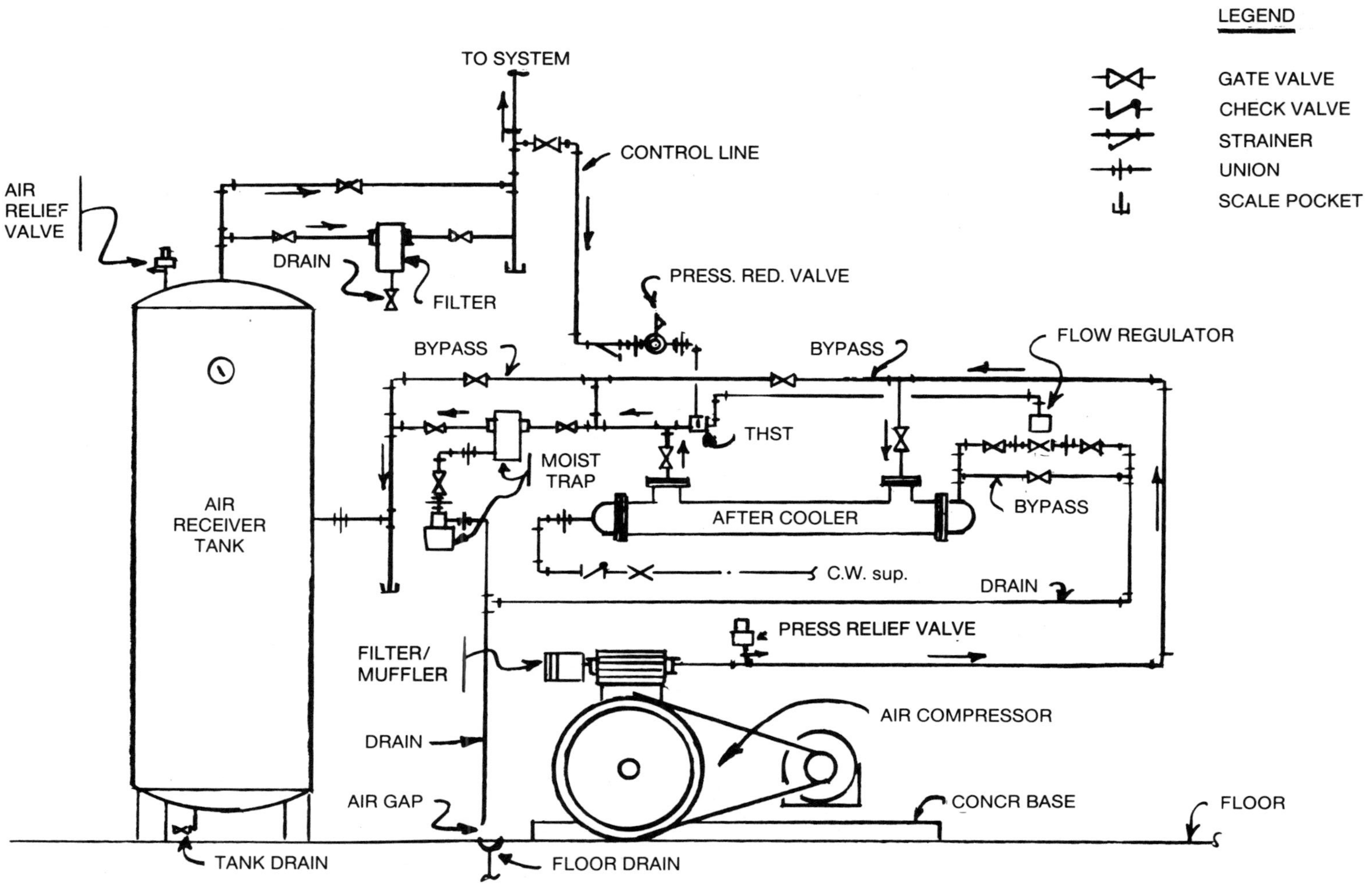

LEGEND

	GATE VALVE
	CHECK VALVE
	STRAINER
	UNION
	SCALE POCKET

FLOW REGULATOR

BYPASS

BYPASS

DRAIN

PRESS. RED. VALVE

CONTROL LINE

THST

AFTER COOLER

C.W. sup.

PRESS RELIEF VALVE

AIR COMPRESSOR

CONCR BASE

FLOOR

TO SYSTEM

FILTER

DRAIN

BYPASS

MOIST TRAP

FILTER/MUFFLER

DRAIN

AIR GAP

FLOOR DRAIN

AIR RELIEF VALVE

AIR RECEIVER TANK

TANK DRAIN

AIR COMPRESSOR PIPING

—— NO SCALE ——

FIGURE 10-14

be wasted in this manner. If the water that passes through the after-cooler cannot be used for some other purpose and is of a large quantity, the after-cooler piping will have to be separated and fed to an exterior cooling tower where it can be cooled and recirculated. On very large systems the compressor is likely to also be water cooled and both compressor and after-cooler water would be circulated through an exterior cooling tower.

Water Cooled Compressors. As we have noted in previous discussions, the work of the plumbing designer and contractor may be limited to the water piping that is related to the compressed air system. Figures 10.15 and 10.16 illustrate that portion of the work.

Figure 10.15 again illustrates the total compressed air system using dual 50-psig water-cooled compressors. As can be seen in this detail, the water piping portion of the work is somewhat difficult to differentiate from the compressed air piping. Thus a composite detail, if used to delineate only water piping, would have to be enlarged to clarify the division of work. If only water piping is involved in the design, Fig. 10.16 would be a superior presentation. The plumbing contractor would see, as is seen in these two typical details, only the work which

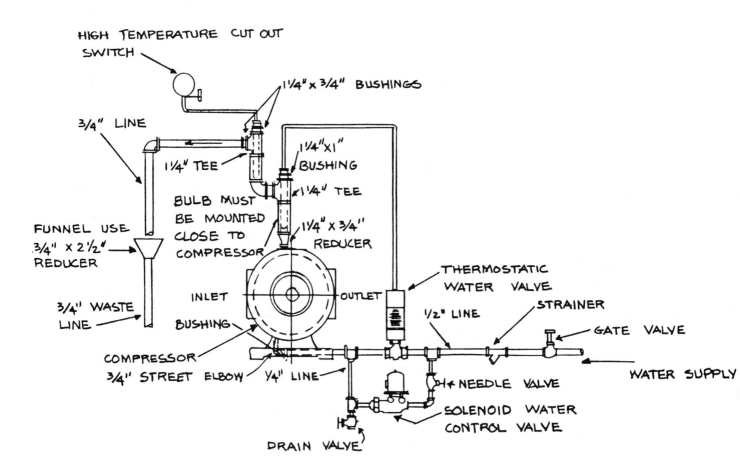

COMPRESSOR & VACUUM PUMP COOLING SYSTEMS

FIGURE 10-15

is his or her responsibility. This detail presumes that the water supply can be wasted. If this is not permitted, the water supply would come from an exterior cooling tower and the ¾-in line would return to the cooling tower and the drain would be eliminated.

High-Pressure Air. Beyond the normal (up to 150 psig) compressed air system, the whole system of compressed air including compressors, piping, and specialties becomes a very specialized area of expertise. As such this high-pressure system design is not part of the plumbing design or within the area covered by this book. How-

ever, situations can and do arise in which only a minor change of air outlets is required and there may be air at more than one pressure involved.

Figures 10.17 and 10.18 depict outlets for a very high-pressure air system in a shipyard. The pressures used in the yard are delivered at various values including one system that delivers air at 5,000 psig. This extreme case is used to illustrate a point. Any system over 150 psig must have double valved outlets and pressure gauges as shown in these two figures. This is a code and a safety requirement. In addition the cap on the end of the line must be attached with a safety chain. While caps are

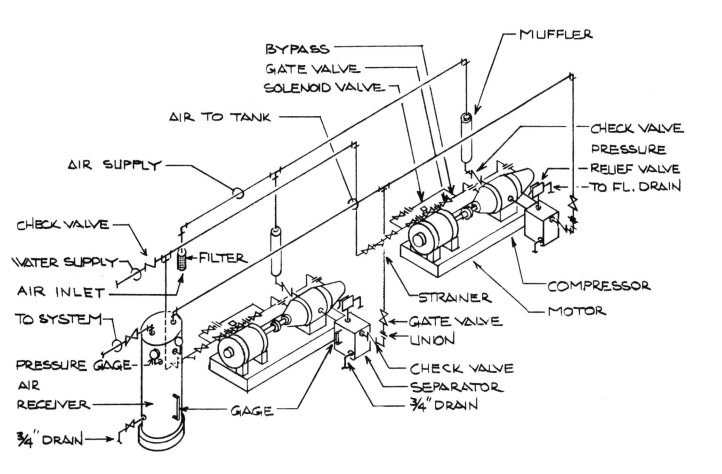

FIGURE 10-16

VALVE OPERATING PROCEDURE
TO REMOVE MOISTURE CLOSE GLOBE
VALVE. VENT DRIP LEG TO ATMOSPHERE
OPENING CAP. REPLACE CAP AND MAKE
AIR TIGHT BEFORE REOPENING VALVE.

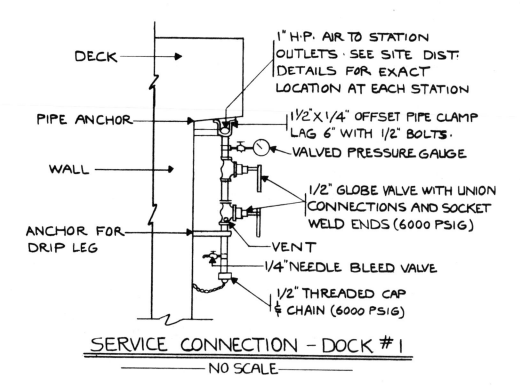

DECK

PIPE ANCHOR

WALL

ANCHOR FOR
DRIP LEG

1" H.P. AIR TO STATION
OUTLETS. SEE SITE DIST.
DETAILS FOR EXACT
LOCATION AT EACH STATION

1½"X¼" OFFSET PIPE CLAMP
LAG 6" WITH ½" BOLTS.
VALVED PRESSURE GAUGE

½" GLOBE VALVE WITH UNION
CONNECTIONS AND SOCKET
WELD ENDS (6000 PSIG)

VENT
¼" NEEDLE BLEED VALVE

½" THREADED CAP
& CHAIN (6000 PSIG)

SERVICE CONNECTION – DOCK #1
——— NO SCALE———

FIGURE 10-17

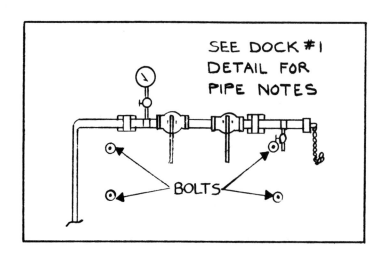

SEE DOCK #1
DETAIL FOR
PIPE NOTES

BOLTS

PLATE DETAIL
——— NO SCALE———

FIGURE 10-18

needed on all compressed air lines, it is very important at pressures over 150 psig that they do not get lost and, when the line is not in use, are properly attached to the end of the pipe.

These high-pressure outlets, and any similar ones, should be firmly attached to safe concrete niches as shown in Fig. 10.17 or to steel plates as shown in Fig. 10.18. The object of the plate, usually ⅜ to ½ in thick, is to provide a secure anchor for the gauge and the dual valves. The plate may be anchored to a building column, to structural concrete, to a masonry wall, or to a specially designed structural support frame. The plate should be painted and marked with cautionary signs as to the pressure contained in the pipe.

Vacuum Systems

Vacuum piping systems have the opposite of the problem of compressed air piping systems in that their problem is preventing air from leaking in rather than from leaking out. This is frequently a more difficult problem, and considerable attention must be paid by the installing contractor and the system specifier to the proper sealing of pipe joints and the testing for leakage.

The requirements for vacuum systems can vary considerably and specialized calculations and equipment may require specialized approaches. More detailed calculations appear in Tyler G. Hicks' *Standard Handbook of Engineering Calculations* and C. M. Van Atta's *Vacuum Science and Engineering,* both published by McGraw-Hill.

In the usual hospital or laboratory installation the minimum vacuum to be maintained at a given outlet is 15 in mercury (Hg) and the minimum value to be maintained in the receiver tank is 10 in Hg. Pipe sizing should be based on the calculated flow rate, and the maximum line loss must not exceed the difference between the vacuum at the receiver tank and the vacuum at the outlet. In the usual situation this is 4 in Hg (19 − 15). Figure 10.19 is a copy of a chart used by government agencies for pipe sizing and can be used for systems limited to 15 in Hg at the outlet.

Outlet sizing depends on usage, and exact data are required. In most hospital applications the outlet requirement varies from 1 to 2 cfm per outlet. Vacuum systems should not be sized smaller than 1 in for mains, ¾ in for laterals and risers, and ½ in for branches to single outlets.

Clinical Vacuum Pump System. As in compressed air systems, the required piping detailing of the vacuum system is at the pump and storage tank. Figure 10.20 is a typical detail of a 19-in Hg system. Only one pump, for the sake of clarity, is shown in this detail. Commonly two pumps are supplied. This would simply mean duplicating the associated piping and tying it to the main line that is shown from the vacuum pump to the vacuum storage tank. A two-pump system is shown in Fig. 10.21.

When two pumps are supplied, each shall be full size for total system requirements. The vacuum pump starting switches shall be tied to a control which will permit both pumps to run simultaneously when a single pump, for any reason, cannot maintain a vacuum of 19-in Hg in the storage (receiver) tank.

In the detail the vent is depicted as discharging to the atmosphere. This discharge line should conform to code requirements. The requirement is similar to that for other vents for other systems and usually is that the discharge terminate in a screened, turned down elbow which is a minimum of 10 ft from any door, window, air intake, or similar item. To avoid back pressure on the pump the line should be increased ½ in over required size for each 50 ft of equivalent length.

The detail depicts a water cooled pump. Unless there is some special local requirement, the cooling system is a one-pass water cooling system using a local domestic water source. To avoid water contamination the vacuum breaker on the water supply and the indirect waste to drain are both mandatory. The drain is, unless otherwise allowed, a vented drain. The tank does require cleaning, and there is a water supply with an air break and funnel to allow for flushing the tank of evacuated liquid buildup.

Laboratory Vacuum System. Figure 10.21 is essentially the same as Fig. 10.20 except that it is intended for use in a much larger central vacuum system. While there are some differences, the basic concept is the same. In the larger system, as in any large device, noise is a problem. In the vacuum system exhaust air noise is a problem. The detail depicts a muffler on each vacuum pump. If the plumbing work merely includes piping connections for client-supplied equipment, the equipment should include, as part of the equipment package, a muffler supplied by the equipment manufacturer. In addition these larger systems should include an air-water separator as depicted in the detail. Finally all of the various devices shown on the tank in Fig. 10.21 should be supplied with the client-supplied equipment. Each manufacturer usually has a series of specialty items related to equipment being purchased. The plumbing-vacuum piping contractor should not be responsible for these items.

Vacuum Cleaning System. Figure 10.22 depicts schematically a central vacuum cleaning system and Fig. 10.23 depicts an enlarged detail of the typical vacuum outlet. The detail representation of Fig. 10.22 is the sort of presentation that should be represented on the drawings. On a schematic detail of this type all of the pipe sizing can readily be depicted.

In essence a vacuum cleaning system should have a minimum of 70 cfm at 2-in Hg vacuum at the end of a 1½-in hose located at the inlet valve farthest away from the vacuum producer. The vacuum piping should be designed to maintain conveying velocities between 3,000

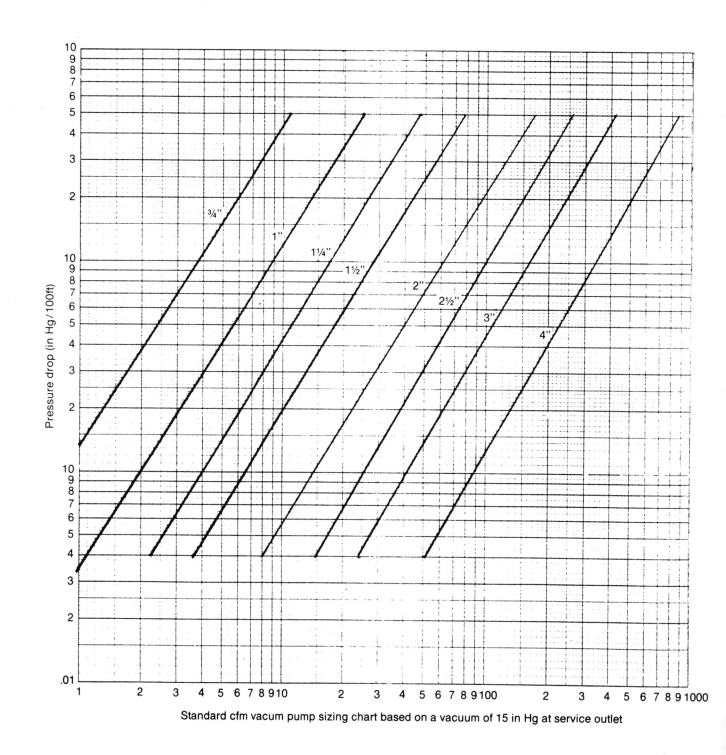

Standard cfm vacum pump sizing chart based on a vacuum of 15 in Hg at service outlet

FIGURE 10-19
Vacuum pump sizing chart.

①	WATER SUPPLY	⑧	VACUUM SWITCH
②	VACUUM GAUGE	⑨	WATER SUPPLY
③	FROM SYSTEM	⑩	VENT - TO ATMOSPHERE
④	VACUUM BREAKER	⑪	OVERFLOW
⑤	FLUSHING CONNECTION	⑫	FUNNEL WITH 2" AIR GAP
⑥	DRAIN CONNECTION	⑬	DRAIN
⑦	LIQUID LEVEL GAUGE	⑭	BACKFLOW PREVENTER OR VACUUM BREAKER (IF REQ'D OR AS REQ'D BY LOCAL PLB CODES)

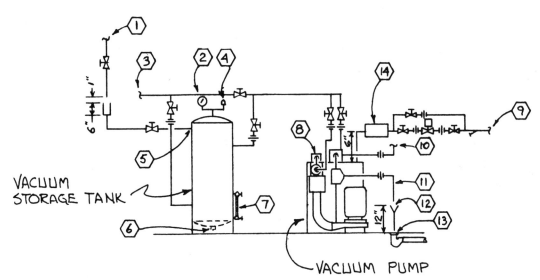

VACUUM STORAGE TANK

VACUUM PUMP

LEGEND

⊣⊦ UNION

GATE VALVE

SOLENOID VALVE

Y "Y" STRAINER

CLINICAL VACUUM PUMP SYSTEM

NO SCALE

FIGURE 10-20

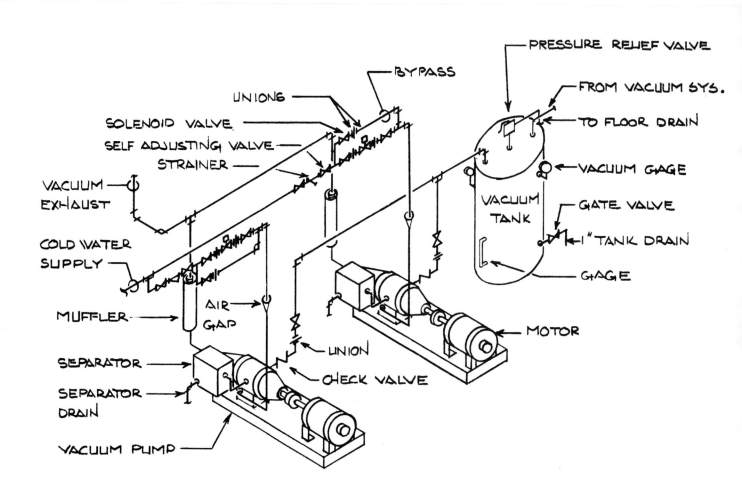

PRESSURE RELIEF VALVE

BYPASS

UNIONS

FROM VACUUM SYS.

SOLENOID VALVE

TO FLOOR DRAIN

SELF ADJUSTING VALVE

STRAINER

VACUUM GAGE

VACUUM
EXHAUST

VACUUM
TANK

GATE VALVE

COLD WATER
SUPPLY

1" TANK DRAIN

GAGE

MUFFLER

AIR
GAP

MOTOR

UNION

SEPARATOR

CHECK VALVE

SEPARATOR
DRAIN

VACUUM PUMP

LABORATORY VACUUM SYSTEM
NOT TO SCALE

FIGURE 10-21

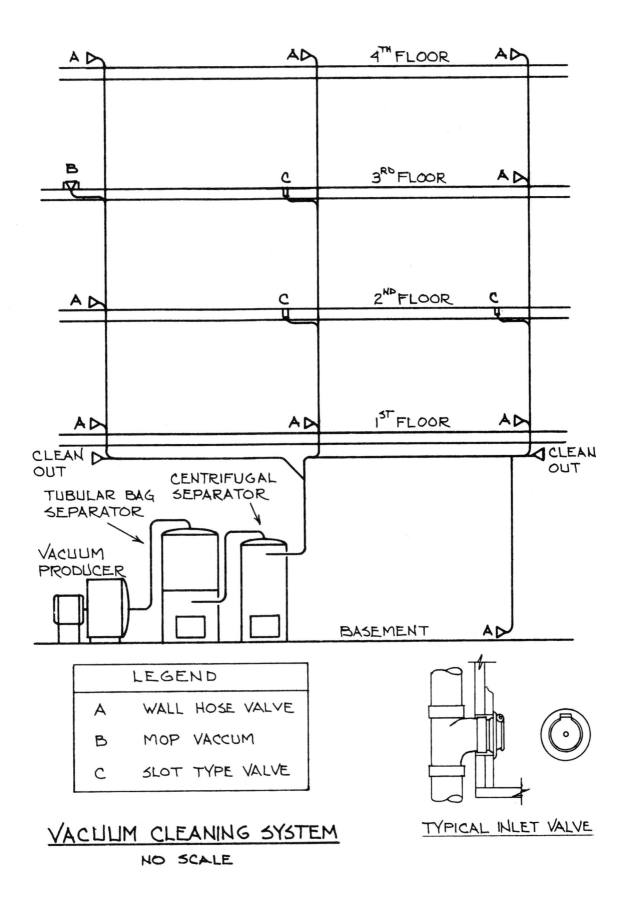

A▷   A▷   4TH FLOOR   A▷

B▽   C   3RD FLOOR   A▷

A▷   C   2ND FLOOR   C

A▷   A▷   1ST FLOOR   A▷

CLEAN OUT ▷   ◁ CLEAN OUT

TUBULAR BAG SEPARATOR

CENTRIFUGAL SEPARATOR

VACUUM PRODUCER

BASEMENT   A▷

LEGEND

A	WALL HOSE VALVE
B	MOP VACCUM
C	SLOT TYPE VALVE

VACUUM CLEANING SYSTEM
NO SCALE

TYPICAL INLET VALVE

FIGURE 10-22

and 4,500 feet per minute (fpm). Since vacuum cleaning pipe sizes are similar to high-velocity air ductwork sizing and are based on the same formulas, we present in Fig. 10.24 part of a common pipe sizing chart in which the friction loss is in inches of mercury, which is the standard of vacuum machinery. The recommended range of velocity is 3,000 to 4,000 fpm. The recommended friction loss is 0.6 in Hg for 100 ft of 2-in line. Thus if this concept is followed and all pipe sizing is to be within the 3,000 to 4,000 fpm range, the larger sized pipe will necessarily have a smaller friction drop per hundred feet. A line carrying 1,000 cfm would fall in the range of 0.10 to 0.20 in Hg/100 ft and would be sized at 8 in. The losses for the system include the basic 2.0-in Hg inlet loss plus the hose piping, separator, and exhaust line losses for the system. Commonly the hose loss will be 1.3 in Hg (standard for hose is 1.5 in). Thus the inlet and hose losses will usually amount to 3.3 in Hg and will represent nearly two-thirds of the total system loss.

Presuming the total system loss is 5.5 in Hg, the proper vacuum exhauster to be selected is one that exceeds this calculated value and the nearest standard size system is one rated at 6 in Hg.

All central vacuum systems require a means to filter out the dust collected. Figure 10.22 includes a set of filters for a typical system. The dust collector passes through a centrifugal separator and a tubular bag separator before reaching the vacuum producer. Normally the pump and filters are located in a remote area such as a basement, equipment room, or outside the building.

The manufacturers of central vacuum systems can provide excellent catalogs that further expand the coverage of all aspects of the system and its related components. Any designer planning to specify this type of system would be well advised to contact the equipment manufacturer in advance for complete design and installation data.

Vacuum Systems Summary. Not depicted in the vacuum detailing in this book are various small wet and dry vacuum systems that may be used in dental facilities, surgical suites, emergency rooms, and operating rooms.

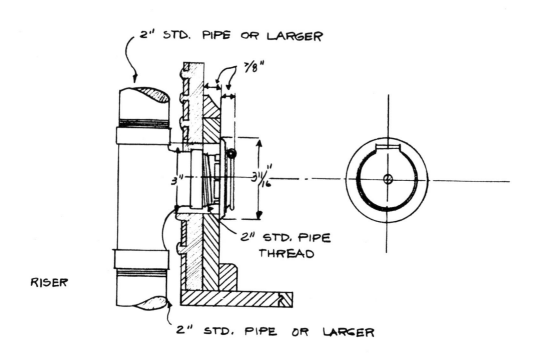

TYPICAL INLET VALVE

CENTRAL VACUUM SYSTEM

FIGURE 10-23

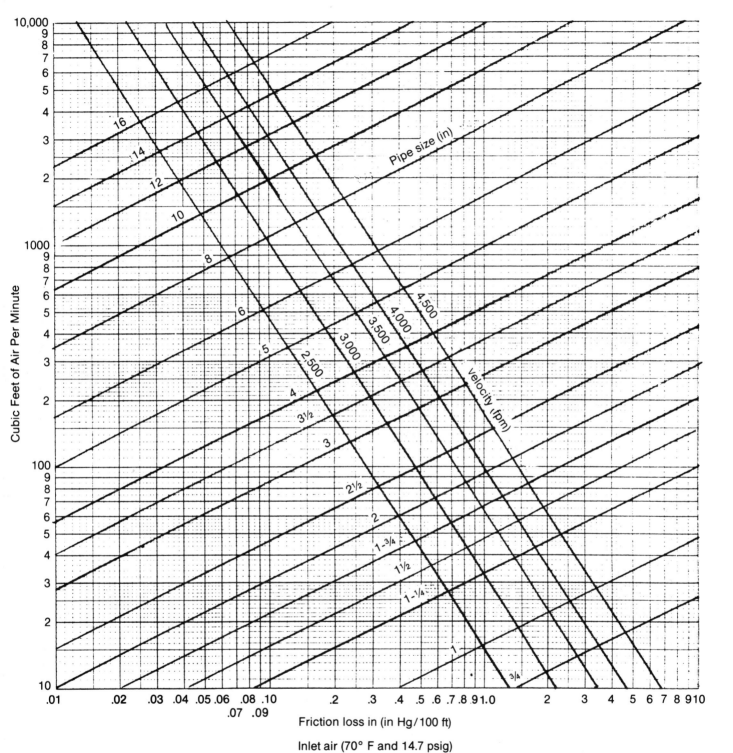

Cubic Feet of Air Per Minute

Pipe size (in)

velocity (fpm)

Friction loss in (in Hg / 100 ft)

Inlet air (70° F and 14.7 psig)

CENTRAL VACUUM PIPE SIZING CHART

FIGURE 10-24

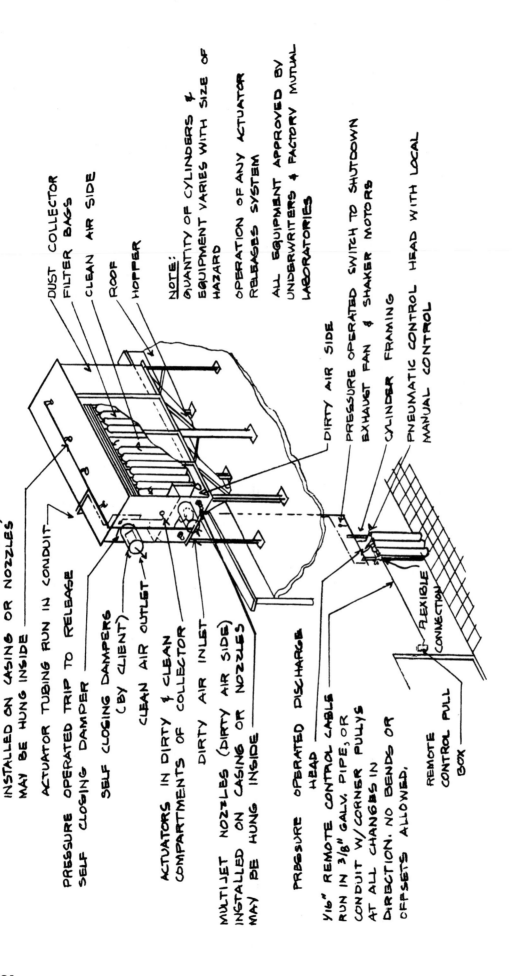

DUST COLLECTORS

CARBON DIOXIDE FIRE EXTINGUISHING SYSTEMS

FIGURE 10-25

224

The principles previously covered apply to these systems as well. The systems differ usually in the amount of vacuum required. Dental systems may be a combination wet-dry system requiring 10 to 12 cfm per location and 5-in to 7-in Hg vacuum with the exhausted air passing through a decontaminating vacuum filter. Wet systems in which liquids enter the vacuum pump are discharged to the sanitary sewer system with solids being intercepted before reaching the pumps. The pumps are the wet rotary type.

The wet system may require vacuum up to 18 in Hg. Each operating room or other similar situation requires an in-line filter and dual full capacity pumps. Piping of wet-dry and wet systems should be PVC with long turn drainage fittings. These systems are commonly small individualized systems with pipe sizing and installation provided as per manufacturers' instructions.

Fire Extinguishing Systems

As considered by the National Fire Protection Association (NFPA), fire protection systems not only include the well-known sprinkler system but also various chemical and foam systems. Each of these chemical and foam systems are specialties. Very commonly they are separately designed by experts in each particular area. This is especially true for large scale foam systems at oil company bulk terminals and large scale carbon dioxide or halon systems.

While a book on plumbing design is not basically involved with these chemical fire treatment systems, the plumbing designer is, on occasion, involved in limited situations, such as we will note in the details that follow, in which a gaseous system using either carbon dioxide or halon is required for fire protection for a limited, local condition.

A complete, detailed presentation of the various chemical fire extinguishing systems is presented in Volume 1 of the NFPA codes. For the reader who seriously wants to expand his or her area of expertise the acquisition of this volume and the complete understanding of the particular requirements of this area of work are a mandatory first step. The concern of this section of this book on plumbing is to cover the detailing required in a small number of the common, limited type of problems facing the sprinkler system designer whose concern is one relatively unique situation.

Under the NFPA codes only halon 1301 carries a specific approval for use in computer areas. The details that follow do not cover a computer room installation since the system is readily and commonly depicted on the building plans. It consists of a collection of tanks, as we typically illustrate in our carbon dioxide details, piped to a series of ceiling outlets in the computer room with a suitable fire detecting device. What is shown in the details that follow are some specialized, not uncommon situations that occur as fire protection design problems.

Dust Collector Protection. Common to the office in which the plumbing designer functions is a total mechanical design group which frequently designs dust collector systems. This is an area which is local, special, and able to be resolved by a total flooding system. Any fire in the duct transport system would be rapidly drawn to the dust collector, which is also the likely source of the fire's origination. Figure 10.25 depicts schematically how the fire protection system, using carbon dioxide as the chemical, is applied.

One of the important elements in this detail is the note in the approximate center of the drawing stating "pressure operated switch to shut down exhaust fan and shaker motors." Filter systems of this type have automatic, timed filter bag shakers to release collected dust, which are driven by integral, operating motors. To preclude accelerating the fire spread, these items should be automatically stopped when the carbon dioxide system is activated.

The detail not only presents a reasonably accurate representation of the dust collector, but it also shows the location of the fire extinguishing system piping outlets that are to be installed and connected. As depicted, this system will operate automatically and can be operated manually. Note that the automatic system actuators are on both sides of the filter intake (dirty) and exhaust (clean).

The quantity of carbon dioxide, its pressure, and the actual pipe size should be obtained from the carbon dioxide supplier as it varies depending on the size of the equipment being protected. While the system is inherently simple, it is the three-dimensional aspect of the detail which clarifies an installation that otherwise could be improperly applied.

Oil Bath Filter Protection. Figure 10.26 is another typical situation that may be faced by the plumbing designer. Your office has designed a special air system utilizing an oil bath air filter. Fire protection is required. To get all of the elements of the fire extinguishing system in one detail, Fig. 10.26 is presented as a composite detail. In an actual installation the room housing the carbon dioxide cylinders may be some distance away from the filter area.

Again the composite detail in three dimensions makes the installation seem very simple. Because the systems are usually in fairly remote locations, the operation of the system through its actuators not only shuts down the fan and puts out the fire, but it also sounds an alarm. Again the system can be manually operated.

The key element in the detail is properly depicting where the carbon dioxide discharge nozzles are located. In this typical detail six nozzles are shown. Consultation should be held with the manufacturer of the carbon dioxide units and equipment not only for pipe and tank sizing and selection but also for nozzle coverage to be

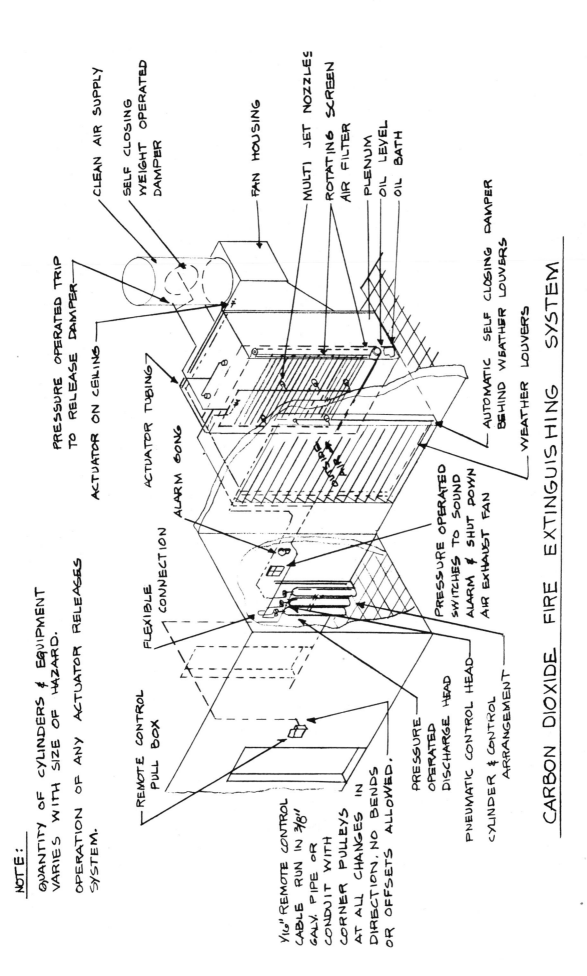

NOTE:

QUANTITY OF CYLINDERS & EQUIPMENT VARIES WITH SIZE OF HAZARD.

OPERATION OF ANY ACTUATOR RELEASES SYSTEM.

CLEAN AIR SUPPLY

SELF CLOSING WEIGHT OPERATED DAMPER

FAN HOUSING

MULTI JET NOZZLES

ROTATING SCREEN AIR FILTER

PLENUM

OIL LEVEL

OIL BATH

PRESSURE OPERATED TRIP TO RELEASE DAMPER

ACTUATOR ON CEILING

ACTUATOR TUBING

ALARM GONG

FLEXIBLE CONNECTION

AUTOMATIC SELF CLOSING DAMPER BEHIND WEATHER LOUVERS

WEATHER LOUVERS

PRESSURE OPERATED SWITCHES TO SOUND ALARM & SHUT DOWN AIR EXHAUST FAN

REMOTE CONTROL PULL BOX

5/16" REMOTE CONTROL CABLE RUN IN 3/8" GALV. PIPE OR CONDUIT WITH CORNER PULLEYS AT ALL CHANGES IN DIRECTION. NO BENDS OR OFFSETS ALLOWED.

PRESSURE OPERATED DISCHARGE HEAD

PNEUMATIC CONTROL HEAD

CYLINDER & CONTROL ARRANGEMENT

CARBON DIOXIDE FIRE EXTINGUISHING SYSTEM

PROTECTION FOR OIL BATH AIR FILTERS

FIGURE 10-26

certain that you have a sufficient number of nozzles to cover your particular filter.

Kitchen Equipment Protection. Plumbing and sprinkler system designers normally think of commercial kitchen protection being part of the sprinkler system design. While this is commonly true, Fig. 10.27 illustrates a commercial kitchen fire extinguishing system that uses carbon dioxide as the source of protection.

Again the advantage of the three-dimensional detail is clear cut, and what could be a cluttered, difficult to interpret plan becomes what it actually is, a very simple system. In essence there are a couple of cylinders of carbon dioxide connected via piping to nozzles at the fryer, in the filter area, and in the central exhaust collecting duct. As in previous cases the system is automatic

with local manual control. The sizing of the piping and the cylinders should be done in consultation with the carbon dioxide equipment supplier. Finally, in this system as in all others, a device to shut off the exhaust fan when the system is actuated is a basic requirement.

Industrial Oven Protection. Figure 10.28 represents two typical carbon dioxide fire extinguishing systems for industrial ovens and is the final presentation in this series of specialized fire protection systems. As in previous details the exact sizing of piping and storage cylinders would be found by consulting with the carbon dioxide supplier and the oven equipment manufacturer.

The detail, as did the previous details, locates the nozzle outlets in their required locations. It is necessary to check with the oven supplier in your project to be

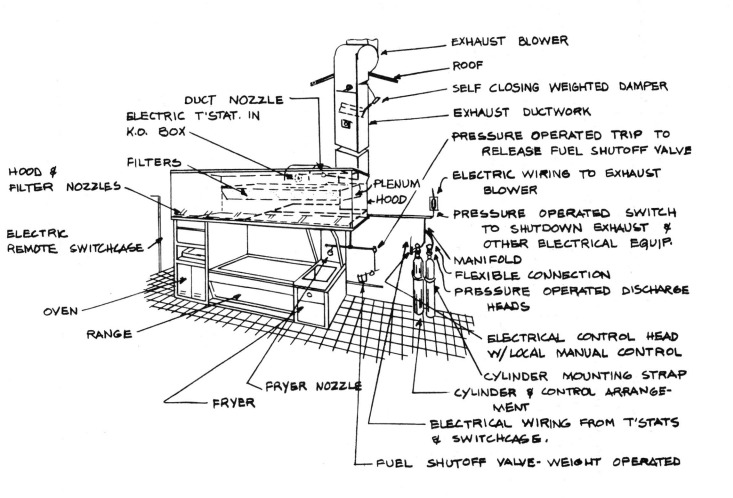

CARBON DIOXIDE FIRE EXTINGUISHING SYSTEMS

PROTECTION FOR COMMERCIAL KITCHEN EQUIPMENT

FIGURE 10-27

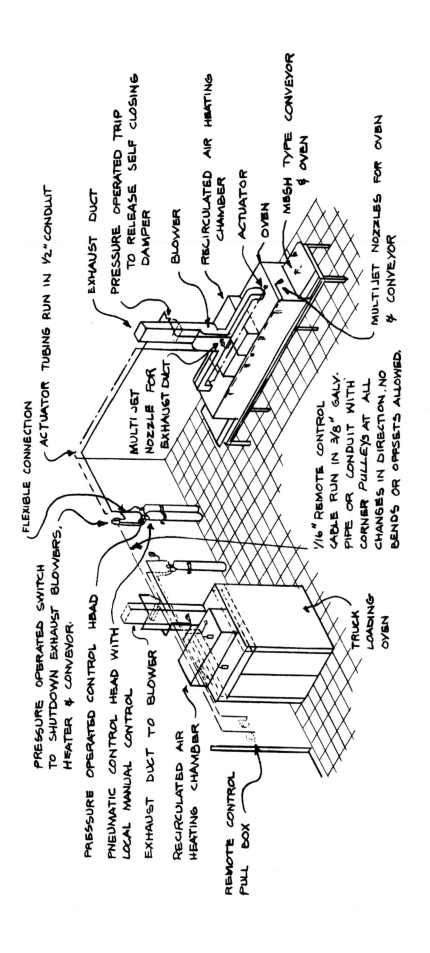

PRESSURE OPERATED SWITCH
TO SHUTDOWN EXHAUST BLOWERS,
HEATER & CONVEYOR.

FLEXIBLE CONNECTION

ACTUATOR TUBING RUN IN ½" CONDUIT

EXHAUST DUCT

PRESSURE OPERATED TRIP
TO RELEASE SELF CLOSING
DAMPER

BLOWER

RECIRCULATED AIR HEATING
CHAMBER

ACTUATOR

OVEN

MESH TYPE CONVEYOR
& OVEN

PRESSURE OPERATED CONTROL HEAD

PNEUMATIC CONTROL HEAD WITH
LOCAL MANUAL CONTROL

EXHAUST DUCT TO BLOWER

RECIRCULATED AIR
HEATING CHAMBER

MULTI JET
NOZZLE FOR
EXHAUST DUCT

MULTIJET NOZZLES FOR OVEN
& CONVEYOR

REMOTE CONTROL
PULL BOX

1/16" REMOTE CONTROL
CABLE RUN IN 3/8" GALV.
PIPE OR CONDUIT WITH
CORNER PULLEYS AT ALL
CHANGES IN DIRECTION. NO
BENDS OR OFFSETS ALLOWED.

TRUCK
LOADING
OVEN

CARBON DIOXIDE FIRE EXTINGUISHING SYSTEM

PROTECTION FOR INDIVIDUAL OVENS

FIGURE 10-28

228

certain there are no other locations required for proper fire extinguishing operation. Another item shown in this detail is the self-closing damper connection. Where these dampers exist, they may be gravity or motor operated. If they are gravity operated, they will close when the exhaust fan stops. If they are motor operated they may or may not be interconnected with the exhaust fan especially if the exhaust blower is not located on top of the oven, as shown in Fig. 10.28, but rather is remotely located and provides exhaust for more than one oven. As a basic rule, anytime there is a motor-operated damper located in the exhaust duct at the item of equipment being protected, the fire extinguishing system should include a means to close this damper when the extinguishing system is actuated. Finally this system, as in the previous systems, has a means of manually starting the extinguishing process. This is to be certain that the employees involved in the process can activate the fire protection system if, for any reason, the automatic actuator fails.

Chemical Treatment Systems

Except for chlorinating the water supply system, the plumbing designer and detailer is not the proper source of information about water treatment. And the designer seldom, if ever, is actually consulted. One of two common procedures are normally employed. In the first procedure the plumbing contractor simply leaves a capped

tee in the water line that is to be protected, and the treatment supplier provides the equipment, piping, and installation labor necessary to install the treatment. In the second procedure the plumbing designer's effort becomes much more frustrating. The project's owner buys the treatment equipment and advises the plumbing designer that the plumbing contractor will have to design the piping necessary to install the treatment equipment. Obviously this a difficult situation since there is no applicable standard for treatment equipment piping. The only solution is for the plumbing designer to consult with the treatment equipment supplier and ascertain precisely how all equipment piping connections are to be made. Three details that follow illustrate this procedure. In no case can they be assumed to be exactly correct for your particular situation. They are, however, correct for the situation in which they were applied.

Injection System. Figure 10.29 represents a pool mounted chlorinator and is similar to previously presented chlorinator details. It is, however, representative of the first situation of treatment requirement. The work of the plumbing contractor can stop at the injector or, because of the simplicity of the treatment system, the designer could ascertain the pipe sizes of the treatment unit and readily depict the treatment system, as was done in Fig. 10.29. Many simple treatment systems are similar to this detail. There is certainly no reason to be

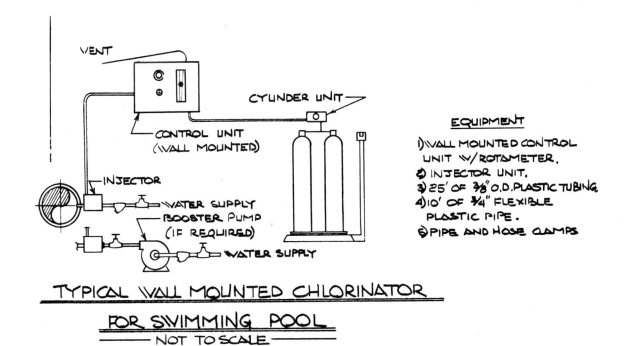

FIGURE 10-29

concerned over the piping design that is being handled by a competent plumbing designer.

Chemical Treatment. To illustrate what is meant in Fig. 10.29 about relatively simple treatment systems and their application, Fig. 10.30 presents a detail of chemical treatment for a high-temperature hot water heating plant.

Not shown in the detail are the piping connections which are simply that, and nothing more, to the specific lines in the boiler plant piping system. In this case the plumbing designer has ascertained from the treatment supplier where, how many, and what size the connections to be detailed are. All of the gate and check valves, unions, and other specialties, if required, are shown at the chemical treatment tank. In actual practice it is not necessary to go to elaborate detail, as was done in Fig. 10.30, to depict the various components that are part of the chemical treatment tank. These were shown in Fig. 10.30 so that the reader, who may be unfamiliar with this device, could relate to tank connection terms that may come up in his or her discussion with the chemical treatment equipment supplier. Note that there is a pump on top of the tank that is capable of delivering treated water at pressures above 250 psig.

Water Softener. Figure 10.31 is an ideal example of the second case in our opening discussion. Here is depicted a complicated water treatment system for which

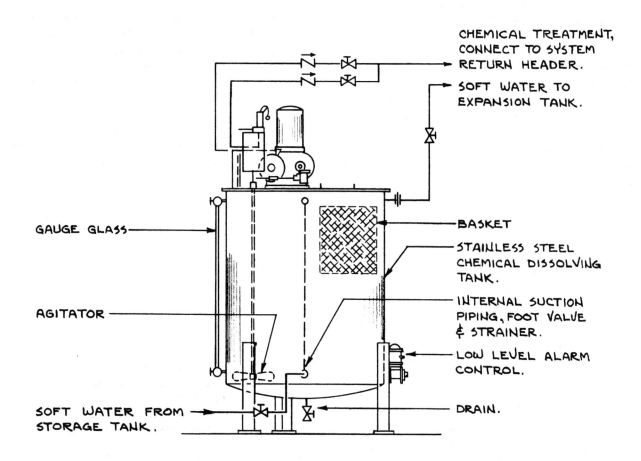

TYPICAL HIGH TEMPERATURE HOT WATER
CHEMICAL TREATMENT SYSTEM

FIGURE 10-30

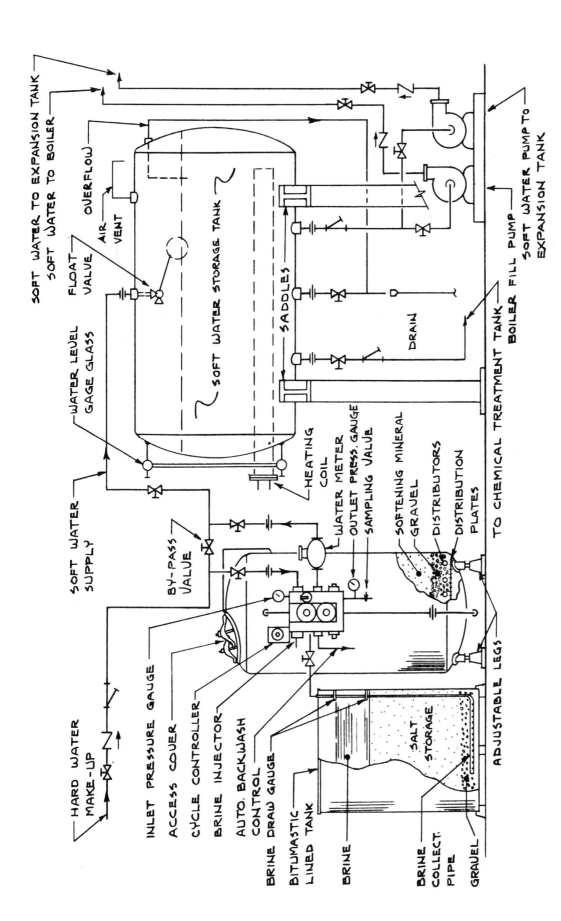

WATER SOFTENING SYSTEM FOR
HIGH TEMPERATURE HOT WATER SYSTEM

FIGURE 10-31

a number of items are not furnished by the water treatment supplier and certainly almost none of the piping is supplied.

In essence what is required is a considerable input of data to the plumbing designer who must size the soft water storage tank and heater, the pump, and all the piping and design all of it to fit the project requirements. Thus the beginning point of data acquisition is how much soft water is required. In most closed heating systems the answer comes from two separate items of information. The first requirement is to fill the system with hard water that has been reduced through treatment to soft water. Usually this means the removal of dissolved iron and calcium in the hard water. Once the heating piping system is properly filled and all air eliminated, the requirement is merely to maintain a reasonable supply of softened water for possible system losses.

The original system volume can be a variety of values, but we have assumed 10,000 gal. Makeup would be no more than 2 or 3 percent of this sum or .02 to .03 × 10,000. Thus the makeup is 200 to 300 gal. The selection of the softener capacity in gallons per minute is different from one that requires continuous treatment. In general, flow requirements for the initial filling of the system are usually the basis for sizing the treatment equipment. Commonly, equipment that can treat 10 to 20 gpm will be sufficient to fill almost all systems in a reasonable length of time.

The flow through the typical softener is usually based on 6 to 8 gpm/sq ft of surface. The correct value must be obtained from the equipment manufacturer you select. The detail depicts one softener for the sake of clarity. Commonly two are provided to allow one to be regenerated when the other is operating. Each softener, to meet design needs of 10 to 20 gpm, has a diameter of 18 to 24 in. Regeneration time usually varies from 12 hr for manually operated units to 8 hr for automatically operated units.

The storage tank depicted in the detail is commonly in the 500-gal to 1,000-gal capacity range. The heater should be sized to heat the 10 to 20 gpm from 40°F to 200 or 205°F since the probability of air or carbon dioxide entrainment is very great and should be avoided. Heating the water to temperatures above 212°F provides little added improvement in air elimination and greatly increases the cost.

The boiler fill pump does not have a large pressure problem in filling the system, but once it begins to function as a makeup pump, it must overcome a system pressure that is at least 250 psig. The actual system pressure should be verified by checking with the heating system designer. Since the pump to the expansion tank is only concerned with hard water supply pressure as increased by temperature, it can be at a lower value. Its size should be worked out in consultation with the expansion tank supplier.

Thus the basic components of the system are two 18-in to 24-in diameter softeners and salt storage units, a 500-gal to 1,000-gal tank with a heater to raise the water temperature to 200°F, a fill pump capable of delivering 20 gpm at a pressure above 250 psig, and a properly sized small expansion tank pump. The source of heat for the tank heating coil is provided by the heating system designer (steam or water). Since most of the piping must handle 20 gpm, the pipe (at normal pressure drop values) would be 1½ in. The drain line from the tank would be the same size as the tank drain opening.

The detail as depicted came from an actual installation and is correct as shown. The storage tank sits on legs or on a supporting cradle, frame, and legs as required by the project. Note that from the bottom of the tank as detailed there is a soft water line going to a chemical treatment control. In Fig. 10.30 soft water was introduced into the chemical treatment system and pumped into the boiler. Since you are working with a system of high pressures, the system and its piping must be rated at 300 psig. While some elements may not normally encounter this pressure, the stopping for any reason of the fill pump would allow boiler system pressure (250 psig) to be imposed on all parts of the softener system. All hot and cold water piping and storage tanks must be insulated in accordance with the normal temperatures they will encounter.

Summary

All of the special systems in this chapter have been approached from the premise that they are not a logical part of the plumbing and sprinkler system designer's area of usual and common knowledge. Rather they are design problems that, on occasion, occur and for a variety of reasons are handed to the plumbing designer as a part of his or her responsibility. Since each of the areas covered has one or more complete texts devoted to the subject, they are obviously much more complex than our limited treatment would indicate.

Our opinion was that they could not be logically ignored in a book on plumbing design. Nor could they be adequately covered. What has been attempted in this chapter is to explain how to design and detail the basic limited application of these systems. We do not believe any plumbing designer or his or her firm should attempt to design large scale, involved applications of any of these systems unless the firm possesses a bona fide expert on the subject or retains a consultant who is an expert in the area involved.

As we noted, there are books on these areas. In addition the manufacturers of equipment in each of these areas have excellent fairly large books on the individual subjects. For the designer wishing to truly become knowledgeable in any of these fields the next step after reading this chapter is to contact specific equipment manufacturers for their literature and design data.

ELEVEN

Schedules and Symbols

Among the practices of plumbing design firms the quantity, quality, and variety of notes, schedules, and symbols, or legends, result in strikingly different presentations of completed plans. Almost all final sets of plans will include a plumbing symbol or legend list. There are plans prepared by firms for large industrial clients that do not show any symbol list. Usually this is the sort of client who constantly does a large amount of building and retrofit work on a given site. Over time the client has developed a standard set of specifications for most, if not all, of the types of work that will be performed at the particular site. These standard client-prepared specifications include sections describing work performance, installation details, and symbol lists. The reasoning behind this sort of plan and specification presentation is that the bidding contractors are primarily selected from a contractor's bid list and are, or soon become, totally familiar with the symbols, specifications, and other construction details. Further, the client's in-house engineers are also totally familiar with all of the specifications, symbols, and details. The system, effectively, is one of "no surprises." A line with a given symbol is on all drawings always a steam line, water line, high-temperature hot water line, etc., wherever shown.

The ASHRAE guide has also standardized on a set of symbols which generally have been adopted by engineers including the author of this book. There is nothing sacred or sacrosanct about any given set of symbols. So long as the plan has a set of symbols and these are used for the work involved, this is all that is really necessary. For many years cold water piping was depicted as a single dashed line with a single dot in between long (2 to 4 in) dashes. The same cold water line can be depicted with the letters CW in lieu of the dot. Or, for that matter, it can be depicted with any other letter so long as that is the way the piping is depicted. The symbol list can be shown either on the plans or in the specifications. Since most completed plans are one-time situations for one client in one location, it is easier for all concerned, especially the installing contractor, to have the symbols on the plans rather than in the specifications.

The location of schedules and notes are subject to much broader interpretation by designers of plumbing systems. As a general statement any specification is a document spelling out standards of material and equipment and, when applicable, standards of equipment performance. It is the area of performance that creates situations of conflict between the designer and the installer. Schedules of equipment can be shown on the plans or in the specifications. These are welcomed by both the installing contractor and the equipment supplier. While either location is acceptable, most suppliers and installers would prefer the schedules to be on the plans since their work requires a count and the location of each type of equipment. It is easier to refer back and forth between the schedule and the items depicted on the plans if the schedule is also on the plans.

Symbol List

Figures 11.1 and 11.2 depict a standard set of symbols for plumbing work in common use today. They have evolved over time and will generally, but not always, cover all of the work typically involved in plumbing and sprinkler design. In the list cold water piping has a different symbol for interior versus exterior lines. We have opted for each older standard for interior water piping which is a line made up of longer dashes separated by one, two, or three dots. Other systems would use the letter symbols, CW, HW, and RHW, for cold, hot, and recirculating lines. In practice, interior piping is frequently crowded and somewhat difficult to read. In a cramped space, where line space is at a premium, the use of the dots is both an accepted standard and easier to draw.

Storm and sanitary piping have been shown as dashed lines with letters. Some offices and systems still use one or more short dashes in lieu of the letters. In the majority of cases using letters for soil and sanitary lines does not create a drawing problem and does clearly differentiate these lines. Vent lines could also be shown as a long, broken dashed line with the letter V. However, vent lines have long been shown as a series of short dashes, and this is a very acceptable symbol.

With the exception of compressed air and gas, which again have very well-known symbols that utilize the letters A and G, we feel all other lines should have two-letter or three-letter symbols such as VAC for vacuum, SW for soft water, and CTW for chemically treated water.

In riser symbols purists will insist that an up riser is an open circle and a down riser is a closed circle. Presuming your drawings have adequate waste and vent details, we feel it is clearer if the cold water riser is always an open circle and the rain leader or sanitary waste or vent is always a closed circle with a letter symbol as we depict.

The individual symbols are not necessarily all the symbols you may need since the piping you are designing may require some specialty item not shown in our list. You will simply have to create these symbols. Many of the symbols are commonly employed for both plumbing and heating specialties. To help a little in differentiating plumbing from heating piping that may appear on a common plan for both contractors, the detail that we use depicts a different gate and globe valve symbol. The common gate and globe valve horizontal X symbol could have been used. The sprinkler symbols are fairly standard. If our plans require an exterior fire hydrant or siamese connection, we always add the words "fire hydrant" or "siamese connection" to the symbol with an arrow pointing to the device and a separate detail for same.

Specifications

Because the specifications and the schedules on the plan must, to be complete, complement each other, we will first present typical specification paragraphs on plumbing fixtures and two types of water heaters. As we have noted in the opening discussion, properly done schedules usually do not create problems. On the contrary they represent solutions to data presentation problems.

Schedules are not the total specification. They generally provide the performance data, size, and model number. While both the plumbing contractor and the material supplier would be delighted to find a line item giving the quantity, we strongly recommend that the designer *never* indicate quantities. Even if the quantity is glaringly obvious. Once the designer indicates the quantity, any deviation, mistake, or whatever is automatically the designer's problem.

Plumbing fixtures can readily and easily be named by manufacturer and type without much in the way of further data. It may be noted that the only item in the specification on fixtures that follows is a phrase that in effect states that the unit shall include all traps and trim. Other items in the schedule (water heaters as noted are a typical example) require much more in the way of written equipment material and construction description. The following three items are from an actual specification as written and used on an actual project:

PART TWO—PRODUCTS

2.1 PLUMBING FIXTURES

A. Fixtures shall be of the type as specified on the fixture schedule on the plans. Each fixture shall be complete with flush valves, stop valves, exposed or concealed traps, faucets, drain valves and other appurtenances as required for a complete, fully operable unit for normal, intended use. Fixtures shall also comply with ANSI A112.11, A112.18 and A112.19 standards as noted under Applicable Codes.

B. The manufacturer whose equipment is named on the plan schedule shall serve as the basic model to which all comparisons of submittals will be made. Where equal products are available the fixtures as manufactured by Crane, Eljer, or Kohler will be considered as meeting the specification.

2.6 GAS WATER HEATER

A. Water heater shall be of glass lined design, gas-fired, equipped to burn natural gas and certified by the American Gas Association, under Volume III tests for commercial heaters for delivery of 180°F water and shall be approved by the National Sanitation Foundation. Heater shall be equipped with a 2½-in × 3¾-in boiler-type hand hole cleanout and shall have a working pressure of 150 psi. Heater shall be provided with an automatic gas shut-off device and 100 percent safety shutoff in event pilot flame is extinguished; a gas pressure regulator set for the type of gas supplied; an approved draft diverter; and an extruded magnesium anode rod rigidly supported for cathodic protection.

B. ASME pressure and temperature relief valve shall be furnished and installed by Contractor.

C. The heater tank shall be insulated with vermin-proof fiberglass insulation or equal. The outer jacket shall have a baked enamel finish over a bonderized undercoating.

D. All internal surfaces of the heater exposed to water shall be glass lined with an alkaline borosilicate composition that has been fused to the steel by firing at a temperature range of 1,400°F. Heater shall be design certified by AGA or CGA and NSF for 180°F service.

2.7 ELECTRIC WATER HEATER

A. Heater shall be of glass lined design with 150 psi working pressure, equipped with extruded high density magnesium anode and stainless steel cold water inlet tube. Electric heating

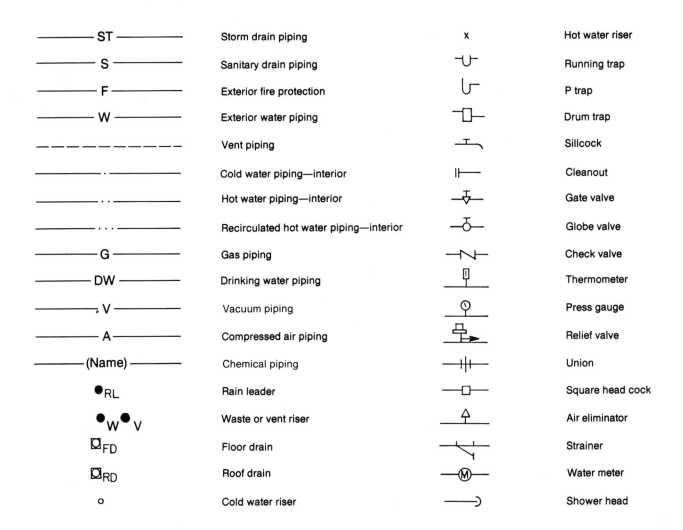

———— ST ————	Storm drain piping	x	Hot water riser
———— S ————	Sanitary drain piping		Running trap
———— F ————	Exterior fire protection		P trap
———— W ————	Exterior water piping		Drum trap
— — — — — — —	Vent piping		Sillcock
———— · ————	Cold water piping—interior		Cleanout
———— · · ————	Hot water piping—interior		Gate valve
———— · · · ————	Recirculated hot water piping—interior		Globe valve
———— G ————	Gas piping		Check valve
———— DW ————	Drinking water piping		Thermometer
———— V ————	Vacuum piping		Press gauge
———— A ————	Compressed air piping		Relief valve
———— (Name) ————	Chemical piping		Union
•RL	Rain leader		Square head cock
•W •V	Waste or vent riser		Air eliminator
▣FD	Floor drain		Strainer
▣RD	Roof drain		Water meter
o	Cold water riser		Shower head

FIGURE 11-1
Plumbing symbol list.

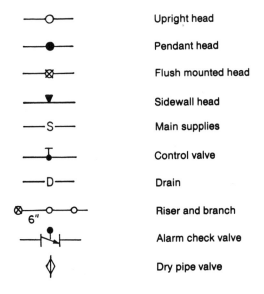

——o——	Upright head
——●——	Pendant head
——⊗——	Flush mounted head
——▼——	Sidewall head
——S——	Main supplies
——丄——	Control valve
——D——	Drain
⊗—o—o (6")	Riser and branch
—N—	Alarm check valve
◇	Dry pipe valve

FIGURE 11-2
Plumbing and sprinkler symbol list.

elements shall be low watt density with zinc inlet tube. Electric heating elements shall be low watt density with zinc plated copper sheath and 2½-in bolt circle flange. Each element shall be controlled by an individually mounted thermostat and high temperature cut-off switch. The outer jacket shall be of baked enamel finish and shall be provided with full size control compartment for performance of service and maintenance through hinged front panel and shall enclose the tank with 2 in of vermin proof fiberglass insulation. Electrical junction box with heavy duty terminal block shall be provided. The drain valve shall be located in the front for ease of servicing. The heater tank shall be guaranteed against leakage due to corrosion for a period of 3 years. Controls and accessories shall be guaranteed for 1 full year. Manufacturer shall supply ASME rated temperature and pressure relief valve. Fully insulated instruction manual shall be included. Heater shall be listed by the Underwriter's Laboratories.

Schedules

Figure 11.3 depicts a typical plumbing fixture schedule and Fig. 11.4 depicts the same schedule with appropriate data which would have been filled out by the engineer, specification writer, or designer. The schedule spacing is designed to fit single-spaced standard typing. Thus the completed schedule may be typed or hand lettered and affixed to the drawing. As we noted earlier, the schedule can also be included in the specifications.

If there are a large number of different items of equipment, there are advantages to hand lettering the schedule data. It is not unusual for many last minute changes to be made. Frequently these are minor in nature. If the schedule is already on the plans, it is usually very simple

PLUMBING FIXTURE SCHEDULE							
FIXTURE	SIZE – TYPE	MANUFACTURER AND MODEL NO.	PIPE SIZE				REMARKS, NOTES ACCESSORIES
			SOIL	VENT	CW	HW	

PLUMBING EQUIPMENT SCHEDULE									
NO.	SIZE – TYPE	MANUFACTURER AND MODEL NO.	CONNECTION SIZE				RATING		NOTES
			SOIL	VENT	CW	HW	INPUT	OUTPUT	

FIGURE 11-3
Fixture and equipment schedules.

to erase and/or alter the hand lettered data. It may not be as easy if the data has been typed on a regular typewriter. If the data have been typed on a computer, the process of change is also simple provided the schedule has not as yet been firmly affixed to the drawing. This is why some firms prefer to include the schedule with the specification. For both contractor convenience and ease of checking shop drawings, a schedule on the plans is far more convenient for everyone.

The data in Fig. 11.4 complement the specification that was described previously. Generally there is no problem with this arrangement. The performance data of the water heaters and other equipment items may also be included in the specification of the particular item and the equipment schedule may be eliminated.

However, the schedule on the plans serves two very useful purposes. First, when the designer and detailer both know there will be a schedule, a second source of specification input is created and will help ensure that the item is covered. Second, while the plans and the specification are both part of the contract, the general tendency of contractors, lawyers, and courts is to put more weight on data shown on the plans even if they are not covered in the specification.

System Installation

Some of the favorite phrases of plumbing specifications are "furnish and install all fixtures as shown on the plumbing and/or architectural plans," "nothing shall preclude the contractor from completing the work in

PLUMBING FIXTURE SCHEDULE							
FIXTURE	SIZE - TYPE	MANUFACTURER AND MODEL NO.	PIPE SIZE				REMARKS, NOTES ACCESSORIES
			SOIL	VENT	CW	HW	
Water Closet	Std - Wall Mtd Siphon jet 1 1/2" top spud	Amer. Std "Glenco" Sloan Royal Flush Valve	4"	2"	1"	—	Olsonite, black open, #10cc Seat Zurk Series 1200 Carrier
Service Sink	Std - Enameled cast iron	Amer. Std "Lakewell" Faucet w/Vacuum Bkr 8340.242	3"	2"	1/2"	1/2"	Wall Hanger Trap Std #8340.242

PLUMBING EQUIPMENT SCHEDULE									
NO.	SIZE - TYPE	MANUFACTURER AND MODEL NO.	CONNECTION SIZE				RATING		NOTES
			SOIL	VENT	CW	HW	INPUT	OUTPUT	
①	100 gal gas fired	A.O. Smith # BT-1971/ GC-166	—	—	3/4"	3/4"	155 MBH input 166 gph @ 40-140 output		incl diverter
②	80 gal electric	A.O. Smith #DRE-80	—	—	3/4"	3/4"	52 MBH input 61 gph @ 40-140 output		3 element 15KW

FIGURE 11-4
Fixture and equipment schedules.

accordance with all federal, state and local plumbing codes," and "furnish all equipment and appurtenances necessary to furnish and install all equipment including closing connections."

The first of the above phases in effect says that if one forgets a fixture on the plans it is still part of the work since the architect showed it on his or her plans. The second says if one did not show how to install the work, or if the installation was depicted incorrectly, the contractor shall install it correctly according to the code. The third statement says that if anything was forgotten the contractor shall include it and make certain that it is connected properly.

Needless to say all installing contractors have heard these songs many many times. And they don't accept them. Some weight can be given to a specification that specifically lists, by drawing number, what drawings are included in the plumbing contract. Although weak in application, both contractually and legally, the inclusion of the architectural plans for the specifically described purpose of only serving as a reference base for plumbing fixtures *may* be sufficient to at least get in the contract the correct number and location of fixtures. But if the fixtures are not shown on the plumbing drawings, no amount of words will get them installed, piped, and connected.

Without proper details the final connection of any complicated piece of equipment, including waste and vent piping, commonly will result in an extra claim by the installing contractor. If the detail of the installation shows more items or larger sizes than the applicable code allows, the contractor will not only happily put it in according to code but will also argue that that is the only way to properly install the item.

There is some debate about whether the word "install" means "install and connect." Specifications frequently use "install, connect, and make closing connections" to ensure that the plumbing system as designed is operable. Too often the third phrase in the opening paragraph of this section is applied to a piece of equipment that has numerous valves, traps, stops, and the like when none of these items are shown on any drawing or detail. This occurs frequently when the plumbing designer is detailing connections to kitchen and laboratory equipment. Here a detail showing all these specialties, or at least most of them, could be reasonably supplemented with the catch-all words of the third phrase. These situations are examples of common design and specifications which illustrate the value of schedules and details and which are one of the major premises of this book.

Index

About the Author

Jerome F. Mueller has been a practicing consulting engineer in the mechanical and electrical disciplines for over thirty years. A Professional Engineer in many states, he has designed mechanical and electrical systems for government departments, universities and colleges, cities and towns, the military, architectural firms, and manufacturing companies. He also has extensive experience in industrial process engineering and process design.

Mr. Mueller is a leader in the design application of computer-controlled installations and in automation control application. He was an early leader in fluorescent applications lighting and a pioneer in Connecticut's usage of electrical heating systems. He also has special expertise in project evaluation, cost estimation, and contract administration. He is the author of *Standard Application of Mechanical Details, Standard Application of Electrical Details,* and *Standard Mechanical and Electrical Details* and coauthor of *Standard Handbook of Consulting Engineering Practice* (all McGraw-Hill).